IET POWER AND ENERGY SERIES 51

Series Editors: Professor A.T. Johns
Professor D.F. Warne

Short-circuit Currents

Other volumes in this series:

Volume 1	**Power circuit breaker theory and design** C.H. Flurscheim (Editor)
Volume 4	**Industrial microwave heating** A.C. Metaxas and R.J. Meredith
Volume 7	**Insulators for high voltages** J.S.T. Looms
Volume 8	**Variable frequency AC motor drive systems** D. Finney
Volume 10	**SF6 switchgear** H.M. Ryan and G.R. Jones
Volume 11	**Conduction and induction heating** E.J. Davies
Volume 13	**Statistical techniques for high voltage engineering** W. Hauschild and W. Mosch
Volume 14	**Uninterruptable power supplies** J. Platts and J.D. St Aubyn (Editors)
Volume 15	**Digital protection for power systems** A.T. Johns and S.K. Salman
Volume 16	**Electricity economics and planning** T.W. Berrie
Volume 18	**Vacuum switchgear** A. Greenwood
Volume 19	**Electrical safety: a guide to causes and prevention of hazards** J. Maxwell Adams
Volume 21	**Electricity distribution network design, 2nd edition** E. Lakervi and E.J. Holmes
Volume 22	**Artificial intelligence techniques in power systems** K. Warwick, A.O. Ekwue and R. Aggarwal (Editors)
Volume 24	**Power system commissioning and maintenance practice** K. Harker
Volume 25	**Engineers' handbook of industrial microwave heating** R.J. Meredith
Volume 26	**Small electric motors** H. Moczala *et al.*
Volume 27	**AC-DC power system analysis** J. Arrill and B.C. Smith
Volume 29	**High voltage direct current transmission, 2nd edition** J. Arrillaga
Volume 30	**Flexible AC Transmission Systems (FACTS)** Y-H. Song (Editor)
Volume 31	**Embedded generation** N. Jenkins *et al.*
Volume 32	**High voltage engineering and testing, 2nd edition** H.M. Ryan (Editor)
Volume 33	**Overvoltage protection of low-voltage systems, revised edition** P. Hasse
Volume 34	**The lightning flash** V. Cooray
Volume 35	**Control techniques drives and controls handbook** W. Drury (Editor)
Volume 36	**Voltage quality in electrical power systems** J. Schlabbach *et al.*
Volume 37	**Electrical steels for rotating machines** P. Beckley
Volume 38	**The electric car: development and future of battery, hybrid and fuel-cell cars** M. Westbrook
Volume 39	**Power systems electromagnetic transients simulation** J. Arrillaga and N. Watson
Volume 40	**Advances in high voltage engineering** M. Haddad and D. Warne
Volume 41	**Electrical operation of electrostatic precipitators** K. Parker
Volume 43	**Thermal power plant simulation and control** D. Flynn
Volume 44	**Economic evaluation of projects in the electricity supply industry** H. Khatib
Volume 45	**Propulsion systems for hybrid vehicles** J. Miller
Volume 46	**Distribution switchgear** S. Stewart
Volume 47	**Protection of electricity distribution networks, 2nd edition** J. Gers and E. Holmes
Volume 48	**Wood pole overhead lines** B. Wareing
Volume 49	**Electric fuses, 3rd edition** A. Wright and G. Newbery
Volume 50	**Wind power integration: connection and system operational aspects** B. Fox *et al.*
Volume 51	**Short circuit currents** J. Schlabbach
Volume 52	**Nuclear power** J. Wood
Volume 53	**Condition assessment of high voltage insulation in power system equipment** R.E. James and Q. Su
Volume 55	**Local energy: distributed generation of heat and power** J. Wood
Volume 56	**Condition monitoring of rotating electrical machines** P. Tavner, L. Ran, J. Penman and H. Sedding
Volume 905	**Power system protection, 4 volumes**

Short-circuit Currents

Jürgen Schlabbach

The Institution of Engineering and Technology

Published by The Institution of Engineering and Technology, London, United Kingdom

First published 2005

The Institution of Engineering and Technology
Michael Faraday House
Six Hills Way, Stevenage
Herts, SG1 2AY, United Kingdom

www.theiet.org

British Library Cataloguing in Publication Data
Schlabbach, J. (Jurgen)
Short circuit currents
1. Short circuits – Mathematical models
I. Title II. Institution of Electrical Engineers
621.3'19

ISBN (10 digit) 0 86341 514 8
ISBN (13 digit) 978-0-86341-514-2

Typeset in India by Newgen Imaging Systems (P) Ltd, Chennai

To my wife Bettina and my children Marina and Tobias

Contents

List of figures xiii

List of tables xxiii

Foreword xxvii

1 Introduction **1**
- 1.1 Objectives 1
- 1.2 Importance of short-circuit currents 1
- 1.3 Maximal and minimal short-circuit currents 3
- 1.4 Norms and standards 4

2 Theoretical background **11**
- 2.1 General 11
- 2.2 Complex calculations, vectors and phasor diagrams 11
- 2.3 System of symmetrical components 14
  - 2.3.1 Transformation matrix 14
  - 2.3.2 Interpretation of the system of symmetrical components 18
  - 2.3.3 Transformation of impedances 19
  - 2.3.4 Measurement of impedances of the symmetrical components 20
- 2.4 Equivalent circuit diagram for short-circuits 24
- 2.5 Series and parallel connection 27
- 2.6 Definitions and terms 30
- 2.7 Ohm-system, p.u.-system and %/MVA-system 32
  - 2.7.1 General 32
  - 2.7.2 Correction factor using %/MVA- or p.u.-system 34
- 2.8 Examples 34
  - 2.8.1 Vector diagram and system of symmetrical components 34

2.8.2 Calculation of impedances of a three-winding transformer in %/MVA 37
2.8.3 Conversion of impedances (Ω; %/MVA; p.u.) 40
2.8.4 Impedances in %/MVA-system based on measurement 41
2.8.5 Representation of a line in the RYB-system and in the system of symmetrical components 42

3 Calculation of impedance of electrical equipment 45
3.1 General 45
3.2 Equipment in a.c. systems 45
3.2.1 General 45
3.2.2 Impedance calculation 46
3.3 Equipment in d.c. systems 50
3.3.1 General 50
3.3.2 Impedance calculation 58
3.4 Examples for calculation 63
3.4.1 a.c. equipment 63
3.4.2 d.c. equipment 64

4 Calculation of short-circuit current in a.c. three-phase HV-systems 67
4.1 Types of short-circuits 67
4.2 Methods of calculation 68
4.3 Calculation of parameters of short-circuit currents 70
4.3.1 General 70
4.3.2 Calculation of short-circuit current parameters according to IEC 60909-0 72
4.4 Influence of motors 84
4.5 Minimal short-circuit currents 85
4.6 Examples 86
4.6.1 Three-phase near-to-generator short-circuit 86
4.6.2 Line-to-earth (single-phase) short-circuit 87
4.6.3 Calculation of peak short-circuit current 88
4.6.4 Short-circuit currents in a meshed 110-kV-system 89
4.6.5 Influence of impedance correction factors on short-circuit currents 91
4.6.6 Short-circuit currents in a.c. auxiliary supply of a power station 94

5 Influence of neutral earthing on single-phase short-circuit currents 97
5.1 General 97
5.2 Power system with low-impedance earthing 98
5.3 Power system having earthing with current limitation 102
5.4 Power system with isolated neutral 105

5.5 Power system with resonance earthing (Petersen-coil) 108
5.5.1 General 108
5.5.2 Calculation of displacement voltage 112
5.5.3 Tuning of the Petersen-coil 115
5.6 Handling of neutrals on HV-side and LV-side of transformers 116
5.7 Examples 119
5.7.1 Increase of displacement voltage for systems with resonance earthing 119
5.7.2 Limitation of single-phase short-circuit current by earthing through impedance 123
5.7.3 Design of an earthing resistor connected to an artificial neutral 124
5.7.4 Resonance earthing in a 20-kV-system 124
5.7.5 Calculation of capacitive earth-fault current and residual current 125
5.7.6 Voltages at neutral of a unit transformer 126

6 Calculation of short-circuit currents in low-voltage systems 131
6.1 General 131
6.2 Types of faults 131
6.3 Method of calculation 132
6.4 Calculation of short-circuit parameters 132
6.4.1 Impedances 132
6.4.2 Symmetrical short-circuit breaking current I_b 133
6.4.3 Steady-state short-circuit current I_k 134
6.5 Minimal short-circuit currents 134
6.6 Examples 135

7 Double earth-fault and short-circuit currents through earth 139
7.1 General 139
7.2 Short-circuit currents during double earth-faults 139
7.2.1 Impedances and initial symmetrical short-circuit current I''_k 139
7.2.2 Power system configurations 140
7.2.3 Peak short-circuit current i_p 143
7.2.4 Symmetrical short-circuit breaking current I_b and steady-state short-circuit current I_k 143
7.3 Short-circuit currents through earth 143
7.3.1 Introduction 143
7.3.2 Short-circuit inside a switchyard 144
7.3.3 Short-circuit at overhead-line tower 145
7.4 Examples 146
7.4.1 Double earth-fault in a 20-kV-system 146
7.4.2 Single-phase short-circuit in a 110-kV-system 148

8 Factors for the calculation of short-circuit currents **151**
8.1 General 151
8.2 Correction using %/MVA- or p.u.-system 152
8.3 Impedance correction factors 154
8.4 Factor κ for peak short-circuit current 156
8.5 Factor μ for symmetrical short-circuit breaking current 158
8.6 Factor λ for steady-state short-circuit current 160
8.7 Factor q for short-circuit breaking current of asynchronous motors 162

9 Calculation of short-circuit currents in d.c. auxiliary installations **165**
9.1 General 165
9.2 Short-circuit currents from capacitors 169
9.3 Short-circuit currents from batteries 170
9.4 Short-circuit currents from rectifiers 172
9.5 Short-circuit currents from d.c. motors with independent excitation 174
9.6 Total short-circuit current 178
9.7 Example 182
9.7.1 Calculation of the impedances of cables and busbar conductors 184
9.7.2 Calculation of the short-circuit currents of the individual equipment 185
9.7.3 Calculation of the correction factors and corrected parameters 190
9.7.4 Calculation of partial short-circuit currents 191
9.7.5 Calculation of total short-circuit current 193

10 Effects of short-circuit currents **195**
10.1 General 195
10.2 a.c. systems 195
10.2.1 Thermal effects and thermal short-circuit strength 195
10.2.2 Mechanical short-circuit strength of rigid conductors 201
10.3 d.c. auxiliary installations 209
10.3.1 Substitute rectangular function 209
10.3.2 Mechanical short-circuit strength of rigid conductors 212
10.3.3 Thermal short-circuit strength 215
10.4 Calculation examples (a.c. system) 216
10.4.1 Calculation of thermal effects 216
10.4.2 Electromagnetic effect 217
10.5 Calculation examples (d.c. system) 218
10.5.1 Thermal effect 218
10.5.2 Electromagnetic effect 220

11 Limitation of short-circuit currents **225**
11.1 General 225
11.2 Measures 226
11.2.1 Measures in power systems 226
11.2.2 Measures in installations and switchgear arrangement 232
11.2.3 Measures concerning equipment 236
11.3 Structures of power systems 240
11.3.1 General 240
11.3.2 Radial system 241
11.3.3 Ring-main system 241
11.3.4 Meshed systems 241

12 Special problems related to short-circuit currents **245**
12.1 Interference of pipelines 245
12.1.1 Introduction 245
12.1.2 Calculation of impedances for inductive interference 247
12.1.3 Calculation of induced voltage 252
12.1.4 Characteristic impedance of the pipeline 253
12.1.5 Voltage pipeline-to-earth 254
12.2 Considerations on earthing 257
12.2.1 General 257
12.2.2 Resistance of human body 257
12.2.3 Soil conditions 258
12.2.4 Relevant currents through earth 259
12.2.5 Earthing impedance 261
12.3 Examples 262
12.3.1 Interference of pipeline from 400-kV-line 262
12.3.2 Calculation of earthing resistances 264

13 Data of equipment **267**
13.1 Three-phase a.c. equipment 267
13.1.1 System feeders 267
13.1.2 Transformers 267
13.1.3 Generators 270
13.1.4 Overhead lines 271
13.1.5 Cables 276
13.1.6 Reactors and resistors 278
13.1.7 Asynchronous motors 281
13.2 d.c. equipment 281
13.2.1 Conductors 281
13.2.2 Capacitors 283
13.2.3 Batteries 283

Symbols, superscripts and subscripts **287**

References **293**

Index **299**

List of figures

Figure 1.1	Importance of short-circuit currents and definition of tasks as per IEC 60781, IEC 60865, IEC 60909 and IEC 61660	2
Figure 2.1	Vector diagram and time course of a.c. voltage	12
Figure 2.2	Definition of vectors for current, voltage and power in three-phase a.c. systems. (a) Power system diagram and (b) electrical diagram for symmetrical conditions (positive-sequence component)	14
Figure 2.3	Vector diagram of current, voltage and power of a three-phase a.c. system represented by the positive-sequence component. (a) Consumer vector system and (b) generator vector system	15
Figure 2.4	Differentially small section of homogeneous three-phase a.c. line	16
Figure 2.5	Vector diagram of voltages in RYB-system and in the zero-sequence component, positive- and negative-sequence components are NIL	18
Figure 2.6	Vector diagram of voltages in RYB-system and positive-sequence component, zero- and negative-sequence components are NIL	19
Figure 2.7	Vector diagram of voltages in RYB-system and negative-sequence component, zero- and positive-sequence components are NIL	19
Figure 2.8	Measurement of impedance in the system of symmetrical components. (a) Positive-sequence component (identical with negative-sequence component) and (b) zero-sequence component	21
Figure 2.9	Measuring of zero-sequence impedance of a two-winding transformer (YNd). Diagram indicates winding arrangement of the transformer: (a) measuring at star-connected winding and (b) measuring at delta-connected winding	22
Figure 2.10	Measurement of positive-sequence impedance of a three-winding transformer (YNyn + d). Diagram indicates winding arrangement of the transformer	22

Figure 2.11 Measurement of zero-sequence impedance of a three-winding transformer (YNyn + d). Diagram indicates winding arrangement of the transformer 23
Figure 2.12 General scheme for the calculation of short-circuit currents in three-phase a.c. systems using the system of symmetrical components 25
Figure 2.13 Equivalent circuit diagram of a single-phase short-circuit in RYB-system 25
Figure 2.14 Equivalent circuit diagram in the system of symmetrical components for a single-phase short-circuit 26
Figure 2.15 Equations for impedance analysis in power systems 30
Figure 2.16 Equivalent circuit diagram of a power system with different voltage levels 34
Figure 2.17 Graphical construction of voltages in the system of symmetrical components: (a) vector diagram RYB, (b) vector diagram of voltage in the zero-sequence component, (c) vector diagram of voltage in the positive-sequence component and (d) vector diagram of voltage in the negative-sequence component 38
Figure 2.18 Simplified equivalent circuit diagram in RYB-components 41
Figure 2.19 Equivalent circuit diagram in the system of symmetrical components 42
Figure 2.20 Equivalent circuit diagram of an overhead line of infinitesimal length with earth return in RYB-system 43
Figure 2.21 Equivalent circuit diagram of an overhead line of infinitesimal length with earth return in 012-system. (a) Positive-sequence component, (b) negative-sequence component and (c) zero-sequence component 43
Figure 4.1 Types of short-circuits and short-circuit currents. (a) Three-phase short-circuit, (b) double-phase short-circuit without earth/ground connection, (c) double-phase short-circuit with earth/ground connection and (d) line-to-earth (line-to-ground) short-circuit 68
Figure 4.2 Time-course of short-circuit currents. (a) Near-to-generator short-circuit (according to Figure 12 of IEC 60909:1988), (b) far-from-generator short-circuit (according to Figure 1 of IEC 60909:1988). I''_k – initial (symmetrical) short-circuit current, i_p – peak short-circuit current, I_k – steady-state short-circuit current and A – initial value of the aperiodic component i_{dc} 69
Figure 4.3 Example for short-circuit current calculation with an equivalent voltage source at s.-c. location. (a) Three-phase a.c. system with three-phase short-circuit, (b) equivalent circuit diagram in 012-system (positive-sequence system), (c) equivalent circuit diagram in 012-system with equivalent voltage source 70

Figure 4.4 Estimate of maximal initial short-circuit current for different types of short-circuit and different impedance ratios Z_1/Z_0 and Z_2/Z_1. Phase angle of $\underline{Z}_0$, $\underline{Z}_1$ and $\underline{Z}_2$ assumed to be identical. Parameter r: ratio of asymmetrical short-circuit current to three-phase short-circuit current 74
Figure 4.5 Equivalent circuit diagram for the calculation of short-circuit currents inside power plant 75
Figure 4.6 Equivalent circuit diagram for single-fed three-phase short-circuit 77
Figure 4.7 Factor κ for the calculation of peak short-circuit current 78
Figure 4.8 Equivalent circuit diagram for three-phase short-circuit fed from non-meshed sources 78
Figure 4.9 Equivalent circuit diagram of a three-phase short-circuit in a meshed system 79
Figure 4.10 Factor μ for calculation of symmetrical short-circuit breaking current 81
Figure 4.11 Factors λ_{max} and λ_{min} for turbine generators (Figure 17 of DIN EN 60909.0 (VDE 0102)). (a) Series one and (b) series two 83
Figure 4.12 Factors λ_{max} and λ_{min} for salient-pole generators (Figure 18 of DIN EN 60909.0 (VDE 0102) 1988). (a) Series one and (b) series two 83
Figure 4.13 Factor q for the calculation of symmetrical short-circuit breaking current 86
Figure 4.14 Equivalent circuit diagram of a 220-kV-system with short-circuit location 87
Figure 4.15 Equivalent circuit diagram of a 110-kV-system with 220-kV-feeder 88
Figure 4.16 Equivalent circuit diagram of a 10-kV system, $f = 50$ Hz 89
Figure 4.17 A 110-kV system with short-circuit location 90
Figure 4.18 System with different voltage levels with short-circuit location 91
Figure 4.19 High-voltage system configuration for the auxiliary supply of a power station 95
Figure 5.1 Equivalent circuit diagram of a single-phase short-circuit (system with low-impedance earthing). (a) Diagram in RYB-system, (b) equivalent circuit diagram in the system of symmetrical components 100
Figure 5.2 Ratio of single-phase to three-phase short-circuit current depending on Z_1/Z_0 and $(\gamma_1 - \gamma_0)$ 101
Figure 5.3 Earth-fault factors in relation to $\underline{Z}_1/\underline{Z}_0$ and $(\gamma_1 - \gamma_0)$. (a) Earth-fault factor δ_Y and (b) earth-fault factor δ_B 103
Figure 5.4 Earth-fault factor δ depending on X_0/X_1 for different ratios R_0/X_0 and $R_1/X_1 = 0.01$ 104
Figure 5.5 Earth-fault factor δ and ratio I''_{k1}/I''_{k3} depending on X_0/X_1 104

Figure 5.6	Power system with isolated neutral with single-phase earth-fault. (a) Equivalent circuit diagram in RYB-system and (b) equivalent circuit diagram in the system of symmetrical components	105
Figure 5.7	Limit for self-extinguishing of capacitive currents in air according to VDE 0228 part 2	106
Figure 5.8	Vector diagram of voltages, power system with isolated neutral. (a) Prior to fault and (b) during earth-fault	107
Figure 5.9	Time course of phase-to-earth voltages, displacement voltage and earth-fault current. System with isolated neutral, earth-fault in phase R	108
Figure 5.10	System with resonance earthing, earth-fault in phase R. (a) Equivalent diagram in RYB-system and (b) equivalent diagram in the system of symmetrical components	109
Figure 5.11	Current limits according to VDE 0228 part 2:12.87 of ohmic currents I_{Res} and capacitive currents I_{CE}	111
Figure 5.12	Equivalent circuit diagram of a power system with asymmetrical phase-to-earth capacitances. (a) Equivalent circuit diagram in the RYB-system and (b) equivalent circuit diagram in the system of symmetrical components	112
Figure 5.13	Polar plot of the displacement voltage in a power system with resonance earthing	114
Figure 5.14	Voltages and residual current in the case of an earth-fault; displacement voltage without earth-fault	115
Figure 5.15	Current–voltage characteristic of a Petersen-coil; $U_r = 20\ \text{kV}/\sqrt{3}$; $I_r = 640$ A. (a) Minimal adjustment (50 A) and (b) maximal adjustment (640 A)	117
Figure 5.16	Displacement voltage in non-faulted operation and residual current under earth-fault conditions; non-linear characteristic of the Petersen-coil	118
Figure 5.17	Transformation of voltage in the zero-sequence component of transformers in the case of single-phase faults. (a) Equivalent circuit diagram in RYB-system and (b) equivalent circuit diagram in the system of symmetrical components	119
Figure 5.18	Alternate earthing of transformer neutrals by Petersen-coils. (a) Two parallel transformers and (b) earthing at artificial neutral with reactor X_{D2}	121
Figure 5.19	Fault current in the MV-system in the case of a short-circuit in the HV-system	121
Figure 5.20	Resonance curve (displacement voltage) for different detuning factors in a 20-kV-system for different conditions	122
Figure 5.21	Voltages in a 20-kV-system with resonance earthing for different tuning factors. (a) Phase-to-earth voltages and (b) displacement voltage (resonance curve)	123

Figure 5.22	Equivalent circuit diagram of a 20-kV-system with resonance earthing	125
Figure 5.23	Connection of a power station to a 220-kV-system with short-circuit location	126
Figure 5.24	Equivalent diagram in the zero-sequence component for fault location F	128
Figure 6.1	Equivalent circuit diagram of a LV-installation	135
Figure 7.1	Equivalent circuit diagram with short-circuit inside switchyard B	144
Figure 7.2	Equivalent circuit diagram with short-circuit at overhead-line tower	146
Figure 7.3	Equivalent circuit diagram of a 20-kV-system	147
Figure 7.4	Equivalent circuit diagram of a 110-kV-system with short-circuit location	148
Figure 8.1	Equivalent circuit diagram of a power system with different voltage levels	152
Figure 8.2	Equivalent circuit diagram for the calculation of impedance correction factor using %/MVA- or p.u.-system	153
Figure 8.3	Generator directly connected to the power system. (a) Equivalent system diagram and (b) equivalent circuit diagram in the positive-sequence component	154
Figure 8.4	Determination of the short-circuit current by superposition	155
Figure 8.5	Equivalent circuit diagram of a power system with three-phase short-circuit. (a) Circuit diagram, (b) simplified diagram of a single-fed three-phase short-circuit and (c) time course of voltage with voltage angle φ_U	157
Figure 8.6	Characteristic saturation curve method for determination of Potier's reactance	161
Figure 8.7	Calculated and measured values of factor q for the calculation of short-circuit breaking current of asynchronous motors; values of q as per Figure 4.13 (According to Figure 20 of IEC 60909-1:1991.)	163
Figure 9.1	Equivalent circuit diagrams of equipment in d.c. auxiliary installations; typical time course of short-circuit current (according to Figure 1 of DIN EN 61660-1 (VDE 0102 Teil 10)). (a) Capacitor, (b) battery, (c) rectifier in three-phase a.c. bridge connection and (d) d.c. motor with independent excitation	166
Figure 9.2	Standard approximation function of the short-circuit current (according to Figure 2 of IEC 61660-1:1997)	167
Figure 9.3	Factor κ_C for the calculation of peak short-circuit current of capacitors (according to Figure 12 of IEC 61660-1:1997)	169
Figure 9.4	Time-to-peak t_{pC} for the calculation of short-circuit currents of capacitors (according to Figure 13 of IEC 61660-1:1997)	170

Figure 9.5	Factor k_{1C} for the calculation of rise-time constant (according to Figure 14 of IEC 61660-1:1997)	171
Figure 9.6	Factor k_{2C} for the calculation of decay-time constant (according to Figure 15 of IEC 61660-1:1997)	171
Figure 9.7	Rise-time constant τ_{1B} and time to peak t_{pB} of short-circuit currents of batteries (according to Figure 10 of IEC 61660-1:1997)	172
Figure 9.8	Factor λ_D for the calculation of quasi steady-state short-circuit current of rectifiers (according to Figure 7 of IEC 61660-1:1997)	173
Figure 9.9	Factor κ_D for the calculation of peak short-circuit currents of rectifiers. Factor: $R^* = (R_N/X_N)(1 + 2R_{DBr}/3R_N)$ (according to Figure 8 of IEC 61660-1:1997)	173
Figure 9.10	Factor κ_M for the calculation of peak short-circuit current of d.c. motors with independent excitation (according to Figure 17 of IEC 61660-1:1997)	176
Figure 9.11	Time to peak of short-circuit currents for d.c. motors with independent excitation and $\tau_{Mec} < 10 * \tau_F$ (according to Figure 19 of IEC 61660-1:1997)	176
Figure 9.12	Factor k_{1M} in the case of d.c. motors with independent excitation and $\tau_{Mec} \geq 10 * \tau_F$ (according to Figure 18 of IEC 61660-1:1997)	177
Figure 9.13	Factor k_{2M} in the case of d.c. motors with independent excitation and $\tau_{Mec} < 10 * \tau_F$ (according to Figure 19 of IEC 61660-1:1997)	177
Figure 9.14	Factor k_{3M} in the case of d.c. motors with independent excitation and $\tau_{Mec} < 10 * \tau_F$ (according to Figure 20 of IEC 61660-1:1997)	178
Figure 9.15	Factor k_{4M} in the case of d.c. motors with independent excitation and $\tau_{Mec} < 10 * \tau_F$ (according to Figure 21 of IEC 61660-1:1997)	178
Figure 9.16	Equivalent circuit diagram of a d.c. auxiliary installation	180
Figure 9.17	Typical time curves of total short-circuit current in d.c. auxiliary installations, e.g., (a) with dominating part of motors, (b) with dominating part of rectifiers, (c) with dominating part of batteries and (d) in the case of low rectifier load (according to Figure 22 of DIN EN 61660-1 (VDE 0102 Teil 10))	181
Figure 9.18	Equivalent circuit diagram of the d.c. auxiliary installation (220 V), e.g., of a power station	182
Figure 9.19	Partial short-circuit currents and total short-circuit current, d.c. auxiliary system as per Figure 9.18	193
Figure 9.20	Total short-circuit current, obtained by superposition of the partial short-circuit currents and approximated short-circuit current, d.c. auxiliary system as per Figure 9.18	194

Figure 10.1 Factor n for the calculation of thermal short-time current (heat dissipation of a.c. component) (according to Figure 22 of IEC 60909-0:2001) 197
Figure 10.2 Factor m for the calculation of thermal short-time current (heat dissipation of d.c. component) (according to Figure 21 of IEC 60909-0:2001) 198
Figure 10.3 Rated short-time current density of conductors. δ_0 is the temperature at beginning of short-circuit and δ_1 is the temperature at end of short-circuit. (a) ——: Copper; - - - -: unalloyed steel and steel cables and (b) Al, aluminium alloy, ACSR 199
Figure 10.4 Maximal permissible thermal short-circuit current for impregnated paper-insulated cables U_n up to 10 kV 200
Figure 10.5 Arrangement of parallel conductors 202
Figure 10.6 Correction factor k_{12} for the calculation of effective distance (according to Figure 1 of IEC 61660-2:1997) 203
Figure 10.7 Factors V_σ and $V_{\sigma s}$ for the calculation of bending stress (according to Figure 4 of IEC 60865-1:1993) 205
Figure 10.8 Factors V_r and V_{rs} for the calculation of bending stress (according to Figure 5 of IEC 60865-1:1993) 205
Figure 10.9 Factor V_F for the calculation of bending stress (according to Figure 4 of IEC 60865-1:1993) 207
Figure 10.10 Calculation of mechanical natural frequency (Factor c). Arrangement of distance elements and calculation equation (according to Figure 3 of IEC 60865-1:1993) 208
Figure 10.11 Standard approximation function (a) and substitute rectangular function (b) (according to Figure 4 of IEC 60660-2:1997). Not to scale 210
Figure 10.12 Factors V_σ and $V_{\sigma s}$ for the calculation of bending stress on conductors (according to Figure 9 of IEC 61660-2:1997) 214
Figure 10.13 Factor V_F for the calculation of forces on supports (according to Figure 9 of IEC 61660-2:1997) 215
Figure 10.14 Equivalent circuit diagram, data of equipment, resistance at 20°C 216
Figure 10.15 Equivalent circuit diagram of a power system with wind power plant 217
Figure 10.16 Arrangement of busbar conductor (data, see text) 219
Figure 10.17 Standardised rectangular function and approximated total short-circuit current 223
Figure 11.1 Selection of suitable voltage level for the connection of power stations 226
Figure 11.2 Schematic diagram of a 400/132-kV-system for urban load; values of short-circuit currents in case of operation as two subsystems 228
Figure 11.3 Schematic diagram of a 132-kV-system with power station 229

Figure 11.4 Equivalent circuit diagram of a 30-kV-system with feeding 132-kV-system: (a) Operation with transformers in parallel and (b) limitation of short-circuit current. Result of three-phase short-circuit current: $S''_{kQ} = 3.2$ GVA; $S_{rT} = 40$ MVA; $u_{krT} = 12\%$; $t_{rT} = 110/32$; OHTL 95Al; $l_{tot} = 56$ km 230
Figure 11.5 Equivalent circuit diagram of a 380-kV-system and results of three-phase short-circuit current calculation: (a) Radial fed system and (b) ring fed system. $S''_{kQ} = 8$ GVA; OHTL ACSR/AW 4 × 282/46; $l_i = 120$ km 231
Figure 11.6 Schematic diagram of a 110-kV-substation fed from the 220-kV-system: (a) Operation with buscoupler closed and (b) operation with buscoupler open. Result of three-phase short-circuit current calculation 232
Figure 11.7 Equivalent circuit diagram of a 6-kV-industrial system. Results of three-phase short-circuit current calculation: (a) Busbar sectionaliser closed and (b) Busbar sectionaliser open 233
Figure 11.8 Equivalent circuit diagram of switchgear with single busbar 234
Figure 11.9 Time course of short-circuit current in installations with and without Ip-limiter 235
Figure 11.10 Cutaway view of an Ip-limiter support: (1) insulating tube, (2) explosive loading, (3) main conductor, (4) fuse element and (5) transducer 236
Figure 11.11 Equivalent circuit diagram of a 10-kV-system with incoming feeder. Results of three-phase short-circuit current calculation: (a) impedance voltage 13% and (b) impedance voltage 17.5% 237
Figure 11.12 Equivalent circuit diagram of a 10-kV-system with short-circuit limiting reactors. Results of three-phase short-circuit current calculation 238
Figure 11.13 Equivalent circuit diagram of 11.5-kV-system fed from the 132-kV-system 239
Figure 11.14 Equivalent circuit diagram of a power station with 132-kV-busbar. Results of three-phase short-circuit current calculation: $S_{rG} = 150$ MVA; $x''_d = 12$–17.8% 240
Figure 11.15 General structure of a radial system with one incoming feeder 241
Figure 11.16 General structures of ring-main systems: (a) Simple structure with one feeding busbar and (b) structure with two feeding busbars (feeding from opposite sides) 242
Figure 11.17 Principal structure of a high voltage system with different voltage levels 243
Figure 11.18 Principal structure of meshed low voltage system: (a) Single-fed meshed system and (b) meshed system with overlapping feeding 244

Figure 12.1	Outline and distances of a high-voltage transmission-line tower. B: counterpoise; P: pipeline; L: conductor nearest to pipeline. E: first earth conductor (nearest to pipeline), also named E1; E2: second earth conductor	250
Figure 12.2	Oblique exposure and crossing of pipeline and overhead line. (a) Plot plan and (b) elevation plan (detail from crossing location)	256
Figure 12.3	Impedance of the human body (hand-to-hand) depending on the touch voltage	258
Figure 12.4	Permissible touch voltage depending on the time of exposure	259
Figure 12.5	Plot plan of the exposure length pipeline and transmission line	263
Figure 12.6	Elevation plan of the overhead transmission tower and the pipeline	264
Figure 12.7	Specific electric field strength (a) and specific induced voltage (b) of the pipeline between towers 2 and 11	265
Figure 12.8	Voltage pipeline-to-earth along the exposure length (0–6400 m)	266
Figure 13.1	Principal structure of a power supply system and typical values of initial short-circuit power of public supply system	268
Figure 13.2	Typical values for the impedance voltage of two-winding transformers	269
Figure 13.3	Typical values for the ohmic losses, no-load losses and no-load current of two-winding transformers	269
Figure 13.4	Tower outline of high-voltage transmission lines. (a) Single-circuit line and (b) double-circuit line	271
Figure 13.5	Capacitances MV-cables ($U_n < 20\,kV$)	276
Figure 13.6	Capacitances C_1' (a) and capacitive loading current I_c' (b) of HV-cables	277
Figure 13.7	Reactance (positive-sequence system) of three-phase cables ($U_n \leq 110\,kV$)	277
Figure 13.8	Arrangement of a short-circuit limiting reactor	278
Figure 13.9	Adjustable Petersen-coil 21 kV/$\sqrt{3}$; 4 MVAr; $I_r = 70.1$–330 A; adjustable in 64 steps, 4.13 A each	279
Figure 13.10	Earthing resistor made from CrNi-alloy steel fabric for indoor installation 3810 Ω, 5 A for 10 s, 170 kV BIL, IP 00	280
Figure 13.11	Earthing resistor made from meandering wire for outdoor installation 16 Ω, 400 A for 10 s, 75 kV BIL, IP 20	281

List of tables

Table 1.1 International documents and norms for short-circuit current calculation 5
Table 1.2 Selection of norms as referred in standards for short-circuit current calculation and as mentioned in this book 6
Table 1.3 Cross-reference list of standards on short-circuit current calculation 9
Table 2.1 Equivalent circuit diagrams and equations to represent short-circuits (single fault location) in three-phase a.c. systems in the RYB-system and in the 012-system 28
Table 2.2 Definitions of quantities in physical, relative and semirelative units 33
Table 2.3 Conversion of quantities between %/MVA-system, Ω-system and p.u.-system (100 MVA-base) 35
Table 3.1 Impedance of system feeder, equivalent circuit diagram, calculation equations and remarks 46
Table 3.2 Impedance of two-winding transformer, equivalent circuit diagram, calculation equations and remarks 47
Table 3.3 Impedance of three-winding transformer, equivalent circuit diagram, calculation equations and remarks 48
Table 3.4 Equivalent circuit diagram of two- and three-winding transformers in the positive- and zero-sequence component 50
Table 3.5 Impedance of synchronous generator, equivalent circuit diagram, calculation equations and remarks 51
Table 3.6 Impedance of power-station unit, equivalent circuit diagram, calculation equations and remarks 52
Table 3.7 Impedance of overhead line (single-circuit), equivalent circuit diagram, calculation equations and remarks 53
Table 3.8 Impedance of short-circuit limiting reactor, equivalent circuit diagram, calculation equations and remarks 54
Table 3.9 Impedance of asynchronous motor, equivalent circuit diagram, calculation equations and remarks 55

Table 3.10	Impedance of static converter fed drive, equivalent circuit diagram, calculation equations and remarks	56
Table 3.11	Impedance of system load, equivalent circuit diagram, calculation equations and remarks	57
Table 3.12	Impedance of a conductor, equivalent circuit diagram, calculation equations and remarks	58
Table 3.13	Impedance of capacitor, equivalent circuit diagram, calculation equations and remarks	59
Table 3.14	Impedance of battery, equivalent circuit diagram, calculation equations and remarks	60
Table 3.15	Impedance of rectifier, equivalent circuit diagram, calculation equations and remarks	61
Table 3.16	Impedance of d.c. motor with independent excitation, equivalent circuit diagram, calculation equations and remarks	62
Table 3.17	Results of calculation of impedance in three-phase a.c. equipment	63
Table 3.18	Results of calculation of impedance of equipment in d.c. installations (without common branch as per IEC 61660-1)	65
Table 4.1	Voltage factor *c* according to IEC 60909-0	71
Table 4.2	Equations for the calculation of initial symmetrical short-circuit currents	73
Table 4.3	Equivalent frequency for the calculation of decaying component	81
Table 4.4	Calculation of short-circuit currents of asynchronous motors	84
Table 5.1	Characteristics of different types of neutral handling in power systems	99
Table 5.2	Voltages in the zero-sequence component U_0 transferred through 110/10-kV-transformer in the case of single-phase fault in the 110-kV-system according to Figure 5.17; U_n: nominal system voltage	120
Table 5.3	Capacitive asymmetry ΔC_E for different parameters in a 20-kV-system	122
Table 5.4	Characteristics of a 20-kV-system with respect to resonance earthing	126
Table 6.1	Voltage factor *c* according to IEC 60909-0. (Voltage factors as per IEC 60781 are of different values. The standard is under review)	132
Table 6.2	Example for the calculation of maximal short-circuit currents in LV-system	136
Table 6.3	Example for the calculation of minimal short-circuit currents in LV-system	137
Table 7.1	System configurations and equations for the calculation of short-circuit currents in the case of double earth-faults	141

Table 8.1	Impedances of equipment and short-circuit current as per Figure 8.1	153
Table 8.2	Impedances of equipment and short-circuit current using correction factor	154
Table 8.3	Assumed values of $u_{\mathrm{f\,max}}$ for the calculation of the factor λ	162
Table 9.1	Resistances R_{ij} and equivalent resistances $R_{\mathrm{res}j}$ for the calculation of correction factors; U: Voltage at short-circuit location prior to the short-circuit	179
Table 10.1	Maximal permissible conductor temperature and rated short-time current density; 1) – Normal operating condition; 2) – Short-circuit condition	196
Table 10.2	Data of materials for screening, armouring and sheaths of cables	201
Table 10.3	Factors α, β and γ for different arrangement of supports (according to Table 3 of IEC 60865-1:1993)	204
Table 10.4	Factors for the calculation of mechanical natural frequency. Swing is at right angle to the area of sub-conductor	209
Table 10.5	Results of calculation of thermal equivalent currents	217
Table 10.6	Results of short-circuit current calculation	218
Table 11.1	Selection of recommended voltage as per IEC 60038:1987	227
Table 11.2	Result of loadflow and short-circuit analysis as per Figure 11.11	237
Table 12.1	Interference between power system, communication circuits and pipelines	246
Table 12.2	Resistivity of soil ρ for different types of soil conditions	248
Table 12.3	Resistance of pipeline coatings	254
Table 12.4	Currents through earth for the design of earthing installations	260
Table 12.5	Reduction factor for typical power system installations; distance of earth conductor to phase conductor $D \approx 20$ m; $\rho = 100\ \Omega\mathrm{m}$	261
Table 12.6	Resistance of earthing installations R_{EI} for different types and arrangement	262
Table 13.1	Data of transformers	270
Table 13.2	Typical data of synchronous generators	271
Table 13.3	Typical values of impedance of the positive-sequence component of MV-overhead lines	274
Table 13.4	Typical values of impedances of the positive- and zero-sequence component of HV-overhead lines ($\rho_{\mathrm{E}} = 100\ \Omega\mathrm{m}$)	274
Table 13.5	Characteristic parameters of resistor elements	280
Table 13.6	Data of asynchronous motors	282
Table 13.7	Typical values of MKP-capacitors; self-healing dry insulation; different make of capacitor can and fuse	283

Table 13.8 Typical values of MKP-capacitors; resin insulation; round can 284
Table 13.9 Resistance of loaded batteries (data from several manufacturers) 284

Foreword

Short-circuit currents are the dominating parameters for the design of equipment and installations, for the operation of power systems and for the analysis of outages and faults. Besides the knowledge about design of equipment in power systems, in auxiliary installations and about system operation constraints, the calculation of short-circuit currents is a central task for power system engineers. The book describes the individual equipment in power systems with respect of the parameters needed for short-circuit current calculation as well as methods for analysing the different types of short-circuits in power systems using the system of symmetrical components. Besides detailed explanation of the calculation methods for short-circuit currents and their thermal and electromagnetic effects on equipment and installations, short-time interference problems and measures for the limitation of short-circuit currents are explained. Detailed calculation procedures for the parameters and typical data of equipment are given in a separate chapter for easy reference. All aspects of the book are explained with examples based on engineering studies carried out by the author.

The preparation of the book was finalised in December 2004 and reflects the actual status of the technique, norms and standards. Carrying out short-circuit studies always requires the application of the latest editions of standards, norms and technical recommendations, which can be obtained from the IEC-secretariat or from the national standard organisation. All comments in this book are given in good faith, based on the comprehensive technical experience of the author.

The author wishes to thank very much his former colleague Dipl.-Ing. Heiner Rofalski for revising the text and improving the book. Most of the drawings were prepared by my students Stefan Drees and Elmar Vogel who spent much effort to obtain clear and understandable presentation. My thanks go also to the staff of IEE-publishers, especially to Ms. Sarah Cramer who encouraged me to write this book.

Comments are highly appreciated.

Bielefeld, December 2004
juergen.schlabbach@fh-bielefeld.de

Professor Dr.-Ing. Jürgen Schlabbach, born 1952, member IEEE and VDI, studied power system engineering at the Technical University of Darmstadt/Germany, from where he got his Ph.D. in 1982. Until 1992 he was working in a consulting engineering company, responsible for planning and design of public and industrial supply systems. Since 1992 he has been professor for 'Power system engineering and utilisation of renewable energy' at the University of Applied Sciences in Bielefeld/Germany. His main working areas are planning of power systems, analysis of faults, power quality, interference problems and connection of renewable energy sources to power systems. He is also a consulting engineer in the mentioned fields.

More information can be found on the author's web-page http://www.fh-bielefeld.de/fb2/labor-ev.

Chapter 1

Introduction

1.1 Objectives

This book deals with the calculation of short-circuit currents in two- and three-phase a.c. systems as well as in d.c. systems, installed as auxiliary installations in power plants and substations. It is not the objective of this book to repeat definitions and rules of norms and standards, but to explain the procedure for calculating short-circuit currents and their effects on installations and equipment. In some cases repetition of equations, tables and diagrams from norms and standards however are deemed necessary for easy understanding. It should be emphasised in this respect that the presentation within this book is mainly concentrated on installations and equipment in high voltage systems, i.e., voltage levels up to 500 kV. Special considerations have to be taken in the case of long transmission lines and in power systems with nominal voltages above 500 kV. The calculation of short-circuit currents and of their effects are based on the procedures and rules defined in the IEC documents 61660, 60909, 60865 and 60781 as outlined in Table 1.1.

1.2 Importance of short-circuit currents

Electrical power systems have to be planned, projected, constructed, commissioned and operated in such a way to enable a safe, reliable and economic supply of the load. The knowledge of the loading of the equipment at the time of commissioning and as foreseeable in the future is necessary for the design and determination of the rating of the individual equipment and of the power system as a whole. Faults, i.e., short-circuits in the power system cannot be avoided despite careful planning and design, good maintenance and thorough operation of the system. This implies influences from outside the system, such as short-circuits following lightning strokes into phase-conductors of overhead lines and damages of cables due to earth construction works as well as internal faults, e.g., due to ageing of insulation materials. Short-circuit currents therefore have an important influence on the design and operation of equipment and power systems.

Switchgear and fuses have to switch-off short-circuit currents in short time and in a safe way; switches and breakers have to be designed to allow even switch-on to an existing short-circuit followed by the normal switch-off operation. Short-circuit currents flowing through earth can induce impermissible voltages in neighbouring metallic pipelines, communication and power circuits. Unsymmetrical short-circuits cause displacement of the voltage neutral-to-earth and are one of the dominating criteria for the design of neutral handling. Short-circuits stimulate mechanical oscillations of generator units which will lead to oscillations of active and reactive power as well, thus causing problems of stability of the power transfer which can finally result in system black-out. Furthermore, equipment and installations must withstand the expected thermal and electromagnetic (mechanical) effects of short-circuit currents.

In Figure 1.1 the typical time course of a short-circuit current is shown, which can be measured at high-voltage installations in the vicinity of power stations with synchronous generators, characterised by decaying a.c. and d.c. components of the

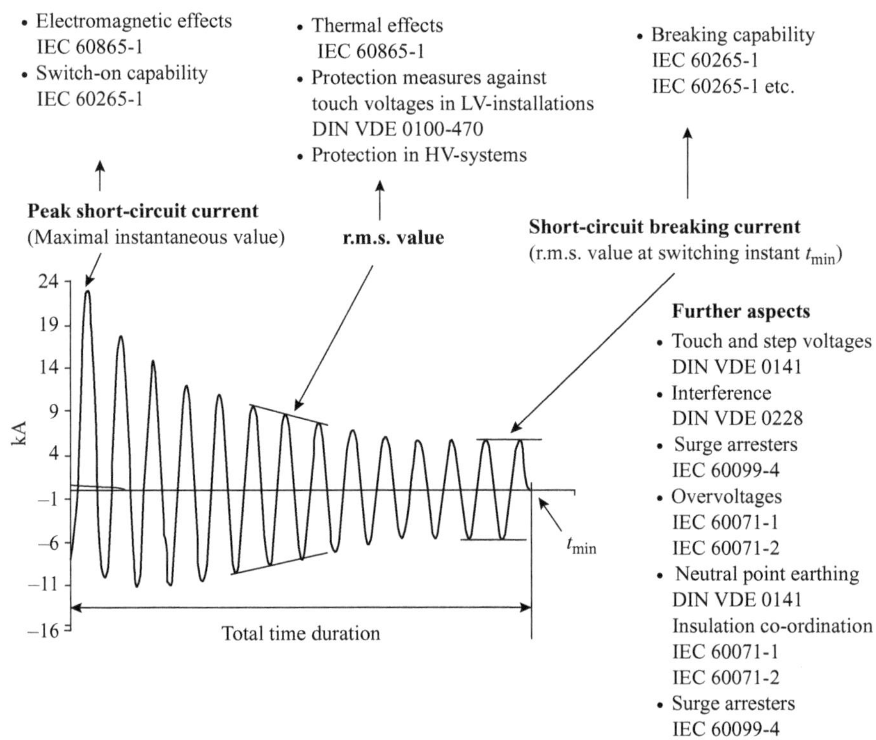

Figure 1.1 Importance of short-circuit currents and definition of tasks as per IEC 60781, IEC 60865, IEC 60909 and IEC 61660

current. It is assumed that the short-circuit is switched-off approximately 14 periods after its initiation, which seems a rather long time, but was chosen for reason of a better visibility in the figure. Attention shall be put on four parameters of the short-circuit current.

- The **total time duration** of the short-circuit current consists of the operating time of the protection devices and the total breaking time of the switchgear.
- The **peak short-circuit current**, which is the maximal instantaneous value of the short-circuit current, occurs approximately a quarter period after the initiation of the short-circuit. As electromagnetic forces are proportional to the instantaneous value of the current, the peak short-circuit current is necessary to know in order to calculate the forces on conductors and construction parts affected by the short-circuit current.
- The **r.m.s.-value** of the short-circuit current is decaying in this example due to the decaying a.c. component. Currents through conductors will heat the conductor due to ohmic losses. The r.m.s. value of the short-circuit current, combined with the total time duration, is a measure for the thermal effects of the short-circuit.
- The **short-circuit breaking current** is the r.m.s.-value of the short-circuit current at switching instant, i.e., at time of operating the circuit-breaker. While opening the contacts of the circuit-breaker, the arc inside the breaker will heat up the installation, which depends obviously on the breaking time as well.

1.3 Maximal and minimal short-circuit currents

Depending on the task of engineering studies, the maximal or minimal short-circuit current has to be calculated. The maximal short-circuit current is the main design criteria for the rating of equipment to withstand the effects of short-circuit currents, i.e., thermal and electromagnetic effects. The minimal short-circuit current is needed for the design of protection systems and the minimal setting of protection relays. The short-circuit current itself depends on various parameters, such as voltage level, actual operating voltage, impedance of the system between any generation unit and the short-circuit location, impedance at the short-circuit location itself, the number of generation units in the system, the temperature of the equipment influencing the resistances and other parameters. The determination of the maximal and the minimal short-circuit current therefore is not as simple as might be seen at this stage [36]. It requires detailed knowledge of the system operation, i.e., which cables, overhead-lines, transformers, generators, machines and reactors are in operation and which are switched-off. The assessment of the results of any calculation of short-circuit currents must take into account these restrictions in order to ensure that the results are on the safe side, i.e., that the safety margin of the calculated maximal short-circuit current is large enough without resulting in an uneconomic high rating of the equipment. The same applies to the minimal short-circuit current for which the safety margin must be assessed in such a way as to distinguish between the highest operating current and any short-circuit current, which has to be switched-off.

1.4 Norms and standards

Technical standards are harmonised on international basis. The international organisation to coordinate the works and strategies is the ISO (International Standards Organisation), whereas IEC (International Electrotechnical Commission) is responsible for the electrotechnical standardisation. The national standard organisations such as CENELEC in Europe, BSI in the United Kingdom, DKE in Germany, ANSI in the United States, JSI in Japan as well as national electrotechnical organisations such as IEE, VDE, IEEE, JES etc. are working in the working groups of IEC to include their sight and knowledge on technical items in the international standards and documents. On national basis, standards are adopted to the widest extent to the internationally agreed standards and documents. In some cases additions to the international standards are included in the national standards, however their status is 'for information only'.

The application of norms and standards has to be based on the latest issues, which can be obtained in Germany from Beuth-Verlag GmbH, Burggrafenstr. 6, D-10787 Berlin or from VDE-Verlag GmbH, Bismarckstr. 33, D-10625 Berlin. English versions are available from British Standards Institution, London/UK, in the United States from American National Standards Institute or any national standard organisation. Standards can also be searched and ordered through the web on the following URLs (appearance in alphabetical order):

American National Standards Institute	http://www.nssn.org/help.html
British Standards Institute	http://www.bsonline.techindex.co.uk
Deutsches Institut für Normung	http://din.de
International Electrotechnical Commission	http://www.iec.ch
VDE-Verlag	http://vde-verlag.de

The structure of standards and norms dealing with short-circuit current calculation as published in IEC or EN-norms are outlined in Table 1.1. The listing should not be understood as a complete catalogue of standards but represents an overview only. Some of the mentioned standards are actually in draft status; others include corrections, additions and appendices. For details reference should be made to the IEC-homepage or the homepage of the national standards committee. The official actual standards catalogue is the only relevant document for any technical application. IEC-documents and national standards refer to other norms and standards. A short overview of these references is outlined in Table 1.2.

With respect to the calculation of short-circuit currents and their effects, the standards are harmonised in most of the countries. The procedures and methods described are identical to those defined in the mentioned IEC-documents. Table 1.3 shows a cross-reference list between IEC, EN and BS.

The classification numbers of the different standards differ in some cases from those of the IEC-documents or the EN-norms, however in most of the cases, classification numbers are similar, e.g.: Australian standard AS 3865, Swedish standard SS-EN 60865-1 and British standard BS EN 60865-1 are identical to IEC 60865-1

(Short-circuit currents – calculation of effects; Part 1: Definitions and calculation methods). The standards' catalogue of American National Standards Institute, to be accessed through the home-page of National Standards Systems Network, directly indicates the IEC- resp. EN-documents under the heading 'short-circuit currents'. It is therefore fully sufficient to apply the IEC-documents for calculation of short-circuit currents and the analysis of their effects.

Table 1.1 International documents and norms (with related VDE-classification) for short-circuit current calculation

IEC (year)	EN (year)	DIN; VDE (year)	Title, contents
61660-1 (1997)	61660-1 (1997)	VDE 0102 Part 10 (1998–06)	Short-circuit currents in d.c. auxiliary installations in power plants and substations Part 1: Calculation of short-circuit currents
61660-2 (1997)	61660-2 (1997)	VDE 0103 Part 10 (1998–05)	Short-circuit currents in d.c. auxiliary installations in power plants and substations Part 2: Calculation of effects
61660-3 (2000)	61660-3	Appendix 1 VDE 0102 Part 10 (2002–11)	Short-circuit currents in d.c. auxiliary installations in power plants and substations Examples of calculation of short-circuit current and effects
60781 (1989)	HD 581 S1 (1991)	Appendix 2 VDE 0102 (1992–09)	Application guide for calculation of short-circuit currents in low-voltage radial systems
60865-1 (1993)	60865-1 (1993)	VDE 0103 (1994–11)	Short-circuit currents – Calculation of effects Part 1: Definitions and calculation methods
—	—	Appendix 1 VDE 103 (1996–06)	Short-circuit currents – Calculation of effects Part 1: Definitions and calculation methods Examples for calculation
60909-0 (2001)	60909-0 (2001)	VDE 0102 (2002–07)	Short-circuit current calculation in a.c. systems

Table 1.1 Continued

IEC (year)	EN (year)	DIN; VDE (year)	Title, contents
60909-1 (1991)	—	Appendix 3 VDE 0102 (1997–05)	Short-circuit current calculation in three-phase a.c. systems Part 1: Factors for the calculation of short-circuit currents in three-phase a.c. systems according to IEC 909
60909-2 (1992)	—	Appendix 4 VDE 0102 (2003–02)	Electrical equipment Data for short-circuit current calculations in accordance with IEC 909 (1998)
60909-3 (1995)	—	VDE 0102 Part 3 (1997–06)	Short-circuit current calculation in three-phase a.c. systems Part 3: Currents during two separate simultaneous single-phase line-to-earth short-circuits and partial short-circuit currents flowing through earth
60909-4 (2000)	—	Appendix 1 VDE 0102 (1992–09)	Examples for the calculation of short-circuit currents

Table 1.2 Selection of norms as referred in standards for short-circuit current calculation and as mentioned in this book

IEC (year)	EN (year)	DIN; VDE (year)	Title, contents
60038 (mod) (2002)	HD472 S1 (1989)	DIN IEC 60038 (2002)	IEC standard voltages
60050(131) (1978)		DIN IEC 60050-131 (1983)	International Electrotechnical Vocabulary (IEV) Chapter 131: Electric and magnetic circuits
60050(151) (2001)		DIN 40200 (1981–12)	International Electrotechnical Vocabulary (IEV) Chapter 151: Electric and magnetic devices *Some parts of DIN 40200 are identical to IEC 60050*

Table 1.2 Continued

IEC (year)	EN (year)	DIN; VDE (year)	Title, contents
60050(195) (1998)			International Electrotechnical Vocabulary (IEV) Chapter 195: Earthing and protection against electric shock
60050(441) (1998)			International Electrotechnical Vocabulary (IEV) Chapter 441: Switchgear, controlgear and fuses
60071-1 (1993)	60071-1 (1995)	VDE 0111 Part 1 (1996–07)	Insulation coordination Definitions, principles and rules
60071-2 (1996)	60071-2 (1997)	VDE 0111 Part 2 (1997–09)	Insulation coordination Application guide
TS 60479-1 (1994)			Effects of currents on human being and livestock Part 1: General aspects
TS 60479-2 (1982)			Effects of currents passing through the human body Part 2: Special aspects
60986 (2000)			Short-circuit temperature limits of electric cables with rated voltages from 6 kV up to 30 kV
60949 (1988)			Calculation of thermally permissible short-circuit currents, taking into account non-adiabatic heating effects
60896-1 (1987)	60896-1 (1991)		Stationary lead-acid batteries – General requirements and methods of test – Part 1: Vented types
61071-1 (mod) (1991)	61071-1 (1996)	VDE 0560 Part 120 (1997–08)	Capacitors for power electronics
60265, 60282, 60298, 60420, 60517, 60644, 60694, etc.		VDE 0670	Switchgear, circuit-breakers, fuses, etc. *Various documents of IEC, parts of VDE 0670*
60099, 61643	60099-1 (1994)	VDE 0675 Part 1 (2000–08)	Surge arresters *Various documents of IEC, parts of VDE 0670*
60949 (1988)			Calculation of thermally permissible short-circuit currents, taking into account non-adiabatic heating effects

Table 1.2 Continued

IEC (year)	EN (year)	DIN; VDE (year)	Title, contents
60986 (2000)			Guide to the short-circuit temperature limits of electrical cables with a rated voltage from 1.8/3 (3.6) kV to 18/30 (36) kV
		VDE 0141 (2000–01)	Earthing of special systems for electrical energy with nominal voltages above 1 kV
		VDE 0228 Part 1 (1987–12)	Proceedings in the case of interference on telecommunication installations by electrical power installations – General
		VDE 0228 Part 2 (1987–12)	Proceedings in the case of interference on telecommunication installations by electrical power installations interference by three-phase installations
		VDE 0228 Part 3 (1988–09)	Proceedings in the case of interference on telecommunication installations by electrical power installations interference by alternating current traction systems
		VDE 0228 Part 4 (1987–12)	Proceedings in the case of interference on telecommunication installations by electrical power installations interference by direct current systems
		VDE 0226 Part 1000 (1995–06)	Current carrying capacity General, conversion factors
		DIN 13321 (1978–04)	Electric power engineering; components in three-phase networks Concepts, quantities and their letter symbols
		DIN 40110-1 (1994)	Quantities used in alternating current theory; two-line circuits
		DIN 40110-2 (2002)	Quantities used in alternating current theory; three-line circuits
	EN 50160 (1999)	DIN EN 50160 (2000–03)	Voltage characteristics of electricity supplied by public distribution systems

Table 1.3 Cross-reference list of standards on short-circuit current calculation

IEC (year)	EN (year)	BS EN (year)	Remarks (title see Table 1.1)
61660-1 (1997)	61660-1 (1997)	61660-1 (1997)	Short-circuit currents in d.c. auxiliary installations in power plants and substations Part 1: Calculation of short-circuit currents
61660-2 (1997)	61660-2 (1997)	61660-2 (1997)	Short-circuit currents in d.c. auxiliary installations in power plants and substations Part 2: Calculation of effects
61660-3 (2000)	61660-1	98/202382 DC	Short-circuit currents in d.c. auxiliary installations in power plants and substations Examples of calculation of short-circuit current and effects
60781 (1989)	HD 581 S1 (1991)	7638 (1993)	Application guide for calculation of short-circuit currents in low-voltage radial systems
60865-1 (1993)	60865-1 (1993)	60865-1 (1994)	Short-circuit currents – Calculation of effects Part 1: Definitions and calculation methods
—	—	PD 6875-2 (1995)	Short-circuit currents – Calculation of effects Part 1: Definitions and calculation methods; Examples for calculation
60909-0 (2001)	60909-0 (2001)	60909-0 (2001)	Short-circuit current calculation in a.c. systems
60909-1 (1991)	—	PD IEC TR 60909-1 (2002)	Short-circuit current calculation in three-phase a.c. systems Part 1: Factors for the calculation of short-circuit currents in three-phase a.c. systems according to IEC 909
60909-2 (1992)	—	PD 7639-2 (1994)	Electrical equipment Data for short-circuit current calculations in accordance with IEC 909 (1988)
60909-3 (1995)	prEN 60909-3	95/203556 DC	Short-circuit current calculation in three-phase a.c. systems Part 3: Currents during two separate simultaneous single-phase line-to-earth short-circuits and partial short-circuit currents flowing through earth
60909-4 (2000)	—	—	Examples for the calculation of short-circuit currents

Chapter 2
Theoretical background

2.1 General

A detailed deduction of the mathematical procedure is not given within the context of this book, but only the final equations are quoted. For further reading, reference is made to [1], [13]. In general, equipment in power systems is represented by equivalent circuits, which are designed for the individual tasks of power system analysis. For the calculation of no-load current and the no-load reactive power of a transformer, the no-load equivalent circuit is sufficient. Regarding the calculation of short-circuits, voltage drops and load characteristic a different equivalent circuit is required. The individual components of the equivalent circuits are resistance, inductive and capacitive reactance (reactor and capacitor), voltage source and ideal transformer. Voltage and currents of the individual components and of the equivalent circuit are linked by Ohm's law.

2.2 Complex calculations, vectors and phasor diagrams

When dealing with two- and three-phase a.c. systems, it should be noted that currents and voltages are generally not in phase. The phase position depends on the amount of inductance, capacitance and resistances of the impedance. The time course, e.g., of a current or voltage in accordance with

$$u(t) = \sqrt{2} * U * \sin(\omega t + \varphi_U) \tag{2.1a}$$

$$i(t) = \sqrt{2} * I * \sin(\omega t + \varphi_I) \tag{2.1b}$$

can in this case be shown as a line diagram as per Figure 2.1. In the case of sinusoidal variables, these can be shown in the complex numerical level by rotating pointers, which rotate in the mathematically positive sense (counterclockwise) with

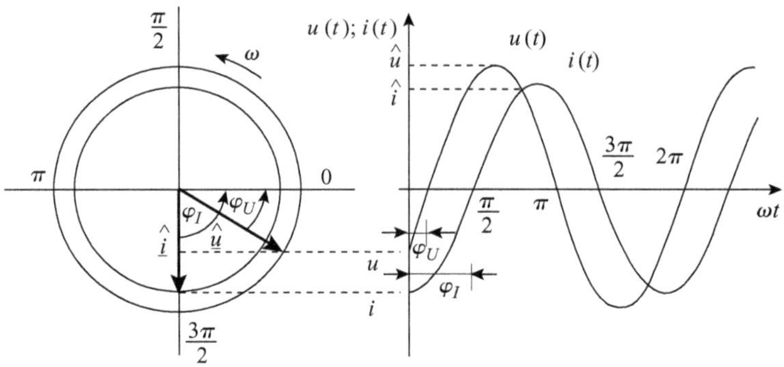

Figure 2.1 Vector diagram and time course of a.c. voltage

angular velocity ω as follows:

$$\underline{U} = \sqrt{2} * U * \mathrm{e}^{(j\omega t + \varphi_U)} \tag{2.2a}$$

$$\underline{I} = \sqrt{2} * I * \mathrm{e}^{(j\omega t + \varphi_I)} \tag{2.2b}$$

where U and I are the r.m.s-values of voltage and current, ω is the angular frequency and φ_U and φ_I are the phase angle of voltage and current. The time course in this case is obtained as a projection on the real axis, see Figure 2.1.

The generic term for an impedance $\underline{Z}$ is given as impedance or apparent resistance

$$\underline{Z} = R + jX \tag{2.3a}$$

The generic term for an admittance $\underline{Y}$ is admittance or apparent admittance

$$\underline{Y} = G + jB \tag{2.3b}$$

where R is the active resistance, X the reactance, G the active conductance and B the susceptance. The terms for the designation of resistances and admittances as per the above are stipulated in DIN 40110 (VDE 0110).

The reactance depends on the particular angular frequency ω under consideration and can be calculated for capacitances C or inductances L from

$$X_C = \frac{1}{\omega C} \tag{2.4a}$$

$$X_L = \omega L \tag{2.4b}$$

For sinusoidal variables, the current $i(t)$ through a capacitor, or the voltage $u(t)$ at an inductance, can be calculated by the first derivative of the voltage, respectively current, as follows:

$$i(t) = C * \frac{\mathrm{d}u(t)}{\mathrm{d}t} \tag{2.5a}$$

$$u(t) = L * \frac{\mathrm{d}i(t)}{\mathrm{d}t} \tag{2.5b}$$

The derivation for sinusoidal currents and voltages at a reactance establishes that the current achieves its maximum value a quarter period after the voltage. When considering the process in the complex level, the pointer of the voltage precedes the pointer of the current by 90°. This corresponds to a multiplication by $+j$.

For a capacitance on the other hand, the voltage does not reach its maximum value until a quarter period after the current; the voltage pointer lags behind the current by 90°, which corresponds to a multiplication by $-j$.

This enables the relationships between current and voltage for inductances and capacitances to be shown in a complex notation

$$\underline{U} = j\omega L * \underline{I} \tag{2.6a}$$

$$\underline{I} = \frac{1}{j\omega C} * \underline{U} \tag{2.6b}$$

The individual explanations of the quantities are given in the text above.

Vectors are used to describe electrical processes. They are therefore used in d.c., a.c. and three-phase systems. Vector systems can, by definition, be chosen as required, but must not be changed during an analysis or calculation. It should also be noted that the appropriate choice of the vector system is of substantial assistance in describing and calculating special tasks. The need for vector systems is clear if one considers the Kirchhoff's laws, for which the positive direction of currents and voltages must be specified. In this way, the positive directions of the active and reactive powers are then also stipulated.

For comparison and transfer reasons, the vector system for the three-phase network (RYB components) is also to be used for other component systems (e.g., system of symmetrical components), which describe the three-phase network.

If vectors are drawn as shown in Figure 2.2, the active and reactive powers generated by a generator in overexcited operation mode are positive. This vector system is designated as a generator vector system. Accordingly, the active and reactive power consumed by the load are positive when choosing the consumer vector system.

When describing electrical systems voltage vectors are drawn from the phase conductor (named L1, L2, L3 or also R, Y, B) to earth (E). In other component systems, for instance in the system of symmetrical components (Section 2.3), the direction of the voltage vector is drawn from the conductor towards the particular reference conductor. On the other hand, vectors in phasor diagrams are shown in the opposite direction. The vector of a voltage conductor to earth is therefore shown in the phasor diagram from earth potential to conductor potential.

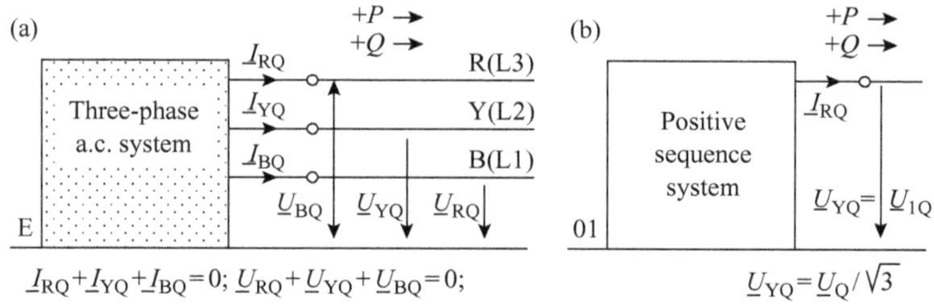

Figure 2.2 *Definition of vectors for current, voltage and power in three-phase a.c. systems. (a) Power system diagram and (b) electrical diagram for symmetrical conditions (positive-sequence component)*

Based on the definition of the vector system, the correlation of voltage and current of an electrical system can be shown in phasor diagrams. Where steady-state or quasi-steady-state operation is shown, r.m.s. value phasors are generally used. Figure 2.3 shows the phasor diagram of an ohmic-inductive load in the generator and in the consumer vector system.

2.3 System of symmetrical components

2.3.1 Transformation matrix

The relationships between voltages and currents of a three-phase system can be represented by a matrix equation, e.g., with the aid of the impedance or admittance matrix. The equivalent circuits created by electrical equipment, such as lines, cables, transformers and machines, in this case have couplings in the three-phase system which are of an inductive, capacitive and galvanic type. This can be explained by using any short element of an overhead line in accordance with Figure 2.4 as an example, see also [1], [7].

The correlation of currents $\underline{I}$ and voltages $\underline{U}$ of the RYB system is as follows:

$$\begin{bmatrix} \underline{U}_R \\ \underline{U}_Y \\ \underline{U}_B \end{bmatrix} = \begin{bmatrix} \underline{Z}_{RR} & \underline{Z}_{RY} & \underline{Z}_{RB} \\ \underline{Z}_{YR} & \underline{Z}_{YY} & \underline{Z}_{YB} \\ \underline{Z}_{BR} & \underline{Z}_{BY} & \underline{Z}_{BB} \end{bmatrix} * \begin{bmatrix} \underline{I}_R \\ \underline{I}_Y \\ \underline{I}_B \end{bmatrix} \tag{2.7}$$

where $\underline{Z}_{RR}$, $\underline{Z}_{YY}$, $\underline{Z}_{BB}$ are the self-impedances of each phase; $\underline{Z}_{RY}$, $\underline{Z}_{RB}$ the coupling (mutual) impedances between phase R and Y, respectively, B; $\underline{Z}_{YR}$, $\underline{Z}_{YB}$ the coupling (mutual) impedances between phase Y and R, respectively, B; and $\underline{Z}_{BR}$, $\underline{Z}_{BY}$ the coupling (mutual) impedances between phase B and R, respectively, Y.

All the values of this impedance matrix can generally be different. Because of the cyclic-symmetrical construction of three-phase systems only the self-impedance and

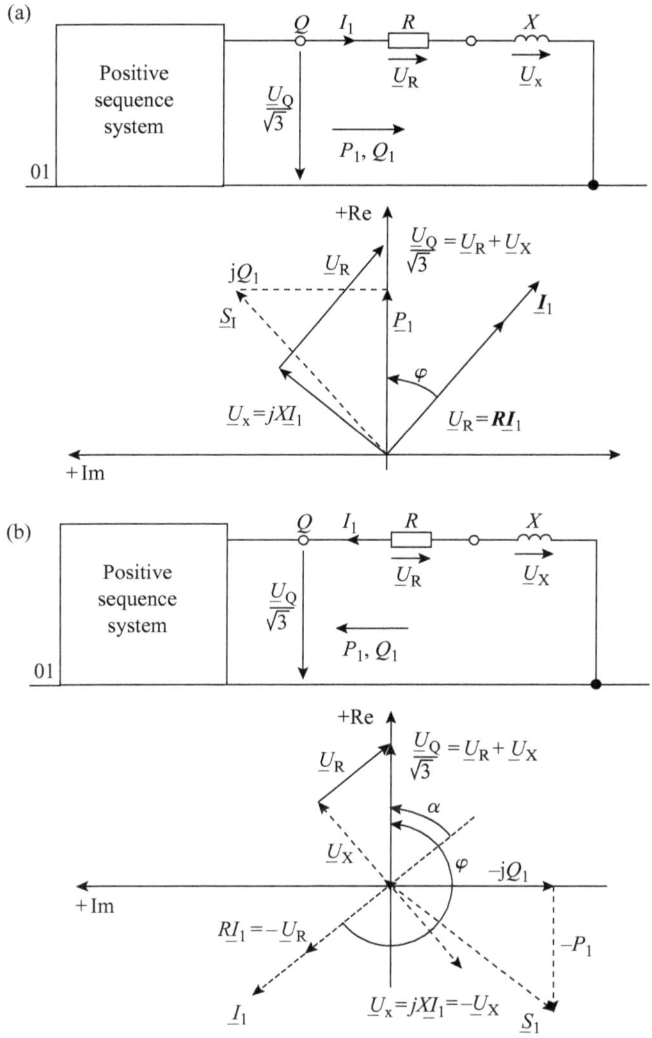

Figure 2.3 Vector diagram of current, voltage and power of a three-phase a.c. system represented by the positive-sequence component. (a) Consumer vector system and (b) generator vector system

two coupling impedances are to be considered. A cyclic-symmetrical matrix is thus obtained.

$$\begin{bmatrix} \underline{U}_R \\ \underline{U}_Y \\ \underline{U}_B \end{bmatrix} = \begin{bmatrix} \underline{Z}_A & \underline{Z}_{M1} & \underline{Z}_{M2} \\ \underline{Z}_{M2} & \underline{Z}_A & \underline{Z}_{M1} \\ \underline{Z}_{M1} & \underline{Z}_{M2} & \underline{Z}_A \end{bmatrix} * \begin{bmatrix} \underline{I}_R \\ \underline{I}_Y \\ \underline{I}_B \end{bmatrix} \tag{2.8}$$

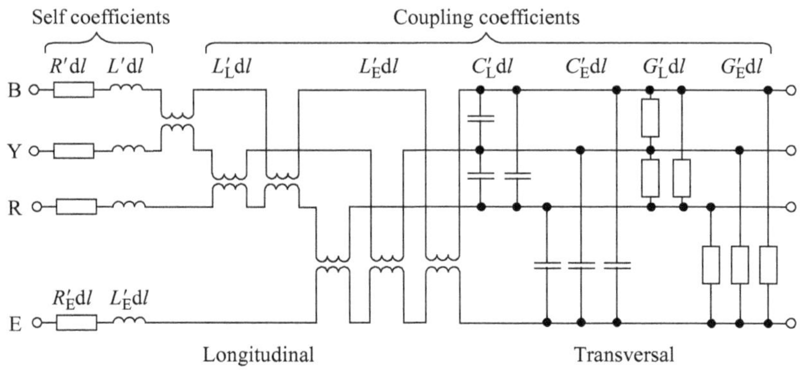

Figure 2.4 Differentially small section of homogeneous three-phase a.c. line

where $\underline{Z}_A$ is the self-impedances of each phase and $\underline{Z}_{M1}$, $\underline{Z}_{M2}$ the coupling (mutual) impedances between the phases.

The multiplicity of couplings between the individual components of three-phase systems complicates the application of the solution methods, particularly when calculating extended networks. For this reason, a mathematical transformation is sought which transfers the RYB-components to a different system. The following conditions should apply for the transformation:

- The transformed voltages should still depend only on one transformed current.
- For symmetrical operation only one component should be unequal to zero.
- The linear relationship between current and voltage should be retained, i.e., the transformation should be linear.
- For symmetrical operation, the current and voltage of the reference component should be retained (reference component invariant).

The desired transformation should in this case enable the three systems to be decoupled in such a way that the three components, named 0, 1 and 2, are decoupled from each other in the following manner:

$$\begin{bmatrix} \underline{U}_0 \\ \underline{U}_1 \\ \underline{U}_2 \end{bmatrix} = \begin{bmatrix} \underline{Z}_0 & 0 & 0 \\ 0 & \underline{Z}_1 & 0 \\ 0 & 0 & \underline{Z}_2 \end{bmatrix} * \begin{bmatrix} \underline{I}_0 \\ \underline{I}_1 \\ \underline{I}_2 \end{bmatrix} \tag{2.9}$$

These requirements are fulfilled by the transformation to the system of symmetrical components [29], which is realised for voltages and currents by the transformation matrix $\underline{\mathbf{T}}$ according to Equations (2.10). It should be noted that the factor $\frac{1}{3}$ is part of the transformation and therefore belongs to the matrix $\underline{\mathbf{T}}$.

$$\underline{U}_{012} = \underline{\boldsymbol{T}} * \underline{U}_{\mathrm{RYB}} \tag{2.10a}$$

$$\underline{I}_{012} = \underline{\boldsymbol{T}} * \underline{I}_{\mathrm{RYB}} \tag{2.10b}$$

$$\begin{bmatrix} \underline{U}_0 \\ \underline{U}_1 \\ \underline{U}_2 \end{bmatrix} = \frac{1}{3} \begin{bmatrix} 1 & 1 & 1 \\ 1 & \underline{a} & \underline{a}^2 \\ 1 & \underline{a}^2 & \underline{a} \end{bmatrix} * \begin{bmatrix} \underline{U}_R \\ \underline{U}_Y \\ \underline{U}_B \end{bmatrix} \tag{2.10c}$$

$$\begin{bmatrix} \underline{I}_0 \\ \underline{I}_1 \\ \underline{I}_2 \end{bmatrix} = \frac{1}{3} \begin{bmatrix} 1 & 1 & 1 \\ 1 & \underline{a} & \underline{a}^2 \\ 1 & \underline{a}^2 & \underline{a} \end{bmatrix} * \begin{bmatrix} \underline{I}_R \\ \underline{I}_Y \\ \underline{I}_B \end{bmatrix} \tag{2.10d}$$

The voltage vector of the 012-system is linearly linked to the voltage vector of the RYB-system (the same applies for the currents). The system of symmetrical components is defined according to DIN 13321.

The reverse transformation of the 012-system to the RYB-system is achieved by the matrix $\underline{\mathbf{T}}^{-1}$ in accordance with

$$\underline{U}_{RYB} = \underline{\boldsymbol{T}}^{-1} * \underline{U}_{012} \tag{2.11a}$$

$$\underline{I}_{RYB} = \underline{\boldsymbol{T}}^{-1} * \underline{I}_{012} \tag{2.11b}$$

$$\begin{bmatrix} \underline{U}_R \\ \underline{U}_Y \\ \underline{U}_B \end{bmatrix} = \begin{bmatrix} 1 & 1 & 1 \\ 1 & \underline{a}^2 & \underline{a} \\ 1 & \underline{a} & \underline{a}^2 \end{bmatrix} * \begin{bmatrix} \underline{U}_0 \\ \underline{U}_1 \\ \underline{U}_2 \end{bmatrix} \tag{2.11c}$$

$$\begin{bmatrix} \underline{I}_R \\ \underline{I}_Y \\ \underline{I}_B \end{bmatrix} = \begin{bmatrix} 1 & 1 & 1 \\ 1 & \underline{a}^2 & \underline{a} \\ 1 & \underline{a} & \underline{a}^2 \end{bmatrix} * \begin{bmatrix} \underline{I}_0 \\ \underline{I}_1 \\ \underline{I}_2 \end{bmatrix} \tag{2.11d}$$

The following applies for both transformation matrices $\underline{\mathbf{T}}$ and $\underline{\mathbf{T}}^{-1}$:

$$\underline{\mathbf{T}} * \underline{\mathbf{T}}^{-1} = \underline{\mathbf{E}} \tag{2.12}$$

with the identity matrix $\underline{\mathbf{E}}$. The complex rotational phasors $\underline{a}$ and $\underline{a}^2$ have the following meanings:

$$\underline{a} = \mathrm{e}^{j120^\circ} = -\tfrac{1}{2} + j\tfrac{1}{2}\sqrt{3} \tag{2.13a}$$

$$\underline{a}^2 = \mathrm{e}^{j240^\circ} = -\tfrac{1}{2} - j\tfrac{1}{2}\sqrt{3} \tag{2.13b}$$

$$1 + \underline{a} + \underline{a}^2 = 0 \tag{2.13c}$$

Multiplication of a vector with either $\underline{a}$ or $\underline{a}^2$ will only change the phase shift of the vector by 120° or 240° but will not change the length (amount) of it.

2.3.2 Interpretation of the system of symmetrical components

If only one zero-sequence component exists, the following applies:

$$\begin{bmatrix} \underline{U}_R \\ \underline{U}_Y \\ \underline{U}_B \end{bmatrix} = \begin{bmatrix} 1 & 1 & 1 \\ 1 & \underline{a}^2 & \underline{a} \\ 1 & \underline{a} & \underline{a}^2 \end{bmatrix} * \begin{bmatrix} \underline{U}_0 \\ 0 \\ 0 \end{bmatrix} = \begin{bmatrix} \underline{U}_0 \\ \underline{U}_0 \\ \underline{U}_0 \end{bmatrix} \quad (2.14)$$

No phase displacement exists between the three a.c. voltages of the RYB-conductors. The zero-sequence component is thus a two-phase a.c. system. Figure 2.5 shows the phasor (vector) diagram of the voltages of the RYB-system and the voltage of the zero-sequence component.

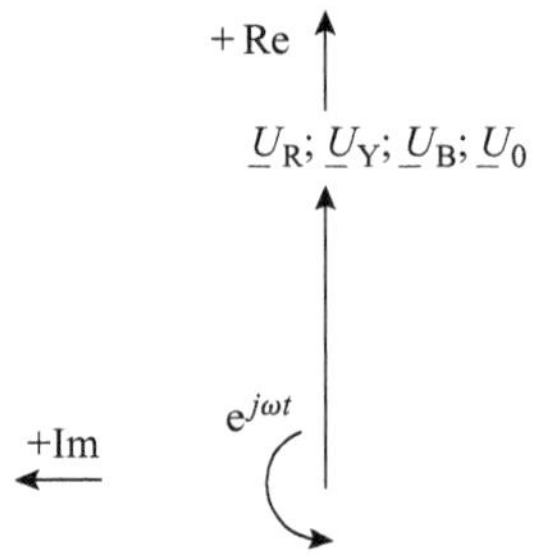

Figure 2.5 Vector diagram of voltages in RYB-system and in the zero-sequence component, positive- and negative-sequence components are NIL

Where only a positive-sequence exists, the following applies:

$$\begin{bmatrix} \underline{U}_R \\ \underline{U}_Y \\ \underline{U}_B \end{bmatrix} = \begin{bmatrix} 1 & 1 & 1 \\ 1 & \underline{a}^2 & \underline{a} \\ 1 & \underline{a} & \underline{a}^2 \end{bmatrix} * \begin{bmatrix} 0 \\ \underline{U}_1 \\ 0 \end{bmatrix} = \begin{bmatrix} \underline{U}_1 \\ \underline{a}^2\,\underline{U}_1 \\ \underline{a}\,\underline{U}_1 \end{bmatrix} \quad (2.15)$$

A three-phase system with a positive rotating phase sequence R, Y, B is represented by the positive-sequence component only. Figure 2.6 shows the phasor (vector) diagram of the voltages of the RYB-system and the voltage of the positive-sequence component.

Where only a negative-sequence component exists, the following applies:

$$\begin{bmatrix} \underline{U}_R \\ \underline{U}_Y \\ \underline{U}_B \end{bmatrix} = \begin{bmatrix} 1 & 1 & 1 \\ 1 & \underline{a}^2 & \underline{a} \\ 1 & \underline{a} & \underline{a}^2 \end{bmatrix} * \begin{bmatrix} 0 \\ 0 \\ \underline{U}_2 \end{bmatrix} = \begin{bmatrix} \underline{U}_2 \\ \underline{a}\,\underline{U}_2 \\ \underline{a}^2\,\underline{U}_2 \end{bmatrix} \quad (2.16)$$

A three-phase system with a positive counter-rotating phase sequence R, B, Y is represented by the negative-sequence component only. Figure 2.7 shows the phasor (vector) diagram of the voltages of the RYB-system and the voltage of the negative-sequence component.

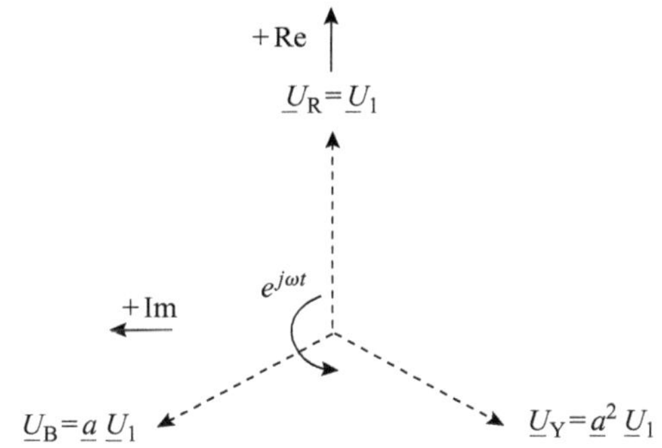

Figure 2.6 Vector diagram of voltages in RYB-system and positive-sequence component, zero- and negative-sequence components are NIL

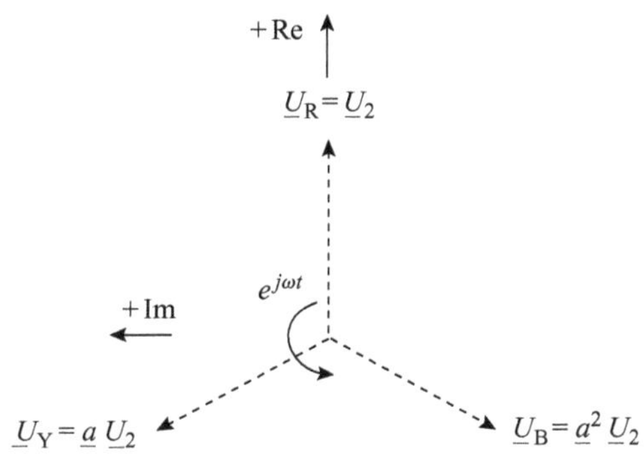

Figure 2.7 Vector diagram of voltages in RYB-system and negative-sequence component, zero- and positive-sequence components are NIL

2.3.3 Transformation of impedances

For the transformation of the impedance matrix, Equations (2.17) apply in accordance with the laws of matrix multiplication, taking account of Equations (2.10) and (2.12)

$$\underline{\mathbf{T}}\,\underline{\mathbf{U}}_{RYB} = \underline{\mathbf{T}}\,\underline{\mathbf{Z}}_{RYB}\underline{\mathbf{T}}^{-1}\quad \underline{\mathbf{T}}\,\underline{\mathbf{I}}_{RYB} \tag{2.17a}$$

$$\underline{\mathbf{U}}_{012} = \underline{\mathbf{Z}}_{012}\qquad\qquad \underline{\mathbf{I}}_{012} \tag{2.17b}$$

As $\underline{\mathbf{T}}\,\underline{\mathbf{U}}_{RYB}$ is equal to $\underline{\mathbf{U}}_{012}$ and $\underline{\mathbf{T}}\,\underline{\mathbf{I}}_{RYB}$ is equal to $\underline{\mathbf{I}}_{012}$ the impedance matrix $\underline{\mathbf{Z}}_{012}$ in the system of symmetrical components is obtained by multiplying the impedance matrix $\underline{\mathbf{Z}}_{RYB}$ in the RYB-system with the matrix $\underline{\mathbf{T}}$ from left and with $\underline{\mathbf{T}}^{-1}$ from right. Based on Equation (2.8) the impedances as per Equations (2.18) for the conversion of the impedances of the three-phase system to the 012-system are obtained.

$$\underline{Z}_0 = \underline{Z}_A + \underline{Z}_{M1} + \underline{Z}_{M2} \tag{2.18a}$$

$$\underline{Z}_1 = \underline{Z}_A + \underline{a}^2\,\underline{Z}_{M1} + \underline{a}\,\underline{Z}_{M2} \tag{2.18b}$$

$$\underline{Z}_2 = \underline{Z}_A + \underline{a}\,\underline{Z}_{M1} + \underline{a}^2\,\underline{Z}_{M2} \tag{2.18c}$$

The impedance values of the positive-sequence and negative-sequence components are generally equal. This applies to all non-rotating equipment. The impedance of the zero-sequence component mainly has a different value from the impedance of the positive-sequence component. If mutual coupling is absent, as perhaps with three single-pole transformers connected together to form a three-phase transformer, the impedance of the zero-sequence component is equal to the impedance of the positive- or (negative-) sequence component.

2.3.4 Measurement of impedances of the symmetrical components

Any equipment can be represented by an equivalent circuit diagram using the system of symmetrical components, which can be determined. To obtain the parameters of the equivalent circuit diagram either short-circuit measurement or no-load measurement has to be carried out in accordance with Figure 2.8 with a voltage system representing the individual component of the system of symmetrical components, i.e.,

- Three-phase voltage system with positive rotating phase sequence RYB to measure the positive-sequence component.
- Three-phase voltage system with positive counter-rotating phase sequence RBY to measure the negative-sequence component.
- a.c. voltage system without any phase displacement of the voltage in the three phases of the equipment to measure the zero-sequence component.

It should be noted in this respect that the type of measurement, i.e., short-circuit measurement or no-load measurement, depends on the type of analysis to be carried out. For short-circuit current calculation or voltage drop estimate, the parameters obtained by short-circuit measurement are needed. Impedance will be calculated based on the measured voltage and current as outlined in Figure 2.8.

Special attention must be taken in case of transformers, as the impedance in the zero-sequence component depends on the type of winding arrangement (star or delta connection) and on the handling of the neutral of the transformer. Figure 2.9 indicates the measurement of the zero-sequence impedance in the case of a two-winding transformer with Yd-arrangement. The Y-connected winding is fed from a single-phase a.c. system; the zero-sequence impedance is present in case the neutral of the transformer is earthed (Figure 2.9(a)). Due to the delta-connection of the second winding a current in the delta winding is present in this case. If the neutral of the transformer remains isolated, the current in the Y-winding is Zero and thus the impedance is infinite. If the zero-sequence impedance of the delta-connected

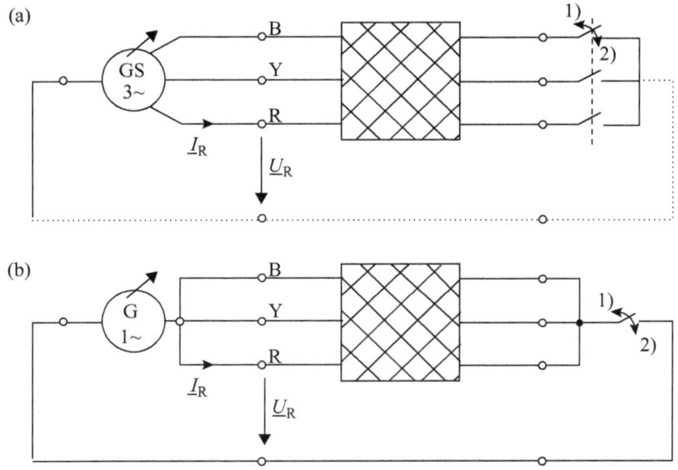

Figure 2.8 Measurement of impedance in the system of symmetrical components. (a) Positive-sequence component (identical with negative-sequence component) and (b) zero-sequence component

winding shall be measured, the current in the delta winding is Zero in any case and the zero-sequence impedance of the delta-connected winding is infinite (Figure 2.9(b)).

To measure the impedance of three-winding transformers in the positive- and zero-sequence component three different measurements have to be carried out. For the positive-sequence component the measurement is outlined in Figure 2.10 and further explained as below:

1. Feeding with three-phase a.c. system at winding 1 (HV-side)
   Short-circuit at winding 2 (MV-side)
   Measurement of u_{k12}, respectively, u_{kHVMV}
2. Feeding with three-phase a.c. system at winding 1 (HV-side)
   Short-circuit at winding 3 (LV-side)
   Measurement of u_{k13}, respectively, u_{kHVLV}
3. Feeding with three-phase a.c. system at winding 2 (MV-side)
   Short-circuit at winding 3 (LV-side)
   Measurement of u_{k23}, respectively, u_{kMVLV}

   The impedances (reactances) as per the equivalent circuit diagram (see Table 3.3) are

$$X_{THV} = 0.5 * \left(\frac{u_{kHVMV}}{S_{rTHVMV}} + \frac{u_{kHVLV}}{S_{rTHVLV}} - \frac{u_{kMVLV}}{S_{rTMVLV}} \right) \tag{2.19a}$$

$$X_{TMV} = 0.5 * \left(\frac{u_{kMVLV}}{S_{rTMVLV}} + \frac{u_{kHVMV}}{S_{rTHVMV}} - \frac{u_{kHVLV}}{S_{rTHVLV}} \right) \tag{2.19b}$$

$$X_{TLV} = 0.5 * \left(\frac{u_{kHVLV}}{S_{rTHVLV}} + \frac{u_{kMVLV}}{S_{rTMVLV}} - \frac{u_{kHVMV}}{S_{rTHVMV}} \right) \tag{2.19c}$$

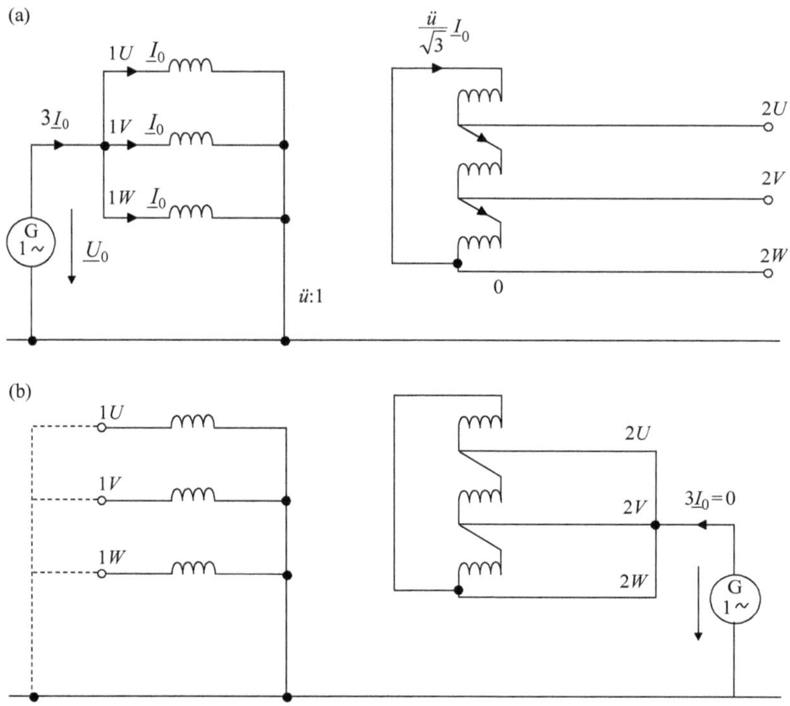

Figure 2.9 *Measuring of zero-sequence impedance of a two-winding transformer (YNd). Diagram indicates winding arrangement of the transformer: (a) measuring at star-connected winding and (b) measuring at delta-connected winding*

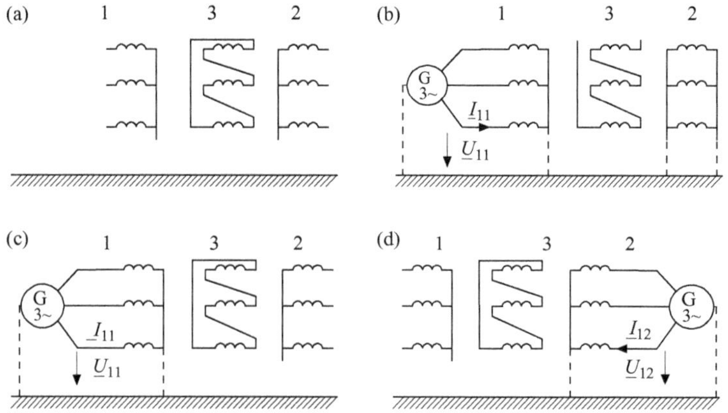

Figure 2.10 *Measurement of positive-sequence impedance of a three-winding transformer (YNyn + d). Diagram indicates winding arrangement of the transformer*

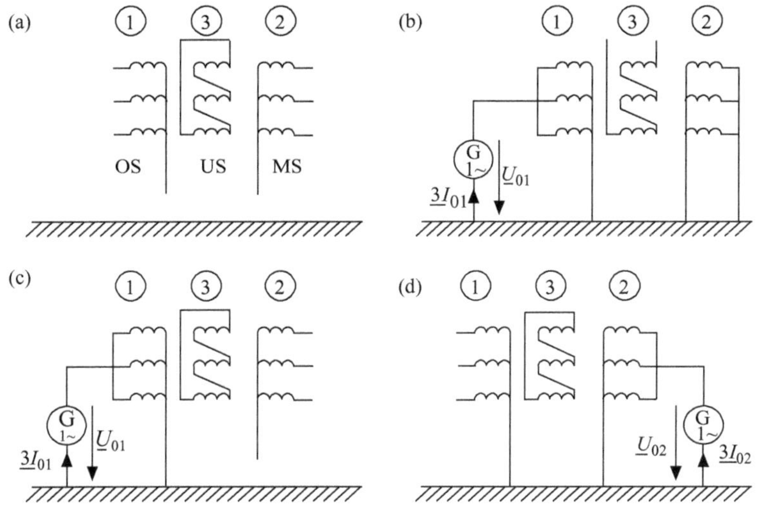

Figure 2.11 *Measurement of zero-sequence impedance of a three-winding transformer (YNyn + d). Diagram indicates winding arrangement of the transformer; for explanations see text*

The measurement in the zero-sequence component is carried out in a similar way as outlined in Figure 2.11.

4. Feeding with single-phase a.c. system at winding 1 (HV-side)
   Short-circuit at winding 2 (MV-side)
   Open-circuit at winding 3 (LV-side)
   Measurement of u_{k012}, respectively, u_{k0HVMV}
5. Feeding with single-phase a.c. system at winding 1 (HV-side)
   Short-circuit at winding 3 (LV-side)
   Open-circuit at winding 2 (MV-side)
   Measurement of u_{k013}, respectively, u_{k0HVLV}
6. Feeding with single-phase a.c. system at winding 2 (MV-side)
   Short-circuit at winding 3 (LV-side)
   Open-circuit at winding 1 (HV-side)
   Measurement of u_{k023}, respectively, u_{k0MVLV}

The impedances (reactances) as per the equivalent circuit diagram (see Table 3.3) are

$$X_{\text{0THV}} = 0.5 * \left(\frac{u_{\text{0kHVMV}}}{S_{\text{rTHVMV}}} + \frac{u_{\text{0kHVLV}}}{S_{\text{rTHVLV}}} - \frac{u_{\text{0kMVLV}}}{S_{\text{rTMVLV}}} \right) \tag{2.20a}$$

$$X_{\text{0TMV}} = 0.5 * \left(\frac{u_{\text{0kMVLV}}}{S_{\text{rTMVLV}}} + \frac{u_{\text{0kHVMV}}}{S_{\text{rTHVMV}}} - \frac{u_{\text{0kHVLV}}}{S_{\text{rTHVLV}}} \right) \tag{2.20b}$$

$$X_{\text{0TLV}} = 0.5 * \left(\frac{u_{\text{0kHVLV}}}{S_{\text{rTHVLV}}} + \frac{u_{\text{0kMVLV}}}{S_{\text{rTMVLV}}} - \frac{u_{\text{0kHVMV}}}{S_{\text{rTHVMV}}} \right) \tag{2.20c}$$

Any impedance in the neutral of the transformer has no effect on the impedance in the positive-sequence component, as the three phase current are summing up to zero at neutral point, and no current flows through the neutral impedance. In the zero-sequence component the neutral impedance will appear with three-times its value in the RYB-system, as the current through the neutral is three-times the phase current during the measurement of zero-sequence impedance.

2.4 Equivalent circuit diagram for short-circuits

The system of symmetrical components can be used for the analysis of symmetrical and asymmetrical operation of power systems. Faults in general and short-circuit currents in particular are the most severe operating conditions in power systems. Each of the different faults, e.g., single-phase-to-ground, three-phase, etc., can be represented by an equivalent circuit diagram in the RYB-system and by this in the 012-system (system of symmetrical components) as well. The calculation of short-circuits in the system of symmetrical components is generally carried out as per Figure 2.12.

1. Draw the equivalent circuit diagram in RYB-components (RYB-system).
2. Draw the short-circuit location at the connection of the RYB-system, the short-circuit should be assumed symmetrical to phase R.
3. Definition of fault equations in RYB-components, equations should be given preference indicating which voltages and/or currents are Zero or are equal to each other.
4. Transformation of fault conditions with the matrices $\underline{\mathbf{T}}$ and $\underline{\mathbf{T}}^{-1}$ into the 012-system (system of symmetrical components). Rearrange the transformed fault equations in such a way that voltages and/or currents are Zero or are equal to each other.
5. Draw the equivalent circuit diagrams in the system of symmetrical components.
6. Draw connection lines between the three components to realise the fault conditions.
7. Calculation of currents and voltages in the system of symmetrical components.
8. Transformation of current and voltages into the RYB-system using transformation matrix $\underline{\mathbf{T}}^{-1}$.

The eight steps as defined above are explained in case of a single-phase short-circuit in a three-phase a.c. system. Figure 2.13 indicates the equivalent circuit diagram in the RYB-system (item 1 above) as well as the type of short-circuit at the short-circuit location (item 2 above).

Any fault in the three-phase a.c. system has to be described by three independent conditions for the voltages of currents of combinations of both. In case of the single-phase short-circuit, the fault equations in the RYB-system (item 3 above) are

$$\underline{U}_{\mathrm{R}} = 0; \qquad \underline{I}_{\mathrm{Y}} = 0; \qquad \underline{I}_{\mathrm{B}} = 0 \tag{2.21}$$

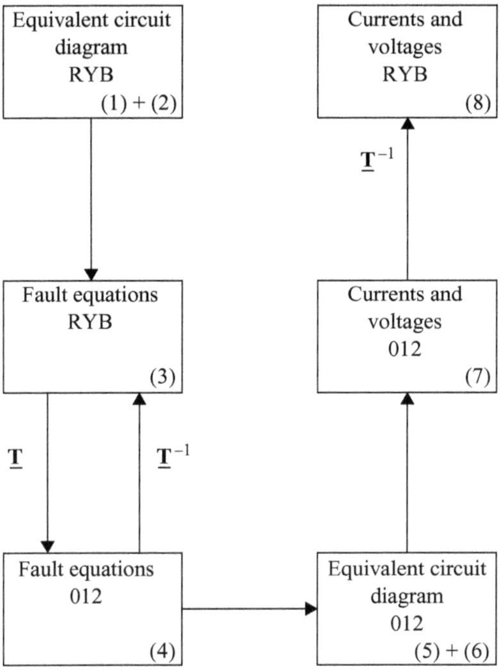

Figure 2.12 *General scheme for the calculation of short-circuit currents in three-phase a.c. systems using the system of symmetrical components. For explanations see text*

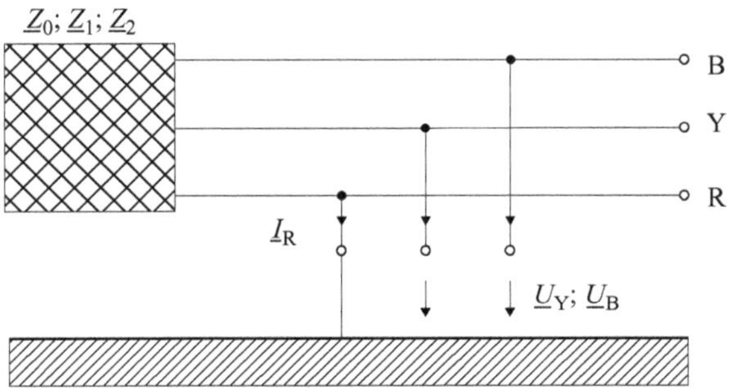

Figure 2.13 *Equivalent circuit diagram of a single-phase short-circuit in RYB-system*

The transformation into the system of symmetrical components (item 4 above) is carried out using the transformation matrices by Equations (2.10) and (2.11). The fault equations for the voltages in the system of symmetrical components are

$$\underline{U}_R = 0 = \underline{U}_0 + \underline{U}_1 + \underline{U}_2 \tag{2.22a}$$

and for the currents

$$\underline{I}_0 = \underline{I}_1 \tag{2.22b}$$

$$\underline{I}_0 = \underline{I}_2 \tag{2.22c}$$

The fault conditions as per Equations (2.22) can only be realised by a series connection of the positive-, negative- and zero-sequence component. The equivalent circuit diagram (item 5 above) in the system of symmetrical components is outlined in Figure 2.14 as well as the connection of the individual components to realise the fault conditions (item 6 above).

The positive-, negative- and zero-sequence component are represented by the impedances $\underline{Z}_1$; $\underline{Z}_2$; $\underline{Z}_0$. The currents and voltages of the system of symmetrical

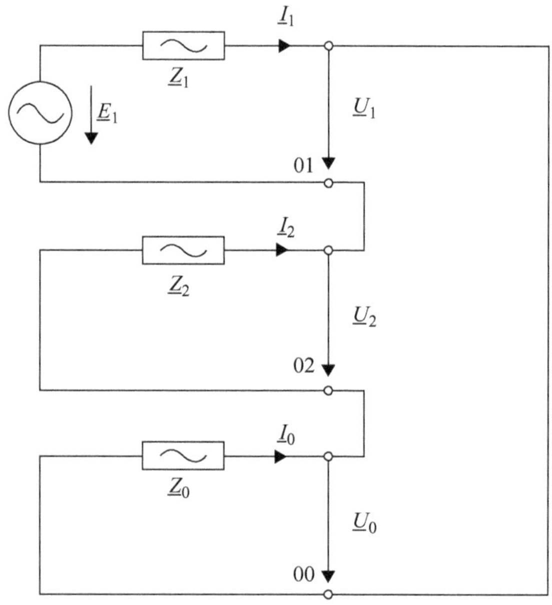

Figure 2.14 Equivalent circuit diagram in the system of symmetrical components for a single-phase short-circuit

components (item 7 above) are then calculated as

$$\underline{I}_0 = \underline{I}_1 = \underline{I}_2 = \frac{\underline{E}_1}{\underline{Z}_0 + \underline{Z}_1 + \underline{Z}_2} \tag{2.23a}$$

$$\underline{U}_0 = -\underline{Z}_0 * \frac{\underline{E}_1}{\underline{Z}_0 + \underline{Z}_1 + \underline{Z}_2} \tag{2.23b}$$

$$\underline{U}_1 = (\underline{Z}_0 + \underline{Z}_2) * \frac{\underline{E}_1}{\underline{Z}_0 + \underline{Z}_1 + \underline{Z}_2} \tag{2.23c}$$

$$\underline{U}_2 = -\underline{Z}_2 * \frac{\underline{E}_1}{\underline{Z}_0 + \underline{Z}_1 + \underline{Z}_2} \tag{2.23d}$$

The currents and voltages of the RYB-components (item 8) are calculated using the transformation matrix, Equation (2.11) and the voltages as per Equations (2.24) are obtained

$$\underline{U}_Y = \underline{E}_1 * \frac{\underline{Z}_0 * (\underline{a}^2 - 1) + \underline{Z}_2 * (\underline{a}^2 - \underline{a})}{\underline{Z}_0 + \underline{Z}_1 + \underline{Z}_2} \tag{2.24a}$$

$$\underline{U}_B = \underline{E}_1 * \frac{\underline{Z}_0 * (\underline{a} - 1) + \underline{Z}_2 * (\underline{a} - \underline{a}^2)}{\underline{Z}_0 + \underline{Z}_1 + \underline{Z}_2} \tag{2.24b}$$

$$\underline{I}_R = \frac{3 * \underline{E}_1}{\underline{Z}_0 + \underline{Z}_1 + \underline{Z}_2} \tag{2.24c}$$

In the case of $\underline{Z}_1 = \underline{Z}_2$ and $|\underline{Z}_0/\underline{Z}_1| = k$ the voltages and the current can be expressed by

$$|\underline{U}_Y| = |\underline{U}_B| = \sqrt{3}E_1 \frac{\sqrt{k^2 + k + 1}}{2 + k} \tag{2.25a}$$

$$I_{k1} = \frac{E_1}{Z_1} * \frac{3}{2 + k} \tag{2.25b}$$

All other faults in three-phase a.c. systems can be analysed in the same manner by the system of symmetrical components. Table 2.1 represents the equivalent circuit diagrams of all short-circuits which can occur in power systems (single fault location only) and the equations to describe the fault both in the RYB-system and in the system of symmetrical components.

2.5 Series and parallel connection

Power systems include numerous equipment, such as lines, transformers, reactors and generators which are connected in series and/or in parallel to other equipment according to their location in the system's topology. The related total impedance at the short-circuit location has to be obtained by mathematical procedures, including

Table 2.1 Equivalent circuit diagrams and equations to represent short-circuits (single fault location) in three-phase a.c. systems in the RYB-system and in the 012-system

RYB-system			012-system		
Equivalent circuit diagram	Fault condition		Fault condition		Equivalent circuit diagram
	Voltages	Currents	Voltages	Currents	
Three-phase 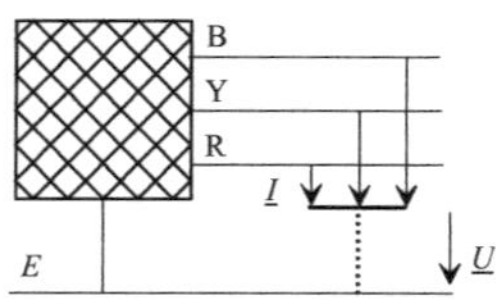	Without earth connection $\underline{U}_R = \underline{U}_Y$ $\underline{U}_Y = \underline{U}_B$ With earth connection $\underline{U}_R = 0$ $\underline{U}_Y = 0$ $\underline{U}_B = 0$	 $\underline{I}_R + \underline{I}_Y + \underline{I}_B = 0$	Without earth connection $\underline{U}_1 = 0$ $\underline{U}_2 = 0$ With earth connection $\underline{U}_1 = 0$ $\underline{U}_2 = 0$ $\underline{U}_0 = 0$	 $\underline{I}_0 = 0$	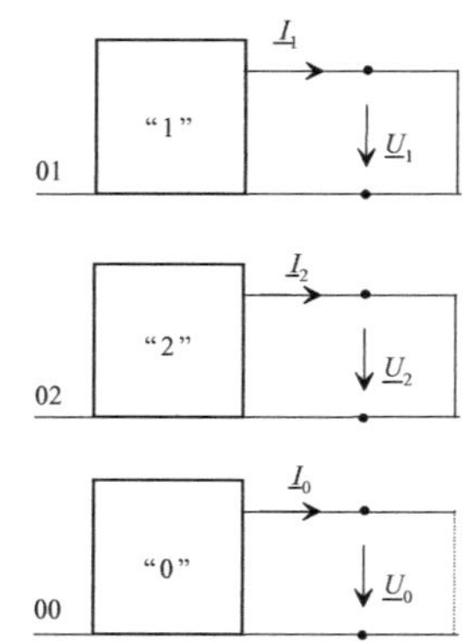

Double-phase

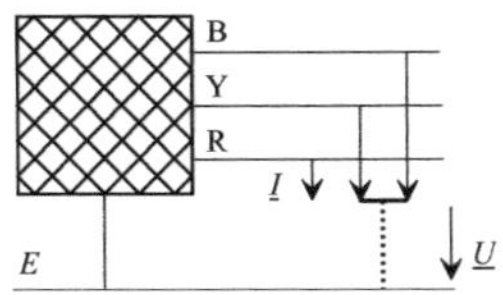

Without earth connection

$\underline{U}_{Y} = \underline{U}_{B}$ $\quad \underline{I}_{R} = 0$

$\underline{I}_{Y} = -\underline{I}_{B}$

With earth connection

$\underline{U}_{Y} = 0$ $\quad \underline{I}_{R} = 0$

$\underline{U}_{B} = 0$

Without earth connection

$\underline{U}_{1} = \underline{U}_{2}$ $\quad \underline{I}_{1} = -\underline{I}_{2}$

$\underline{I}_{0} = 0$

With earth connection

$\underline{U}_{1} = \underline{U}_{2}$ $\quad \underline{I}_{1} + \underline{I}_{2} + \underline{I}_{0} = 0$

$\underline{U}_{1} = \underline{U}_{0}$

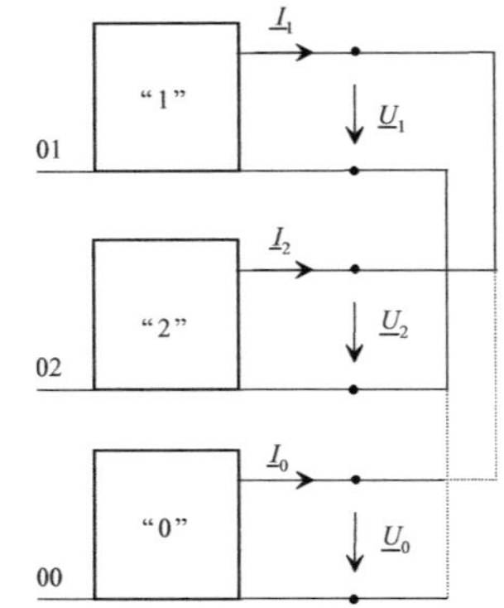

Line-to-ground

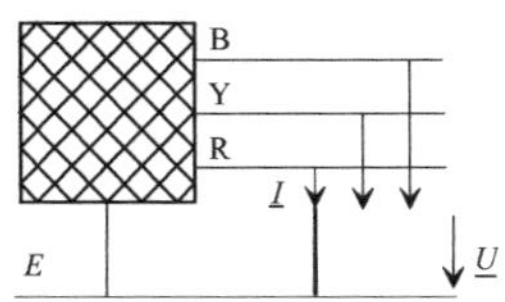

$\underline{U}_{R} = 0$ $\quad \underline{I}_{Y} = 0$

$\underline{I}_{B} = 0$

$\underline{U}_{1} + \underline{U}_{2} + \underline{U}_{0} = 0$ $\quad \underline{I}_{1} = \underline{I}_{2}$

$\underline{I}_{1} = \underline{U}_{0}$

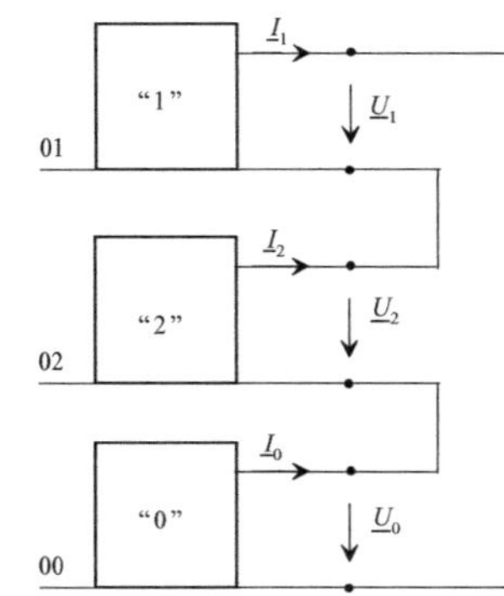

star-delta- and delta-star-transformation. The equations to calculate total impedance within a given system topology are outlined in Figure 2.15.

2.6 Definitions and terms

A clear usage of terms defined in standards and norms is essential in all areas of technique. The knowledge of IEC-documents, national standards and norms therefore

Diagram	Impedance–admittance	Current–voltages
Serial connection		
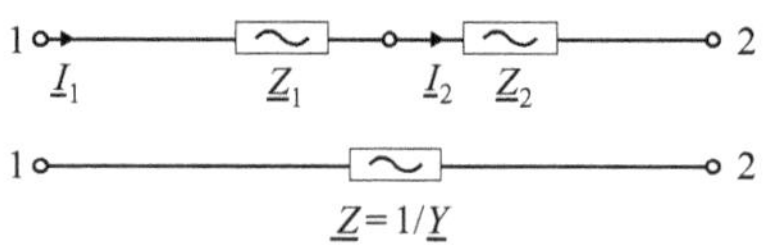	$\underline{Z} = \underline{Z}_1 + \underline{Z}_2$	$\underline{I} = \underline{I}_1 = \underline{I}_2$
	$\underline{Y} = \dfrac{\underline{Y}_1 \underline{Y}_2}{\underline{Y}_1 + \underline{Y}_2}$	$\underline{U} = \underline{U}_1 + \underline{U}_2$
Parallel connection		
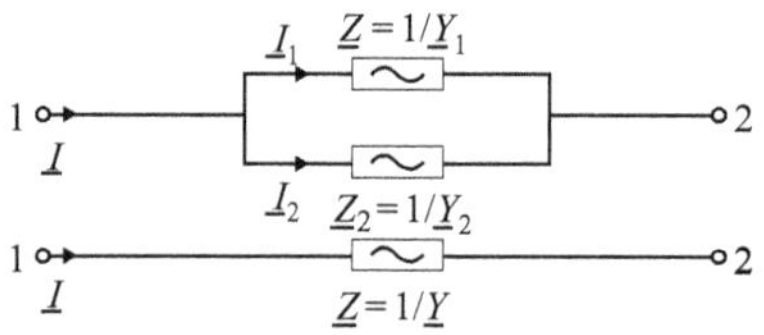	$\underline{Z} = \dfrac{\underline{Z}_1 \underline{Z}_2}{\underline{Z}_1 + \underline{Z}_2}$	$\underline{I} = \underline{I}_1 + \underline{I}_2$
	$\underline{Y} = \underline{Y}_1 + \underline{Y}_2$	$\underline{U} = \underline{U}_1 = \underline{U}_2$
Star-delta-transformation		
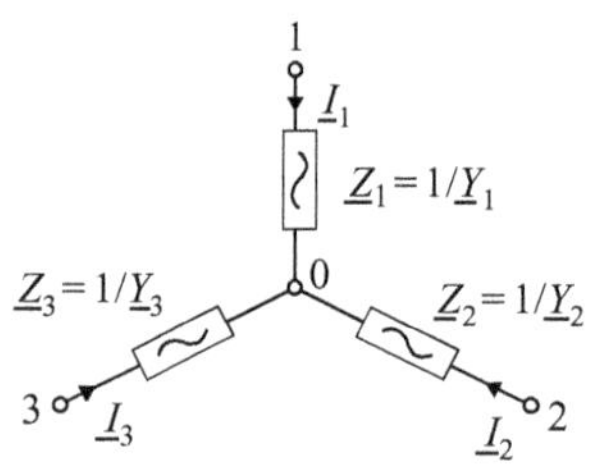	$\underline{Z}_{12} = \dfrac{\underline{Z}_1\underline{Z}_2 + \underline{Z}_1\underline{Z}_3 + \underline{Z}_2\underline{Z}_3}{\underline{Z}_3}$	$\underline{I}_1 + \underline{I}_2 + \underline{I}_3 = 0$
	$\underline{Z}_{13} = \dfrac{\underline{Z}_1\underline{Z}_2 + \underline{Z}_1\underline{Z}_3 + \underline{Z}_2\underline{Z}_3}{\underline{Z}_2}$	
	$\underline{Z}_{23} = \dfrac{\underline{Z}_1\underline{Z}_2 + \underline{Z}_1\underline{Z}_3 + \underline{Z}_2\underline{Z}_3}{\underline{Z}_1}$	
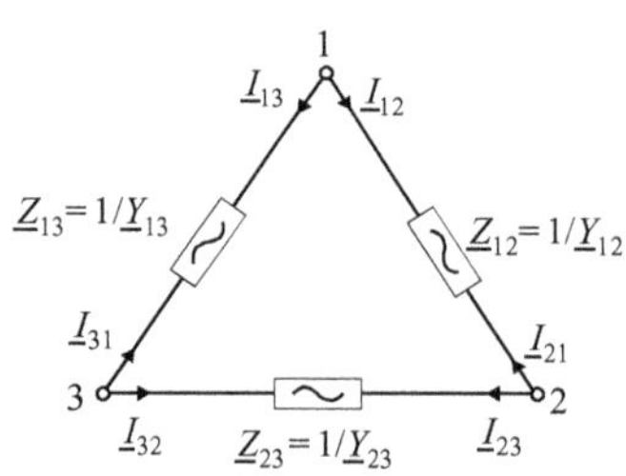	$\underline{Y}_{12} = \dfrac{\underline{Y}_1\underline{Y}_2}{\underline{Y}_1 + \underline{Y}_2 + \underline{Y}_3}$	$\underline{I}_{12} = \dfrac{\underline{I}_1\underline{Z}_1 - \underline{I}_2\underline{Z}_2}{\underline{Z}_{12}}$
	$\underline{Y}_{13} = \dfrac{\underline{Y}_1\underline{Y}_3}{\underline{Y}_1 + \underline{Y}_2 + \underline{Y}_3}$	$\underline{I}_{13} = \dfrac{\underline{I}_1\underline{Z}_1 - \underline{I}_3\underline{Z}_3}{\underline{Z}_{13}}$
	$\underline{Y}_{23} = \dfrac{\underline{Y}_2\underline{Y}_3}{\underline{Y}_1 + \underline{Y}_2 + \underline{Y}_3}$	$\underline{I}_{23} = \dfrac{\underline{I}_2\underline{Z}_2 - \underline{I}_3\underline{Z}_3}{\underline{Z}_{23}}$
Impedance: $\underline{Z} = \underline{R} + j\underline{X}$		

Figure 2.15 Equations for impedance analysis in power systems [30,31]

Diagram	Impedance–admittance	Current–voltages
Delta-star-transformation		
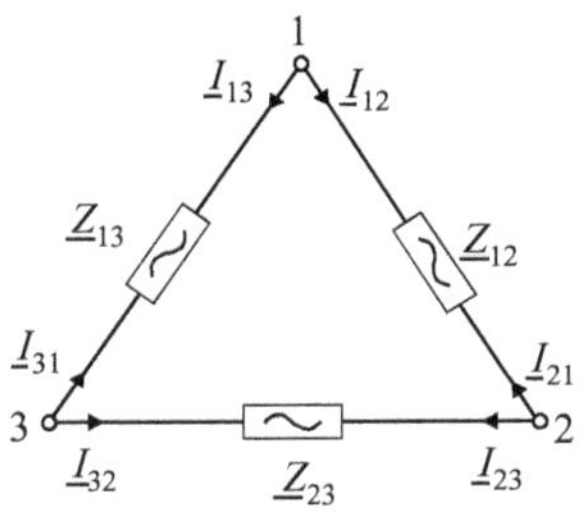	$\underline{Z}_1 = \dfrac{\underline{Z}_{12}\underline{Z}_{13}}{\underline{Z}_{12}+\underline{Z}_{23}+\underline{Z}_{13}}$	$\underline{I}_1 = \underline{I}_{12} + \underline{I}_{13}$
	$\underline{Z}_2 = \dfrac{\underline{Z}_{12}\underline{Z}_{23}}{\underline{Z}_{12}+\underline{Z}_{23}+\underline{Z}_{13}}$	$\underline{I}_2 = \underline{I}_{21} + \underline{I}_{23}$
	$\underline{Z}_3 = \dfrac{\underline{Z}_{13}\underline{Z}_{23}}{\underline{Z}_{12}+\underline{Z}_{23}+\underline{Z}_{13}}$	$\underline{I}_3 = \underline{I}_{31} + \underline{I}_{32}$
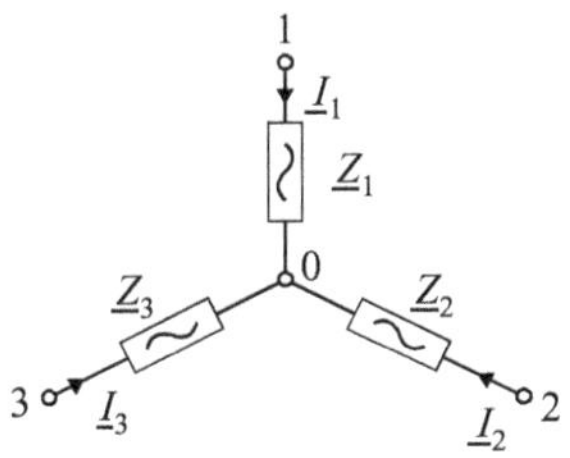 Admittance: $\overline{Y} = \overline{G} + j\overline{B}$		

Figure 2.15 Continued

is absolutely necessary. Some definitions of terms as related to short-circuit currents are based on the German standard DIN 40110 and IEC 60050 as stated below. Further reference is made to the IEC-documents as per Tables 1.1 and 1.2.

Nominal value A suitable rounded value of a physical quantity to define or identify an element, a group of elements or an installation.

Example The nominal value 110 kV defines a voltage level for an electrical power system. Actual voltages in the system are different from the nominal voltage 110 kV.

Limiting value A defined minimal or maximal value of a physical quantity.

Example The minimal value of a current setting has to be defined in order to guarantee the suitable operation of a protection or control device.

Rated value The value of a physical unit for operating conditions as defined for the element, group of elements or installation by the manufacturer.

Example The rated apparent power of a transformer should not be exceeded at the defined operating conditions, in order to protect the transformer from damage by overheating.

Rated data	Summary of rated data and operating conditions.
Example	The definition of the rated current of a cable is not sufficient, as the thermal constraints are fixed besides others by ambient temperature, thermal resistance of the soil, duration of current loading and pre-load conditions.

2.7 Ohm-system, p.u.-system and %/MVA-system

2.7.1 General

It is necessary to calculate the values of the equipment of electrical supply systems in order, for instance, to examine the behaviour of the supply system during normal operation and in the state of disturbed operation. In this connection, equipment such as generators, transformers, lines, motors and capacitors are of interest. Simulation of consumers is only necessary in special cases. It may also be possible to determine the equipment data from name plate rating or tabulated data. Various systems of units are available for calculation.

2.7.1.1 Physical quantities

To describe the steady-state conditions of equipment and of the system, four basic unit quantities are required, i.e., voltage U, current I, impedance Z and power S with the units Volt, Ampere, Ohm and Watt. Other units have to be converted into the ISO-standard unit system [32]. The units are linked to each other by Ohm's law and the power equation.

$$\underline{Z} = \frac{\underline{U}}{\underline{I}} \tag{2.26a}$$

$$\underline{S} = \underline{U} * \underline{I}^{*} \tag{2.26b}$$

where $\underline{U}$ is the voltage across the impedance $\underline{Z}$ and, $\underline{I}$ the current through the impedance ($\underline{I}^{*}$ is the conjugate-complex value of $\underline{I}$).

In the case of a three-phase system the apparent power $\underline{S}$ is calculated as per Equation (2.26c) with the voltage $\underline{U}$ being the phase-to-phase voltage, e.g., the rated voltage of an equipment.

$$\underline{S} = \sqrt{3} * \underline{U} * \underline{I}^{*} \tag{2.26c}$$

If physical quantities are taken to be measurable properties of physical objects, occurrences and states from which useful sums and differences can be formed, the following then applies:

$$\text{Physical quantity} = \text{numerical value} \times \text{unit}$$

2.7.1.2 Relative quantities

On the contrary, the unit of a relative quantity is One or 1 p.u. by definition, i.e.,

Relative quantity = quantity/reference quantity

Because the four quantities voltage, current, impedance and power required for system calculations are linked to each other, two reference quantities only are required to specify a relative system of units. Voltage and power are usually chosen for this purpose. This system is called the per-unit system. As reference voltage either the phase-to-phase or the phase-to-earth voltage can be chosen. If the power of 100 MVA is selected as reference quantity, the system is called the p.u.-system on 100 MVA-base. Table 2.2 gives the definitions in the p.u.-system. It should be observed that the phase-to-phase voltage is chosen as reference voltage. In case the phase-to-earth voltage is selected as reference voltage, the current i' has to be calculated as per [28] as indicated by **(∗1)** in Table 2.2.

2.7.1.3 Semirelative quantities

In the semirelative system of units only one quantity is freely chosen as the reference quantity. If the voltage is chosen, the %/MVA system is obtained, which is outstandingly suitable for network calculations because the values of the equipment

Table 2.2 Definitions of quantities in physical, relative and semirelative units

Ohm-system Physical units	%/MVA-system Semirelative units	p.u.-system Relative units
No reference quantity	One reference quantity	Two reference quantities
Voltage U	$u = \dfrac{U}{U_B} = \dfrac{\{U\}}{\{U_B\}} * 100\%$	$u' = \dfrac{U}{U_B} = \dfrac{\{U\}}{\{U_B\}} * 1$
Current I	$i = I * U_B = \{I\} * \{U_B\}$ $* \mathrm{MVA}$	$i' = I * U_B/S_B = \{I\} * \{U_B\}/\{S_B\} * 1$ **(∗1)** remark see text $i' = I * U_B/S_B = \{I\}$ $*\{\sqrt{3}U_B\}/\{S_B\} * 1$
Impedance Z	$z = Z/U_B^2 = \{Z\} * 100/\{U_B^2\}$ $* \%/\mathrm{MVA}$	$z' = Z * S_B/U_B^2 = \{Z\} * \{S_B\}/U_B^2 * 1$
Power S applies also to P and Q	$s = S = \{S\} * 100\% * \mathrm{MVA}$	$s' = S/S_B = \{S\}/\{S_B\} * 1$

can be calculated very easily. Table 2.2 gives the definitions for the %/MVA-system. The reference voltage U_B in the %/MVA-system should be equal to the rated voltage of equipment U_r or to the nominal system voltage U_n, i.e., it should be a phase-to-phase voltage.

The impedances or reactance of electrical equipment are determined from the data of the respective rating plate (name plate) or from geometrical dimensions. The reactance, resistance or impedance should generally be calculated based on the nominal apparent power or on the nominal voltage of the system in which the equipment is used. Conversion between the different unit systems is made using the data in Table 2.3.

2.7.2 *Correction factor using %/MVA- or p.u.-system*

In case the nominal voltage of the system is unequal to the rated voltages of the equipment connected to this system (most often the case with transformers) a correction factor for the impedances must be applied [33]. The derivation of the correction factor is explained based on Figure 2.16.

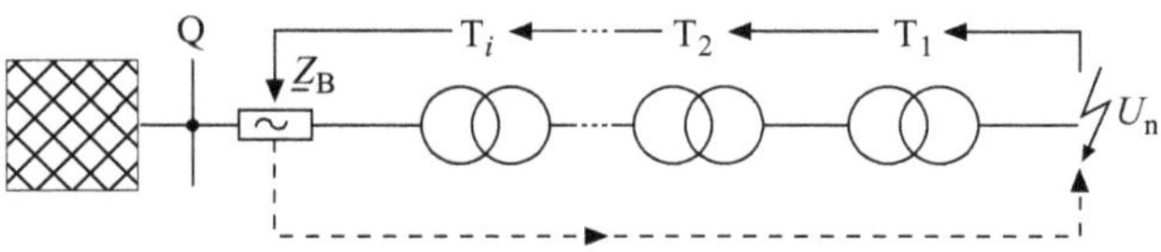

Figure 2.16 Equivalent circuit diagram of a power system with different voltage levels

The impedance correction factor K_B for equipment B is calculated by

$$K_B = \left(\frac{U_{rT1E}}{U_{rT1A}} * \frac{U_{rT2E}}{U_{rT2A}} * \frac{U_{rT3E}}{U_{rT3A}} * \cdots\right)^2 * \left(\frac{U_{rB}}{U_n}\right)^2 \qquad (2.27)$$

where $U_{rT..}$ are the rated voltages of the transformers at side A or E, U_{rB} the rated voltage of the equipment B and, U_n the nominal system voltage at short-circuit location.

The impedance correction factor while using the %/MVA- or the p.u.-system must be applied for any equipment except power station units for which special correction factors (see Table 3.6) are valid.

2.8 Examples

2.8.1 *Vector diagram and system of symmetrical components*

The voltages $\underline{U}_R$, $\underline{U}_Y$ and $\underline{U}_B$ are measured in a power system with nominal voltage $U_n = 10$ kV.

$$\underline{U}_R = 6.64\ \text{kV} * e^{j0^0}; \quad \underline{U}_Y = 6.64\ \text{kV} * e^{j250^0}; \quad \underline{U}_B = 6.64\ \text{kV} * e^{j110^0}$$

Table 2.3 Conversion of quantities between %/MVA-system, Ω-system and p.u.-system (100 MVA-base)

%/MVA-system → Ω-system	Ω-system → %/MVA-system	Ω-system → p.u.-system (100 MVA)	%/MVA-system → p.u.-system (100 MVA)
$\frac{U}{\mathrm{kV}} = \frac{u}{\%} * \frac{1}{100} * \frac{U_\mathrm{B}}{\mathrm{kV}}$	$\frac{u}{\%} = \frac{U}{\mathrm{kV}} * 100 * \frac{1}{U_\mathrm{B}/\mathrm{kV}}$	$\frac{u'}{\mathrm{p.u.}} = \frac{U}{\mathrm{kV}} * \frac{1}{U_\mathrm{B}/\mathrm{kV}}$	$\frac{u'}{\mathrm{p.u.}} = \frac{u}{\%} * \frac{1}{100}$
$\frac{I}{\mathrm{kA}} = \frac{i}{\mathrm{MVA}} * \frac{1}{U_\mathrm{B}/\mathrm{kV}}$	$\frac{i}{\mathrm{MVA}} = \frac{I}{\mathrm{kA}} * \frac{U_\mathrm{B}}{\mathrm{kV}}$	$\frac{i'}{\mathrm{p.u.}} = \frac{I}{\mathrm{kA}} * \frac{U_\mathrm{B}}{\mathrm{kV}} * \frac{1}{S_\mathrm{B}/\mathrm{MVA}}$	$\frac{i'}{\mathrm{p.u.}} = \frac{i}{\mathrm{MVA}}$
$\frac{Z}{\Omega} = \frac{z}{\%/\mathrm{MVA}} * \frac{1}{100} * \left(\frac{U_\mathrm{B}}{\mathrm{kV}}\right)^2$	$\frac{z}{\%/\mathrm{MVA}} = \frac{Z}{\Omega} * \frac{100}{(U_\mathrm{B}/\mathrm{kV})^2}$	$\frac{z'}{\mathrm{p.u.}} = \frac{Z}{\Omega} * \frac{S_\mathrm{B}}{\mathrm{MVA}} * \frac{1}{(U_\mathrm{B}/\mathrm{kV})^2}$	$\frac{z'}{\mathrm{p.u.}} = \frac{z}{\%/\mathrm{MVA}} * \frac{1}{S_\mathrm{B}/\mathrm{MVA}}$
$\frac{S}{\mathrm{MVA}} = \frac{s}{\%\mathrm{MVA}} * \frac{1}{100}$	$\frac{s}{\%\mathrm{MVA}} = \frac{S}{\mathrm{MVA}} * 100$	$\frac{s'}{\mathrm{p.u.}} = \frac{S}{\mathrm{MVA}} * \frac{1}{S_\mathrm{B}/\mathrm{MVA}}$	$\frac{s'}{\mathrm{p.u.}} = \frac{s}{\%\mathrm{MVA}} * \frac{0.01}{S_\mathrm{B}/\mathrm{MVA}}$

The voltages in the system of symmetrical components $\underline{U}_0$, $\underline{U}_1$ and $\underline{U}_2$ have to be calculated by using the transformation equations. The individual voltages in units of kV are obtained by using Equation (2.10). The voltage in the zero-sequence component is

$$\underline{U}_0 = \tfrac{1}{3} * (\underline{U}_R + \underline{U}_Y + \underline{U}_B)$$

$$\underline{U}_0 = \tfrac{1}{3} * (6.64 * e^{j0^0} + 6.64 * e^{j250^0} + 6.64 * e^{j110^0})$$

$$\underline{U}_0 = \tfrac{1}{3}[(6.64 + 6.64 * \cos 250^0 + 6.64 * \cos 110^0) + j(6.64 * \sin 0^0 + 6.64 * \sin 250^0 + 6.64 * \sin 110^0)]$$

$$\underline{U}_0 = \tfrac{1}{3}[(6.64 - 6.64 * 0.342 - 6.64 * 0.342) + j(6.64 * 0 - 6.64 * 0.94 + 6.64 * 0.94)]$$

$$\underline{U}_0 = (0.699 + j0.0)\text{kV}$$

The voltage in the positive-sequence component is

$$\underline{U}_1 = \tfrac{1}{3} * (\underline{U}_R + \underline{a} * \underline{U}_Y + \underline{a}^2 * \underline{U}_B)$$

$$\underline{U}_1 = \tfrac{1}{3} * (6.64 * e^{j0^0} + e^{j120^0} * 6.64 * e^{j250^0} + e^{j240^0} * 6.64 * e^{j110^0})$$

$$\underline{U}_1 = \tfrac{1}{3} * (6.64 * e^{j0^0} + 6.64 * e^{j370^0} + 6.64 * e^{j350^0})$$

$$\underline{U}_1 = \tfrac{1}{3}[(6.64 + 6.64 * \cos 370^0 + 6.64 * \cos 350^0) + j(6.64 * \sin 0^0 + 6.64 * \sin 370^0 + 6.64 * \sin 350^0)]$$

$$\underline{U}_1 = \tfrac{1}{3}[(6.64 + 6.64 * 0.985 + 6.64 * 0.985) + j(6.64 * 0 + 6.64 * 0.174 - 6.64 * 0.174)]$$

$$\underline{U}_1 = (6.57 + j0.0)\ \text{kV}$$

The voltage in the negative-sequence component is

$$\underline{U}_2 = \tfrac{1}{3} * (\underline{U}_R + \underline{a}^2 * \underline{U}_Y + \underline{a} * \underline{U}_B)$$

$$\underline{U}_2 = \tfrac{1}{3} * (6.64 * e^{j0^0} + e^{j240^0} * 6.64 * e^{j250^0} + e^{j120^0} * 6.64 * e^{j110^0})$$

$$\underline{U}_2 = \tfrac{1}{3} * (6.64 * e^{j0^0} + 6.64 * e^{j490^0} + 6.64 * e^{j230^0})$$

$$\underline{U}_2 = \tfrac{1}{3}[(6.64 + 6.64 * \cos 490^0 + 6.64 * \cos 230^0) + j(6.64 * \sin 0^0 + 6.64 * \sin 490^0 + 6.64 * \sin 230^0)]$$

$$\underline{U}_2 = \tfrac{1}{3}[(6.64 - 6.64 * 0.643 - 6.64 * 0.643) + j(6.64 * 0 + 6.64 * 0.766 - 6.64 * 0.766)]$$

$$\underline{U}_2 = (-0.633 + j0.0)\ \text{kV}$$

The voltages in the system of symmetrical components can be obtained by graphical construction as well as outlined in Figure 2.17. The voltages are assumed to have the same value and phase displacement as mentioned above.

2.8.2 Calculation of impedances of a three-winding transformer in %/MVA

The impedances of a three-winding transformer are calculated in units of %/MVA by using the Equations (2.19). The data as mentioned below are taken from the name-plate:

Rated voltages	$U_{rHV}/U_{rMV}/U_{rTLV} = 110$ kV/30 kV/10 kV
Rated apparent power	$S_{rHV} = 30$ MVA; $S_{rMV} = 20$ MVA; $S_{rLV} = 10$ MVA
Rated impedance voltage	$u_{krHVMV} = 10\%$; $u_{krHVLV} = 4.5\%$; $u_{krMVLV} = 10.2\%$

It should be noted that the rated apparent power S_r as per above indicates the apparent power of the individual winding. The maximal permissible apparent power to be transferred between the windings involved for measurement of the impedance voltage had to be taken for the calculation of impedances, i.e., $S_{rHVMV} = \mathrm{MAX}\{S_{rHV}; S_{rMV}\} = 20$ MVA. The impedance of the high-voltage winding is calculated by

$$X_{THV} = 0.5 * \left(\frac{u_{kHVMV}}{S_{rTHVMV}} + \frac{u_{kHVLV}}{S_{rTHVLV}} - \frac{u_{kMVLV}}{S_{rTMVLV}}\right)$$

$$X_{THV} = 0.5 * \left(\frac{10\%}{20\ \text{MVA}} + \frac{4.5\%}{10\ \text{MVA}} - \frac{10.2\%}{10\ \text{MVA}}\right)$$

$$X_{THV} = -0.035\frac{\%}{\text{MVA}}$$

The impedance of the medium-voltage winding is calculated by

$$X_{\mathrm{TMV}} = 0.5 * \left(\frac{u_{\mathrm{kMVLV}}}{S_{\mathrm{rTMVLV}}} + \frac{u_{\mathrm{kHVMV}}}{S_{\mathrm{rTHVMV}}} - \frac{u_{\mathrm{kHVLV}}}{S_{\mathrm{rTHVLV}}} \right)$$

$$X_{\mathrm{TMV}} = 0.5 * \left(\frac{10.2\%}{10\ \mathrm{MVA}} + \frac{10\%}{20\ \mathrm{MVA}} - \frac{4.5\%}{10\ \mathrm{MVA}} \right)$$

$$X_{\mathrm{TMV}} = 0.535 \frac{\%}{\mathrm{MVA}}$$

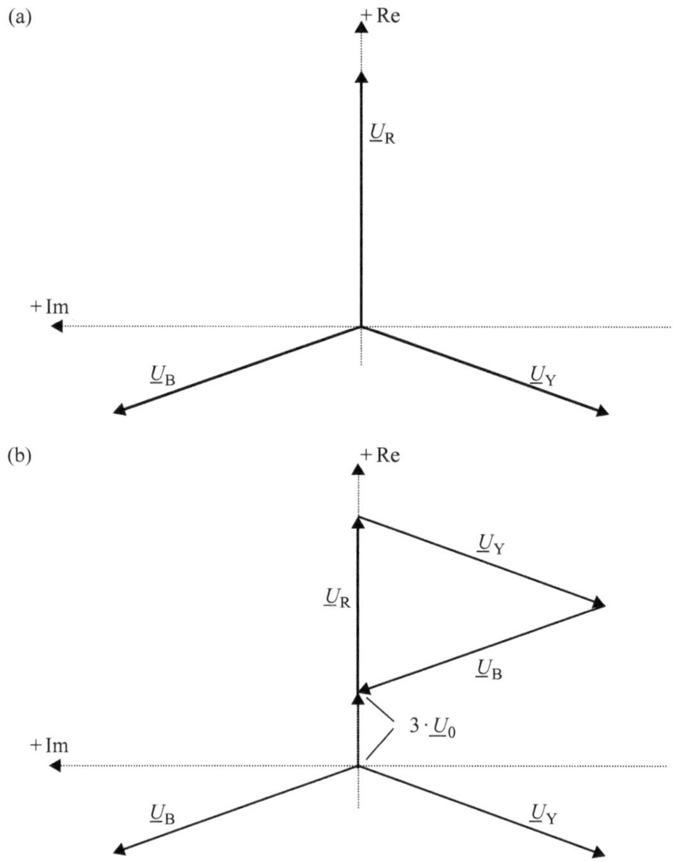

Figure 2.17 Graphical construction of voltages in the system of symmetrical components: (a) vector diagram RYB, (b) vector diagram of voltage in the zero-sequence component, (c) vector diagram of voltage in the positive-sequence component and (d) vector diagram of voltage in the negative-sequence component

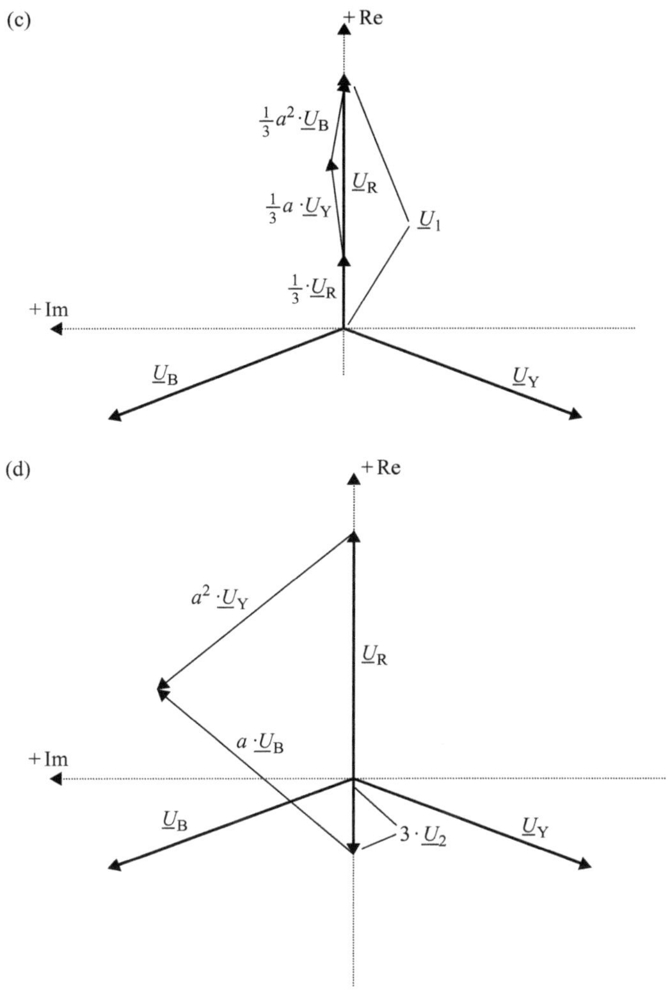

Figure 2.17 Continued

The impedance of the low-voltage winding is calculated by

$$X_{\mathrm{TLV}} = 0.5 * \left(\frac{u_{\mathrm{kHVLV}}}{S_{\mathrm{rTHVLV}}} + \frac{u_{\mathrm{kMVLV}}}{S_{\mathrm{rTMVLV}}} - \frac{u_{\mathrm{kHVMV}}}{S_{\mathrm{rTHVMV}}} \right)$$

$$X_{\mathrm{TLV}} = 0.5 * \left(\frac{4.5\%}{10\ \mathrm{MVA}} + \frac{10.2\%}{10\ \mathrm{MVA}} - \frac{10\%}{20\ \mathrm{MVA}} \right)$$

$$X_{\mathrm{TLV}} = 0.485 \frac{\%}{\mathrm{MVA}}$$

It should be noted that the negative value of X_{THV} shall not be interpreted to have a physical meaning. The model (see Table 3.4) describes the performance of the three-winding transformer at the connection point HV, MV and LV only. The impedance between any two connection points appears physically correct with a positive value.

2.8.3 *Conversion of impedances (Ω; %/MVA; p.u.)*

Impedance values shall be converted from one unit system into the other unit systems (%/MVA-system and p.u.-system on 100 MVA-base) by using equations as per Tables 2.2 and 2.3.

- Overhead line $\underline{Z} = (6.2 + j53.4)\ \Omega$ at $U_n = 380$ kV
  The impedance in the %/MVA-system is calculated by

$$\underline{z} = (6.2 + j53.4)\ \Omega * \frac{100\%}{(380\ \text{kV})^2} = (0.0043 + 0.037)\ \frac{\%}{\text{MVA}}$$

  The impedance in the p.u.-system on 100 MVA-base is calculated by

$$\underline{z}' = (6.2 + j53.4)\ \Omega * \frac{100\ \text{MVA}}{(380\ \text{kV})^2} = (0.0043 + 0.037)\ \text{p.u.}$$

- Apparent power of a transformer $\underline{S} = (1.22 + j300)$ MVA
  The apparent power in the %/MVA-system is calculated by

$$\underline{s} = (1.22 + j300)\ \text{MVA} * 100\% = (122 + j30000)\%/\text{MVA}$$

  The apparent power in the p.u.-system on 100 MVA-base is calculated by

$$\underline{s}' = \frac{(1.22 + j300)\ \text{MVA}}{100\ \text{MVA}} = (0.0122 + j3)\ \text{p.u.}$$

- Voltage drop $\Delta U = 12.5$ kV in a 132 kV-system
  The voltage drop in the %/MVA-system is calculated by

$$\Delta u = 12.5\ \text{kV} * \frac{100\%}{132\ \text{kV}} = 9.47\%$$

  The voltage drop in the p.u.-system on 100 MVA-base is calculated by

$$\Delta u' = 12.5\ \text{kV} * \frac{1}{132\ \text{kV}} = 0.0947\ \text{p.u.}$$

- Rated current of a 320 MVA-transformer on 115 kV-side $I_r = 1.607$ kA
  The current in the %/MVA-system is calculated by

$$i_r = 1.607\ \text{kA} * 115\ \text{kV} = 184.81\ \text{MVA}$$

  The current in the p.u.-system on 100 MVA-base is calculated by

$$i_r' = 1.607\ \text{kA} * \frac{115\ \text{kV}}{100\ \text{MVA}} = 0.185\ \text{p.u.}$$

2.8.4 Impedances in %/MVA-system based on measurement

A simplified example for the application of the system of symmetrical components in representing electrical equipment is outlined in Figure 2.18 by using a short element of an overhead line in accordance to Figure 2.4. The impedance matrix (RYB-system) is thus obtained

$$\underline{\mathbf{Z}}_{\mathrm{RYB}} = \begin{bmatrix} R + jX & jX_{\mathrm{M}} & jX_{\mathrm{M}} \\ jX_{\mathrm{M}} & R + jX & jX_{M} \\ jX_{\mathrm{M}} & jX_{\mathrm{M}} & R + jX \end{bmatrix} \tag{2.28}$$

where R is the resistance of the line, X the reactance of the line and X_{M} the mutual reactance between the individual phases.

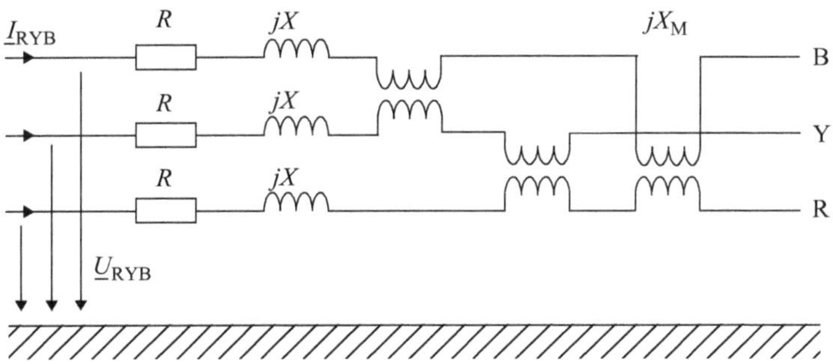

Figure 2.18 Simplified equivalent circuit diagram in RYB-components

Voltages at the individual components R, Y and B are obtained from the current and the impedance by Ohm's law.

The voltage of phase R (component R) is given as follows:

$$\underline{U}_{\mathrm{R}} = (R + jX) * \underline{I}_{\mathrm{R}} + jX_{\mathrm{M}} * \underline{I}_{\mathrm{Y}} + jX_{\mathrm{M}} * \underline{I}_{\mathrm{B}} \tag{2.29}$$

and depends on the currents of all three phases (components) $\underline{I}_{\mathrm{R}}$, $\underline{I}_{\mathrm{Y}}$ and $\underline{I}_{\mathrm{B}}$. Using the system of symmetrical components, see Equation (2.18), the impedance matrix is simplified in such a way that the three components are decoupled from each other in the following manner:

$$\underline{\mathbf{Z}}_{012} = \begin{bmatrix} R + j(X + 2X_{\mathrm{M}}) & 0 & 0 \\ 0 & R + j(X - X_{\mathrm{M}}) & 0 \\ 0 & 0 & R + j(X - X_{\mathrm{M}}) \end{bmatrix} \tag{2.30}$$

The equivalent circuit diagram in the system of symmetrical components is outlined in Figure 2.19.

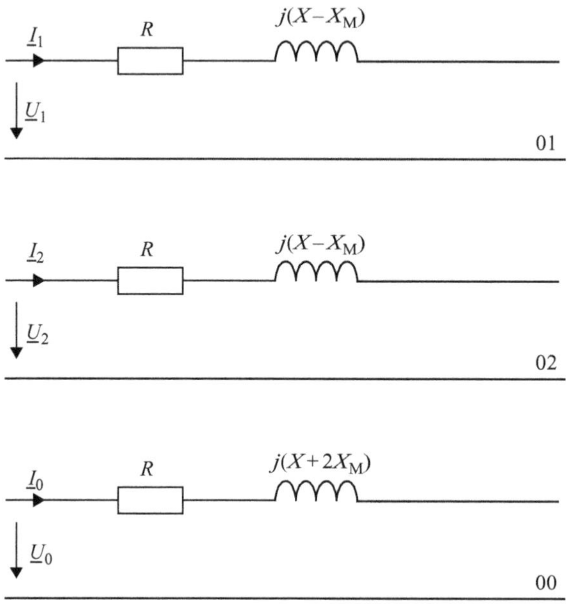

Figure 2.19 Equivalent circuit diagram in the system of symmetrical components

The voltage $\underline{U}_1$ of the component 1 (positive-sequence component) is given as below:

$$\underline{U}_1 = (R + j(X - X_M)) * \underline{I}_1 \tag{2.31a}$$

and depends only on the current $\underline{I}_1$ of the same component. Impedance of the components 1 and 2 (positive- and negative-sequence components) is identical. Voltage of the negative-sequence system therefore is given by

$$\underline{U}_2 = (R + j(X - X_M)) * \underline{I}_2 \tag{2.31b}$$

The impedance of the component 0 (zero-sequence component) is different from the others and the voltage in the zero-sequence component is given by

$$\underline{U}_0 = (R + j(X + 2X_M)) * \underline{I}_0 \tag{2.31c}$$

2.8.5 Representation of a line in the RYB-system and in the system of symmetrical components

Overhead lines are represented in the RYB-system by lumped elements, indicating the inductive and capacitive coupling as well as the impedance of earth return path. The equivalent circuit diagram of an infinitesimal small section of a line is outlined in Figure 2.20.

The representation in the system of symmetrical components (012-system) can either be done by transformation of the individual elements using the transformation

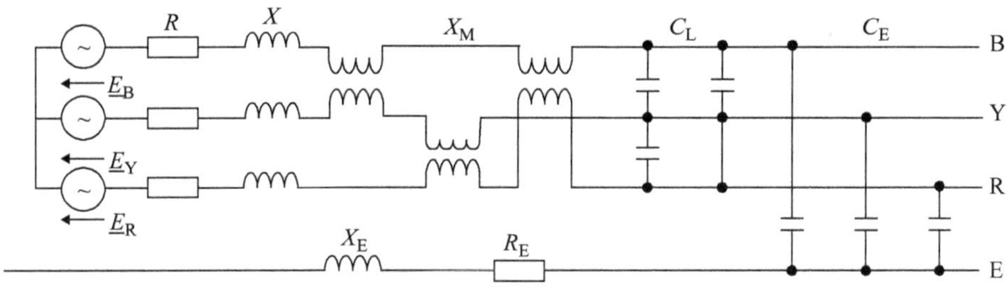

Figure 2.20 Equivalent circuit diagram of an overhead line of infinitesimal length with earth return in RYB-system

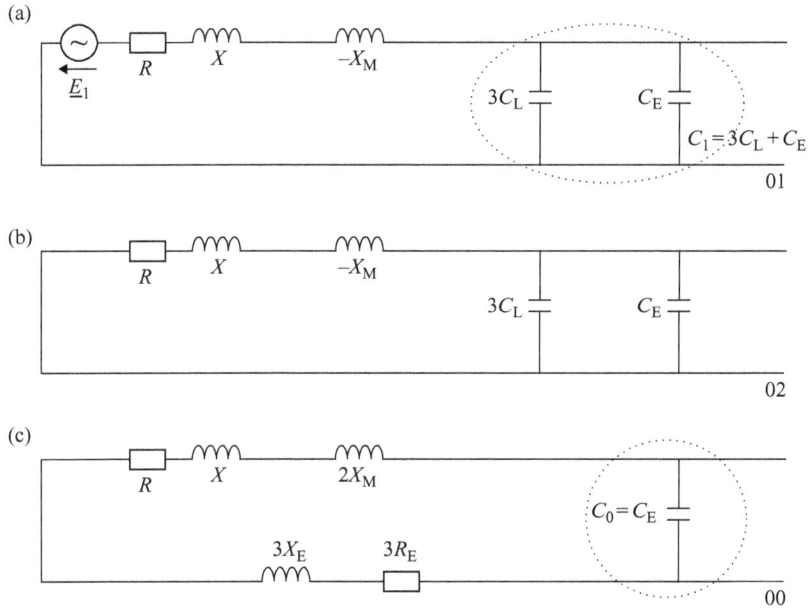

Figure 2.21 Equivalent circuit diagram of an overhead line of infinitesimal length with earth return in 012-system. (a) Positive-sequence component, (b) negative-sequence component and (c) zero-sequence component

matrix as per Equations (2.18) or by measurement according to the procedure as explained in Section 2.3.4. The line is represented in the system of symmetrical components as outlined in Figure 2.21. The reactance X_E and resistance R_E of the earth return path appear with its threefold value in the zero-sequence component, as the current through the neutral is three-times the phase current during the measurement of zero-sequence impedance. The capacitances between the phases C_L are represented with three times its capacitance in the positive- and negative-sequence component, as they appear in the RYB-system as delta-connected and have to be transferred to a star-connection for representation in the 012-system. The capacitance line-to-earth C_E is represented in the three components with the same value.

Please observe that the mutual reactance X_M appears with different values in the positive- (negative-) and zero-sequence component as outlined in Section 2.3.3. The negative value of the reactance in the positive-, respectively, negative-sequence system has no physical meaning, as the total of the mutual reactance X_M and the self-reactance X still remains positive and therefore physically correct.

The equivalent voltage source $\underline{E}_1$, which is the internal voltage of a generator, is only present in the positive-sequence component as the symmetrical operation of the system is represented by this component.

Chapter 3
Calculation of impedance of electrical equipment

3.1 General

In general, equipment in power systems are represented by equivalent circuits, which are designed for the individual tasks of power system analysis, e.g., for the calculation of no-load current and the no-load reactive power of a transformer, the no-load equivalent circuit is sufficient. Regarding the calculation of short-circuits, voltage drops and load characteristic a different equivalent circuit is required. The individual components of the equivalent circuits are resistance, inductive and capacitive reactance (reactor and capacitor), voltage source and ideal transformer. Voltage and currents of the individual components and of the equivalent circuit are interlaced by Ohm's law, which is valid for the three-phase system (RYB-system) as well as for the system of symmetrical components (012-system). A detailed deduction of the mathematical methods and equations is not given within the context of this section of the book, but only the final equations are quoted. For further reading, reference is made to [1,13].

3.2 Equipment in a.c. systems

3.2.1 General

Impedances of equipment are calculated based on name plate data, from manufacturer's data or from geometrical arrangement. For the calculation of impedances of generators, power plants, step-up and step-down transformers, correction factors are necessary. The calculation equations as per Tables 3.1–3.11 are given in the Ohm-system only. For conversion to %/MVA-system, respectively, p.u.-system Tables 2.2 and 2.3 can be used. If not marked by index '1', e.g., $\underline{Z}_{1Q}$, in a different way, impedances are given for the positive-sequence component. The impedance in the zero-sequence system is marked with index '0', e.g., $\underline{Z}_{0Q}$.

3.2.2 *Impedance calculation*

Table 3.1 Impedance of system feeder, equivalent circuit diagram, calculation equations and remarks

Figure	Impedance	Remarks
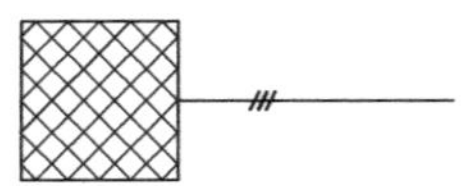 Symbol 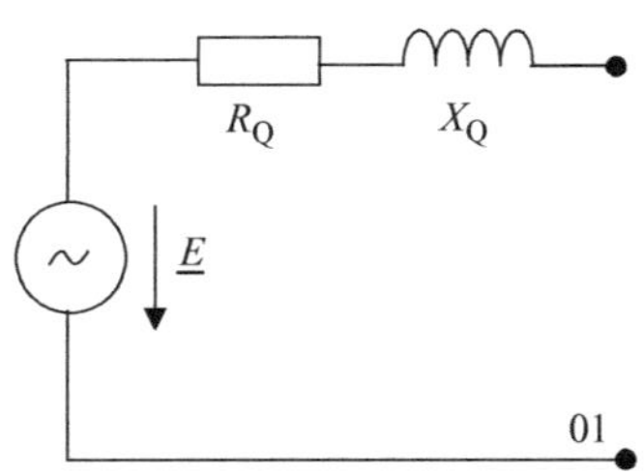 Equivalent circuit diagram	Positive-sequence system $Z_{1Q} = \frac{c * U_{nQ}^2}{S''_{kQ}}$ $Z_{1Q} = \frac{c * U_{nQ}}{\sqrt{3} * I''_{k3Q}}$ $X_{1Q} = \frac{Z_{1Q}}{\sqrt{1 + (R_{1Q}/X_{1Q})^2}}$ Zero-sequence component $Z_{0Q} = 3 * Z_{1Q} * \frac{I''_{k3Q}}{I''_{k1Q}} - 2 * Z_{1Q}$	c Voltage factor according to Table 4.1 I''_{k1Q} Initial short-circuit current (single-phase short-circuit) I''_{k3Q} Initial short-circuit current (three-phase short-circuit) S''_{kQ} Initial short-circuit power (three-phase short-circuit) U_{nQ} Nominal system voltage at connection point Assumptions for resistance: • $R_Q \approx 0$ if $U_{nQ} \geq 35$ kV; system with overhead transmission lines • $R_Q = 0.1X_Q$ respectively $X_Q = 0.995Z_Q$ • The impedance in the zero-sequence component normally is given as a ratio of the impedance in the positive-sequence component. If the single-phase short-circuit current from the system feeder is known, the zero-sequence impedance can be calculated. • Reference Item 3.2 of IEC 60909

Table 3.2 Impedance of two-winding transformer, equivalent circuit diagram, calculation equations and remarks

Figure	Impedance	Remarks
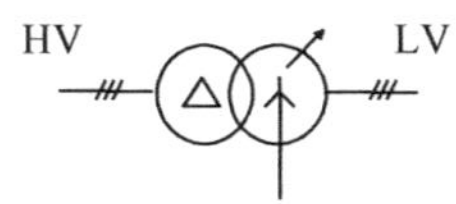 Symbol 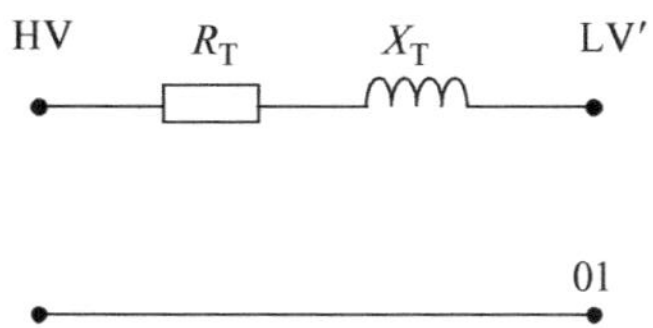 Equivalent circuit diagram	$Z_T = \frac{u_{kr} * U_{rT}^2}{100\% * S_{rT}} * K_T$ $R_T = \frac{u_{Rr} * U_{rT}^2}{100\% * S_{rT}} * K_T$ $R_T = \frac{P_{krT}}{3 * I_{rT}^2} * K_T$ $X_T = \sqrt{Z_T^2 - R_T^2}$ Correction factor K_T: $K_T = \frac{U_{nQ}}{U_{b\max}} * \frac{c_{\max}}{1 + x_T(I_{b\max T}/I_{rT}) \sin \varphi_{bT}}$ Approximation: $K_T = 0.95 * \frac{c_{\max}}{1 + 0.6x_T}$	u_{kr} Impedance voltage U_{rT} Rated voltage (HV- or LV-side) S_{rT} Rated apparent power $c_{\max}$ Voltage factor according to Table 4.1 at LV-side $I_{b\max T}$ Maximal current prior to short-circuit I_{rT} Rated current (HV- or LV-side) P_{krT} Ohmique losses u_{Rr} Ohmique part of impedance voltage $U_{b\max}$ Maximal voltage prior to short-circuit U_{nQ} Nominal system voltage φ_{bT} Phase angle of current prior to short-circuit – Transformer with high rated power: $X_T \approx Z_T$ – Correction factor for positive-, negative- and zero-sequence component [35] – Impedances as per this table are related to HV-side – The impedance in the zero-sequence component can either be given as a ratio of the impedance in the positive-sequence component (see Chapter 13) or can be calculated from the impedance voltage and the losses in the zero-sequence component. – The impedance in the zero-sequence component depends on the handling of transformer neutral, see Table 3.6 – Reference Item 3.3.1 of IEC 60909

Table 3.3 Impedance of three-winding transformer, equivalent circuit diagram, calculation equations and remarks

Figure	Impedance	Remarks
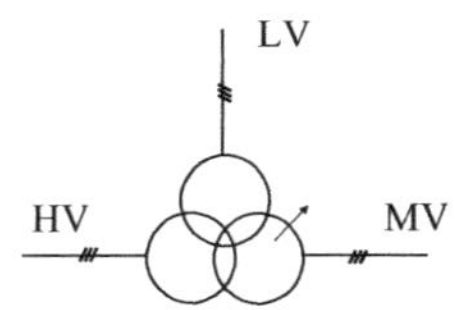 Symbol 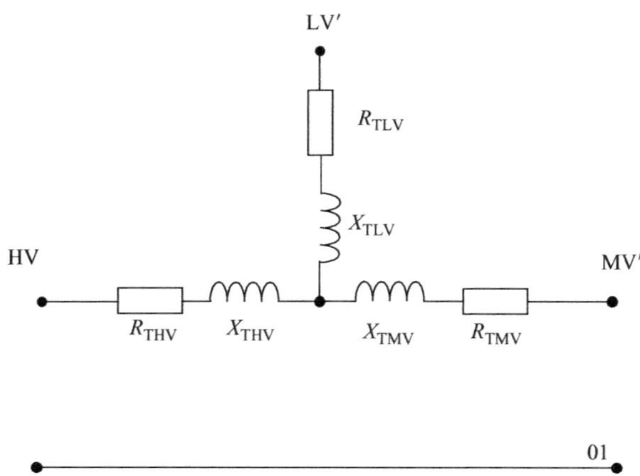 Equivalent circuit diagram	$\lvert \underline{Z}_{THVMV} \rvert = u_{krHVMV} * \frac{U^2_{rTHV}}{S_{rtHVMV}} * K$ $\lvert \underline{Z}_{THVLV} \rvert = u_{krHVLV} * \frac{U^2_{rTHV}}{S_{rtHVLV}} * K$ $\lvert \underline{Z}_{TLVMV} \rvert = u_{krLVMV} * \frac{U^2_{rTHV}}{S_{rtHVMV}} * K$ $Z_{THV} = 0.5 * (\underline{Z}_{THVMV} + \underline{Z}_{THVLV} - \underline{Z}_{TLVMV}) * K_{THV}$ $Z_{TMV} = 0.5 * (\underline{Z}_{TLVMV} + \underline{Z}_{THVMV} - \underline{Z}_{THVLV}) * K_{TMV}$ $Z_{TLV} = 0.5 * (\underline{Z}_{THVLV} + \underline{Z}_{TLVMV} - \underline{Z}_{THVMV}) * K_{TLV}$ $R_T = \frac{u_{Rr} * U^2_{rT}}{100\% * S_{rT}} * K_T$ $R_T = \frac{P_{krT}}{3 * I^2_{rT}} * K_{T**}$ $X_T = \sqrt{Z^2_T - R^2_T}$	c_{max} Voltage factor according to Table 4.1 I_{rT} Rated current P_{krT} Ohmique losses S_{rT} Rated apparent power U_{rT} Rated voltage u_{Rr} Ohmique part of impedance voltage u_{kr} Impedance voltage – HV, MV, LV related to voltage levels – Calculation of R_T and X_T similar to Z_T for HV, MV and LV – Transformers with high rating: $X_T \approx Z_T$ – Correction factor for positive, negative and zero-sequence component [35] – Impedances as per this table are related to HV-side – Correction factor K not defined herewith – Correction factor K_{T**} equal K_{THV}; K_{TMV}; K_{TLV}

Correction factor K_T:

$$K_{THVMV} = 0.95\frac{c_{max}}{1 + 0.6x_{THVMV}}$$

$$K_{THVLV} = 0.95\frac{c_{max}}{1 + 0.6x_{THVLV}}$$

$$K_{TMVLV} = 0.95\frac{c_{max}}{1 + 0.6x_{TMVLV}}$$

- The impedance in the zero-sequence component can either be given as a ratio of the impedance in the positive-sequence component (see Chapter 13) or can be calculated from the impedance voltage and the losses in the zero-sequence component.
- The impedance in the zero-sequence component depends on the handling of transformer neutral, see Table 3.4
- Reference Item 3.3.2 of IEC 60909

Table 3.4 Equivalent circuit diagram of two- and three-winding transformers in the positive- and zero-sequence component

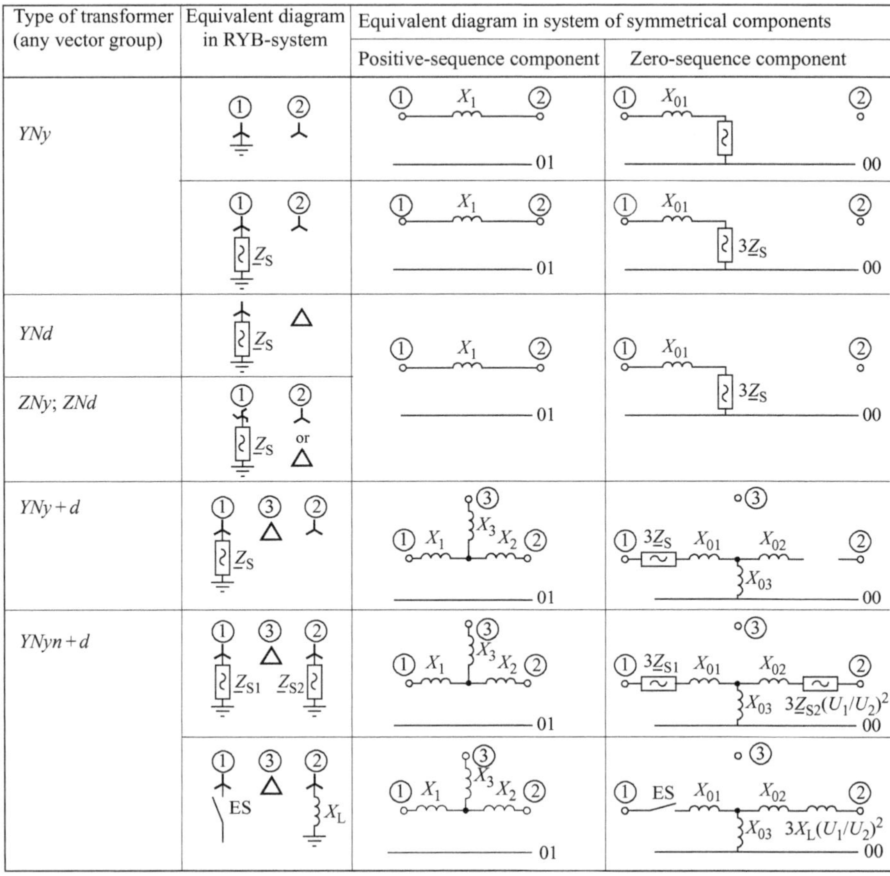

3.3 Equipment in d.c. systems

3.3.1 General

For the calculation of short-circuit currents in d.c. systems, the parameters of equipment contributing to the short-circuit current, i.e., capacitor, battery, rectifier and d.c. motor need to be known besides the parameter of conductors. The calculation equations as per Tables 3.12 to 3.16 are given in the Ohm-system only. For conversion to %/MVA system, p.u. system, respectively, Tables 2.2 and 2.3 can be used.

Tables 3.12 to 3.16 mention the term 'common branch'. The common branch in d.c. systems is the branch (conductor) leading parts of the short-circuit current from several different sources (capacitor, battery, rectifier and d.c. motor) according to IEC 61660-1.

Table 3.5 Impedance of synchronous generator, equivalent circuit diagram, calculation equations and remarks

Figure	Impedance	Remarks
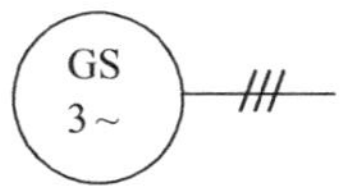 Symbol 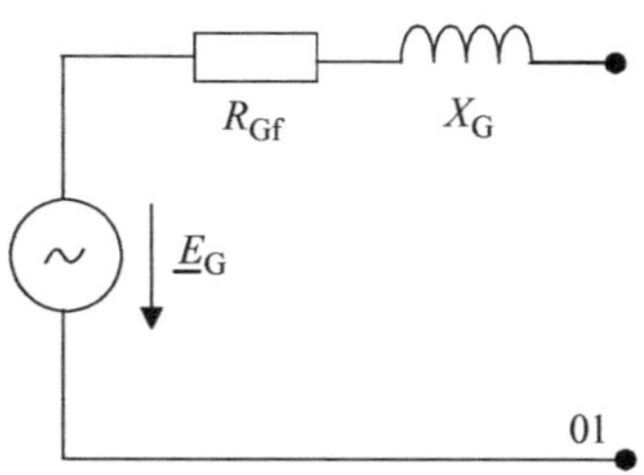 Equivalent circuit diagram	$X_{1G} = \frac{x''_d * U^2_{rG}}{100\% * S_{rG}} * K_G$ Salient pole generators: $X_{2G} = 0.5 * (X''_d + X''_q) * K_G$ Zero-sequence component $X_{0G} \approx (0.4\text{–}0.8) X''_d * K_G$ Correction factor K_G: $K_G = \frac{U_{nQ}}{U_{rG}(1 + p_G)} * \frac{c_{max}}{1 + x''_d * \sin\varphi_{rG}}$	c_{max} Voltage factor according to Table 4.1 p_G Voltage control range: $U_{rG} = \text{const} \Rightarrow p_G = 0$ S_{rG} Rated apparent power U_{rG} Rated voltage x''_d Subtransient reactance φ_{rG} Phase angle between U_{rG} and I_{rG} Fictitious resistance of stator R_{Gf} (for peak s.-c. current) HV: $R_{Gf} = 0.05 * X_G : S_{rG} \geq 100$ MVA $R_{Gf} = 0.07 * X_G : S_{rG} < 100$ MVA LV: $R_{Gf} = 0.15 * X_G$ – Correction factor for positive-, negative- and zero-sequence component – Synchronous motors identical to synchronous generators – Calculation of decaying d.c. component with real resistance of stator – Impedance in the zero-sequence component depends on the type of winding of the generator – Reference Item 3.6 of IEC 60909

Table 3.6 Impedance of power-station unit, equivalent circuit diagram, calculation equations and remarks

Figure	Impedance	Remarks
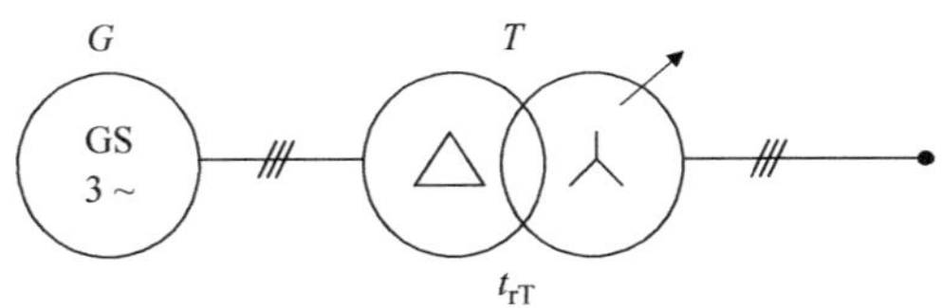 Symbol 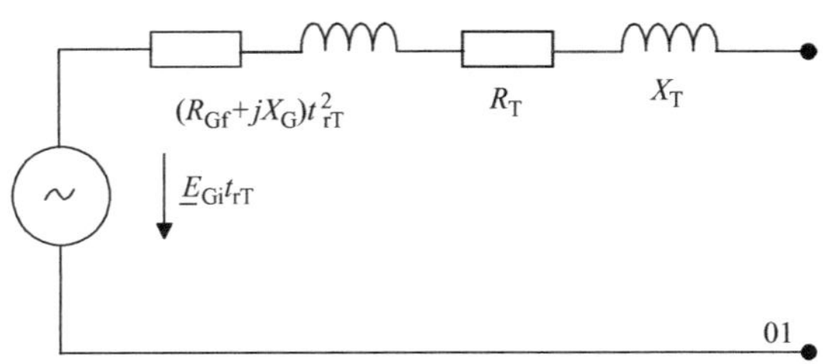 Equivalent circuit diagram	$\underline{Z}_{KW} = (\underline{Z}_G * t_{rT}^2 + \underline{Z}_{THV}) * K_{KWi}$ Correction factor K_{KWi} with tap-changer: $K_{KWs} = \frac{U_{nQ}^2}{(U_{rG}(1+p_G))^2} * \frac{U_{rTLV}^2}{U_{rTHV}^2} * \frac{c_{max}}{1+\lvert x_d'' - x_T\rvert * \sin\varphi_{rG}}$ Correction factor K_{KWi} without tap-changer: $K_{KWo} = \frac{U_{nQ}}{U_{rG}(1+p_G)} * \frac{U_{rTLV}}{U_{rTHV}} * (1 \pm p_T) * \frac{c_{max}}{1+x_d'' \sin\varphi_{rG}}$	U_{nG} Nominal system voltage Generator: p_G Control range of voltage: $U_{rG} = \text{const} \Rightarrow p_G = 0$ U_{rG} Rated voltage x_d'' Subtransient reactance Z_G Impedance according to Table 3.5 φ_{rG} Phase angle between U_{rG} and I_{rG} Unit transformer: c_{max} Voltage factor according to Table 4.1 p_T Permanent setting of winding t_{rT} Rated transformation ratio U_{rTLV} Rated voltage LV-side U_{rTHV} Rated voltage HV-side x_T Reactance (equal impedance voltage) Z_T Impedance according to Table 3.2 (two-winding transformer) • Correction factor for positive-, negative- and zero-sequence component [37] • Reference Item 3.7 of IEC 60909

Table 3.7 Impedance of overhead line (single-circuit), equivalent circuit diagram, calculation equations and remarks

Figure	Impedance	Remarks
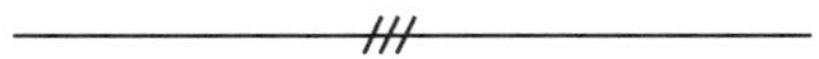 Symbol	$R'_{1L} = \frac{\rho}{n * q_n}$	D Geometric mean distance between conductors $d_{RY}; d_{YB}; d_{BR}$ Distances of conductors R, Y, B
	$R'_{L\delta} = \frac{(1 + \alpha * (\delta - 20°C)) * R'_{L20}}{n}$	n Number of conductors per phase
		R'_{1L} Resistance per length (pos. seq.)
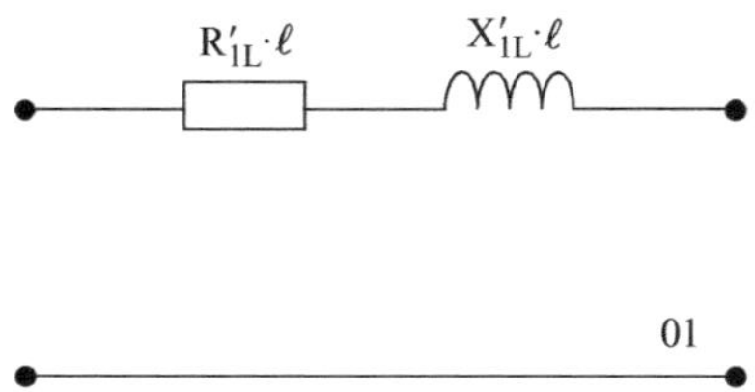 Equivalent circuit diagram	Single-circuit overhead line:	r Radius of conductor
	$X'_{1L} = \omega\frac{\mu_0}{2\pi}\left(\ln\frac{D}{r_B} + \frac{\mu_r}{4n}\right)$	r_t Radius of conductor arrangement per phase
		q_n Nominal cross-section of conductor
		X'_{1L} Reactance per length (pos. seq.)
	Zero-sequence component:	α Temperature coefficient $\alpha = 0.004\ K^{-1}$
	$R'_{0L} = \frac{R'_{1L}}{n} + 3\omega\frac{\mu_0}{8}$	δ Equivalent depth of earth conductor
		ρ Resistivity in $\Omega mm^2/m$
	$X'_{0L} = \omega\frac{\mu_0}{2\pi}\left(3\ln\frac{\delta}{\sqrt[3]{r_B * D^2}} + \frac{\mu_r}{4n}\right)$	ρ_E Specific earth resistance
		μ_0 Permeability of vacuum
		μ_r Relative permeability
	with	Reactance:
	$D = \sqrt[3]{d_{RY} * d_{YB} * d_{BR}}$	– Calculation from geometrical arrangement, see Section 13.1.4
	$r_B = \sqrt[n]{n * r * r_t^{(n-1)}}$	
	$\delta = \frac{1.85}{\sqrt{\mu_0 * \omega / \rho_E}}$	– Specific earth resistance $\rho_E = 30\ \Omega m$ (swamp soil)–3000 Ωm (stone)
		– Reference Item 3.4 of IEC 60909

Note: Impedances of other arrangements of overhead lines needed for special technical problems are dealt with in Section 12.1 and Section 13.1.4. Impedances of cables can be calculated from geometrical data only in a very time consuming manner. It is recommended to use manufacturer's data. Tables and diagrams can be found in [1,2,8,9].

Table 3.8 Impedance of short-circuit limiting reactor, equivalent circuit diagram, calculation equations and remarks

Figure	Impedance	Remarks
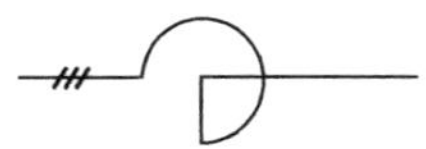 Symbol	$Z_D = \frac{u_{kr}}{100\%} * \frac{U_n}{\sqrt{3} * I_{rD}}$ $S_{rD} = \sqrt{3} * U_n * I_{rD}$	I_{rD} rated current S_{rD} Rated apparent power U_n Nominal system voltage u_{kr} Rated voltage drop (impedance voltage)
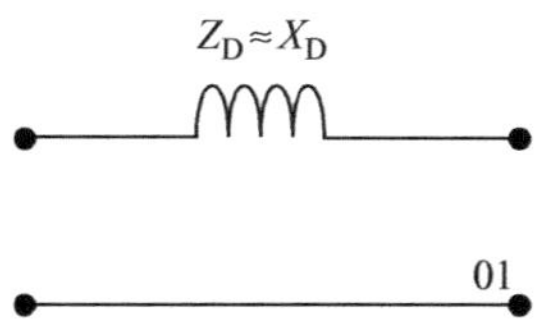 Equivalent circuit diagram		– Impedances in positive-, negative- and zero-sequence component identical in case of symmetrical construction – $R_D \approx 0$ – Impedance in the zero-sequence component equal to the impedance in the positive-sequence component in case three single-phase reactors are used – Reference Item 3.5 of IEC 60909

Table 3.9 Impedance of asynchronous motor, equivalent circuit diagram, calculation equations and remarks

Figure	Impedance	Remarks
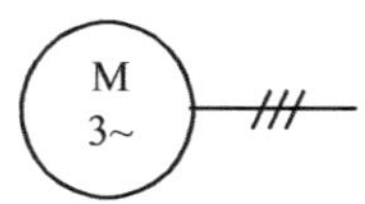 Symbol 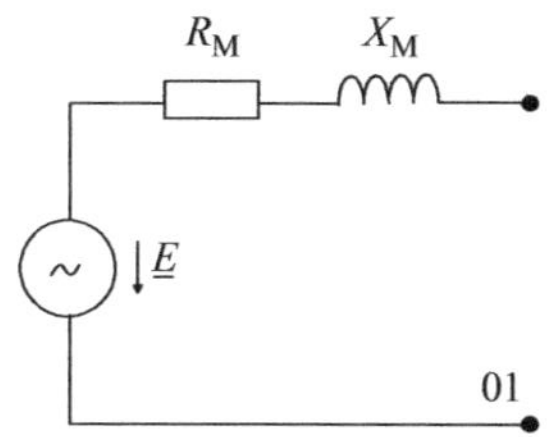 Equivalent circuit diagram	$Z_M = \frac{I_{rM}}{I_{anM}} * \frac{U_{rM}^2}{S_{rM}}$ $X_M = \frac{Z_M}{\sqrt{1 + (R_M/X_M)^2}}$ $S_{rM} = \frac{P_{rM}}{\eta_{rM} * \cos\varphi_{rM}}$	I_{anM} Locked rotor current I_{rM} Rated current P_{rM} Rated active power S_{rM} Rated apparent power φ_{rM} Phase angle at rated power η_{rM} Rated power factor MV: $R_M = 0.1 * X_M$ with $P_{rMp} \geq 1$ MW $R_M = 0.15 * X_M$ with $P_{rMp} < 1$ MW P_{rMp} Rated active power per pole pair LV: $R_M = 0.42 * X_M$ including connection cable – Asynchronous motors are normally operated with isolated neutrals, zero-sequence impedance therefore can be neglected – Reference Item 3.8 of IEC 60909

Table 3.10 Impedance of static converter fed drive, equivalent circuit diagram, calculation equations and remarks

Figure	Impedance	Remarks
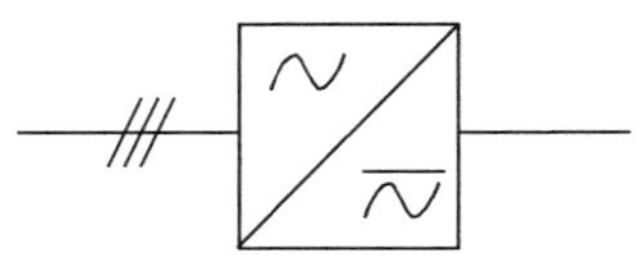 Symbol 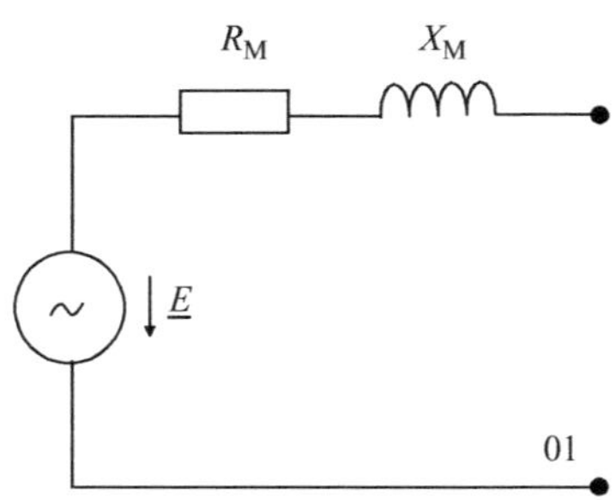 Equivalent circuit diagram	$Z_M = \frac{I_{rM}}{I_{anM}} * \frac{U_{rM}^2}{S_{rM}}$ $\frac{R_M}{X_M} = 0.1$	I_{anM} Locked rotor current I_{rM} Rated current S_{rM} Rated apparent power U_{rM} Rated voltage – $I_{anM}/I_{rM} = 3$ – Only for rectifiers, able to transfer energy for deceleration during the duration of short-circuit – Static converters for photovoltaic generators or fuel cells contribute to short-circuit currents only with their rated current – Reference Item 3.9 of IEC 60909

Table 3.11 Impedance of system load, equivalent circuit diagram, calculation equations and remarks

Figure	Impedance	Remarks
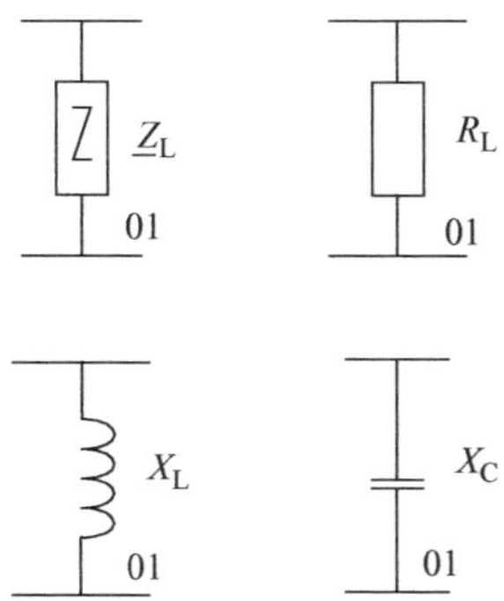 Symbol, Equivalent circuit diagram	$Z_L = \frac{U_{rL}^2}{S_{rL}}$ $R_L = \frac{U_{rL}^2}{P_{rL}}$ $\omega L_L = \frac{U_{rL}^2}{Q_{rL}}$ $\frac{1}{\omega C_L} = \frac{U_{rL}^2}{Q_{rL}}$	P_{rL} Rated active power Q_{rL} Rated reactive power S_{rL} Rated apparent power U_{rL} Rated voltage • In case U_{rL} is unknown, the nominal system voltage U_n is to be used • Reference Item 3.10 of IEC 60909

3.3.2 Impedance calculation

Table 3.12 Impedance of a conductor, equivalent circuit diagram, calculation equations and remarks

Figure	Impedance	Remarks
— Symbol	$R'_{\mathrm{L}} = \frac{\rho}{q_{\mathrm{n}}}$ $R'_{\mathrm{L}\delta} = (1 + \alpha * (\delta - 20°\mathrm{C})) * R'_{\mathrm{L20}}$	a Distance of conductors b Height of conductor (busbar) d Thickness of conductor R_{J} Resistance of bolted joint R'_{L} Specific resistance R_{L20} Resistance at 20°C
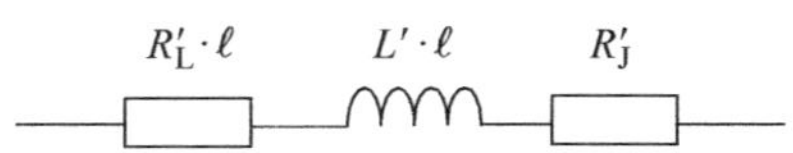 Equivalent circuit diagram	Loop by single cables $L' = \frac{\mu_0}{\pi} * \left(0.25 + \ln\left(\frac{a}{r}\right)\right)$ Loop by busbars $L' = \frac{\mu_0}{\pi} * \left(1.5 + \ln\left(\frac{a}{d+b}\right)\right)$	$R_{\mathrm{L}\delta}$ Resistance at temperature δ r Radius of conductor q_{n} Cross-section of conductor α Temperature coefficient (for Al, Cu) $\alpha = 0.004\ \mathrm{K}^{-1}$ δ Temperature in °C ρ Resistivity in $\Omega\mathrm{mm}^2/\mathrm{m}$ μ_0 Permeability
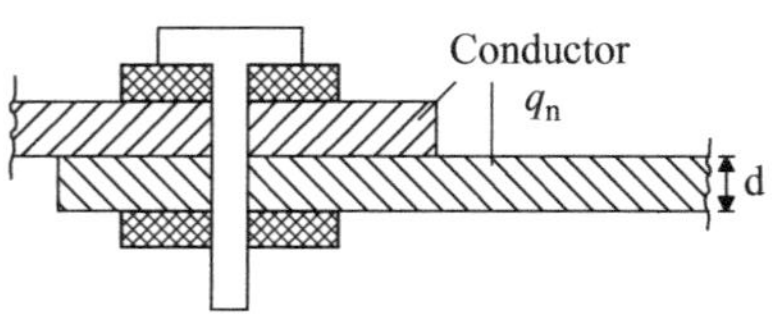 Conductor joint	Resistance of bolted joint: $R_{\mathrm{J}} = \frac{14 * \rho * d}{q_{\mathrm{n}}}$	– R_{J} for calculation of minimal s.-c. currents only – Calculation of L' for simple arrangement only – Reference Item 2.3 of IEC 61660-1

Table 3.13 Impedance of capacitor, equivalent circuit diagram, calculation equations and remarks

Figure	Impedance	Remarks
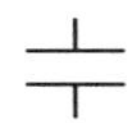 Symbol 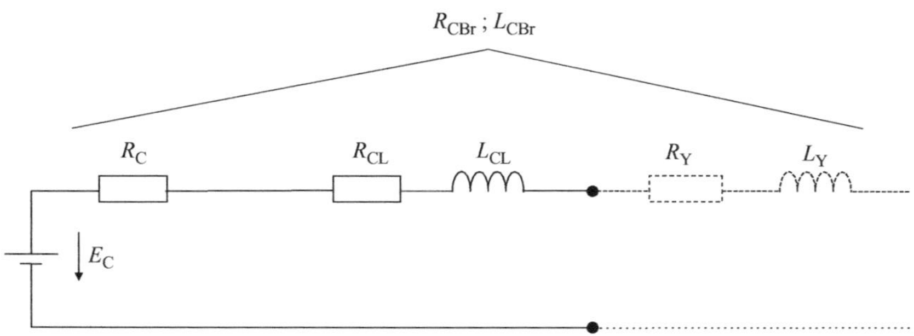 Equivalent circuit diagram	$R_{CBr} = R_C + R_{CL} + R_Y$ $L_{CBr} = L_{CL} + L_Y$ Approximation: $C \approx 1.2 * C_{a.c.}$	Conductor for connection: R_{CL}; L_{CL} Resistance; inductance Common branch: R_Y; L_Y Resistance; inductance Capacitor: C d.c.-capacitance $C_{a.c.}$ a.c.-capacitance at 100 Hz R_C Equivalent series d.c. resistance. If unknown: maximal a.c. resistance – Reference Item 2.6 of IEC 61660-1

Table 3.14 Impedance of battery, equivalent circuit diagram, calculation equations and remarks

Figure	Impedance	Remarks
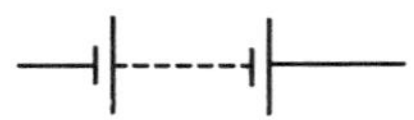 Symbol 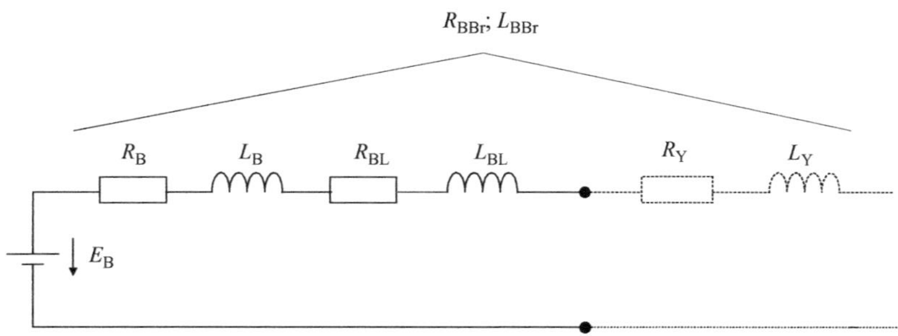 Equivalent circuit diagram	$R_{BBr} = 0.9 * R_B + R_{BL} + R_Y$ $L_{BBr} = L_B + L_{BL} + L_Y$ Approximation: $R_{Bun} = 1.7 * R_B$ $L_{BZ} = 0.2\ \mu H$ $E_{Bge} = 1.05 * U_{nB}$ $E_{Bun} = 0.9 * U_{nB}$ $E_{Bge} = 1.05 * U_{nB}$	Conductor for connection: R_{BL}; L_{BL} Resistance; inductance Common branch: R_Y; L_Y Resistance; inductance Battery: E_{Bge} Voltage of charged battery E_{Bun} Voltage of discharged battery L_{BZ} Inductance of one cell including connection R_B; L_B Resistance; inductance of a charged battery R_{Bun} Resistance of discharged battery U_{nB} Nominal voltage of battery Lead-acid battery: $U_{nB} = 2.0$ V per cell – Reference Item 2.5 of IEC 61660-1

Table 3.15 Impedance of rectifier, equivalent circuit diagram, calculation equations and remarks

Figure	Impedance	Remarks
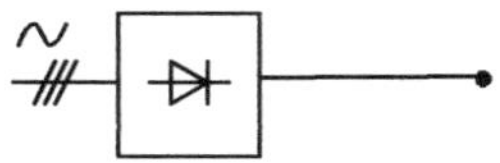 Symbol 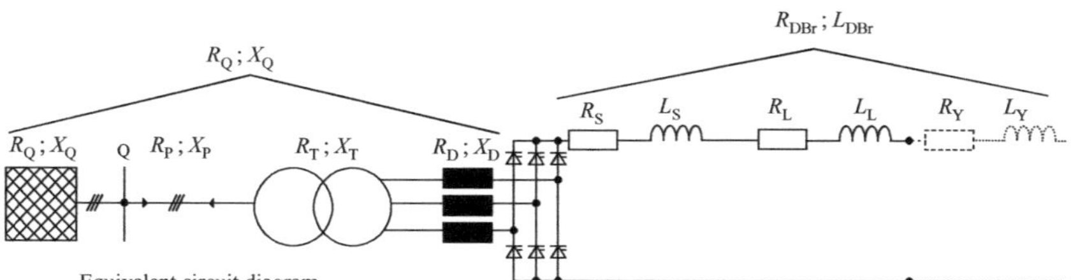 Equivalent circuit diagram	ac-side $R_N = R_Q + R_P + R_T + R_D$ $X_N = X_Q + X_P + X_T + X_D$	Indices for ac-side: *D* Commutating reactor *N* Total *Q* System feeder *P* Connection at secondary side of transformer – Z_{Qmin} (R_Q and X_Q) for calculation of max. short-circuit current – Z_{Qmax} (R_Q and X_Q) for calculation of min. short-circuit current
	dc-side $R_{DBr} = R_S + R_L + R_Y$ $L_{DBr} = L_S + L_L + L_Y$	Indices for dc-side: *L* Power supply cable *S* DC saturated smoothing reactor Coupling branch: L_S saturated inductance R_Y; L_Y Resistance; inductance – Reference Item 2.4 of IEC 61660-1

Table 3.16 Impedance of d.c. motor with independent excitation, equivalent circuit diagram, calculation equations and remarks

Figure	Impedance	Remarks
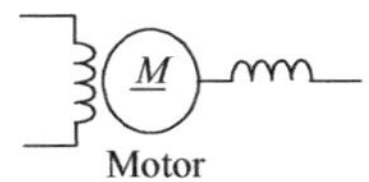 Symbol 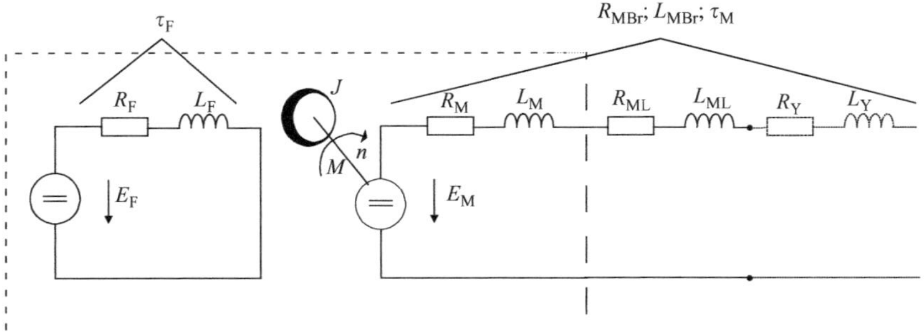 Equivalent circuit diagram	$R_{MBr} = R_M + R_{ML} + R_Y$ $L_{MBr} = L_M + L_{ML} + L_Y$ Time constants: $\tau_M = \frac{L_{MZw}}{R_{MZw}}$ $\tau_{Mec} = \frac{2\pi * n_0 * J * R_{MBr} * I_{rM}}{M_r * U_{rM}}$ $\tau_F = \frac{L_F}{R_F}$	Conductor for connection: R_{ML} Resistance L_{ML} Inductance Common branch: R_Y Resistance L_Y inductance Motor: I_{rM} Rated current J Moment of inertia of whole rotating part M_r Rated torque n_0 No-load speed R_M; L_M Resistance; inductance of armature circuit including brushes U_{rM} Rated voltage τ_F Field circuit time constant τ_M Time constant of armature circuit up to s.-c. location τ_{Mec} Mechanical time constant – Reference Item 2.7 of IEC 61660-1

3.4 Examples for calculation

3.4.1 a.c. equipment

The impedance (resistance and reactance) of equipment in a.c. three-phase power systems has to be calculated based on the data as below. Results are summarised in Table 3.17.

Power system feeder Q:	$S''_{kQ} = 3000$ MVA; $U_{nQ} = 110$ kV
Two-winding transformer:	$S_{rT} = 70$ MVA; $U_{rTHV}/U_{rTLV} = 115$ kV/10.5 kV; $u_{kr} = 12\%$; $u_{Rr} = 0.5\%$
Three-winding transformer:	$U_{rT} = 110$ kV/30 kV/10 kV; $S_{rT} = 30$ MVA/20 MVA/10 MVA $u_{krHVMV} = 10\%$; $u_{krHVLV} = 4.5\%$; $u_{krMVLV} = 10.2\%$ $u_{RrHVMV} = 0.5\%$; $u_{RrHVLV} = 0.6\%$; $u_{RrMVLV} = 0.65\%$
Synchronous machine:	$S_{rG} = 70$ MVA; $U_{rG} = 10.5$ kV; $\cos\varphi_{rG} = 0.9$; $x''_d = 17\%$; $p_G = \pm 10\%$
Power plant consisting of synchronous machine and two-winding transformer:	$S_{rG} = 70$ MVA; $U_{rG} = 10.5$ kV; $x''_d = 17\%$; $p_G = \pm 10\%$ $S_{rT} = 70$ MVA; $U_{rTHV}/U_{rTLV} = 115$ kV/10.5 kV; $u_{kr} = 12\%$; $U_{Rr} = 0.5\%$; without tap-changer
Overhead transmission line:	Al/St 240/40; $r = 10.9$ mm; Line length 10 km Flat arrangement, distance between phase wires 4 m
s.-c. limiting reactor:	$u_{kr} = 6\%$; $I_{rD} = 630$ A; $U_n = 10$ kV
Asynchronous motor:	$P_{rM} = 1.2$ MW; $U_{rM} = 6$ kV; $\cos\varphi_{rM} = 0.84$; $\eta_{rM} = 0.93$; $I_{an}/I_{rM} = 5.6$; 2 pairs of poles
Rectifier:	$S_{rM} = 4$ MVA; $U_{rM} = 6.2$ kV
System load:	$S_{rL} = 6$ MVA; $\cos\varphi_L = 0.87$; $U_n = 10$ kV

Table 3.17 Results of calculation of impedance in three-phase a.c. equipment

Equipment	*Z* (Ω)	*R* (Ω)	*X* (Ω)	Remark	Tab.
System feeder	4.437	0.441	4.414	R_Q/X_Q not defined $X_Q \approx 0.995 \times Z_Q$	3.1
Two-winding transformer	22.76	0.945	22.74	Without correction factor Impedance related to 110 kV	3.2
	22.19	0.921	22.17	$K_T = 0.975$ Impedance related to 110 kV	

Table 3.17 Continued

Equipment	Z (Ω)	R (Ω)	X (Ω)	Remark	Tables
Three-winding transformer	−3.24 62.89 58.65	1.31 1.67 6.08	−2.97 62.87 58.33	Impedance related to 110 kV including correction factors $K_T = 0.986–1.018–0.985$ Values from top: HV ∗ MV ∗ LV	3.3
Synchronous machine	0.268 0.238	0.019 0.017	0.267 0.237	Without correction factor $K_G = 0.887$; $U_{nQ} = 10$ kV	3.5
Power plant	65.14	3.22	54.77	$K_{KWo} = 0.891$; $U_{nQ} = 110$ kV	3.6
Overhead transmission line	4.19	1.23	4.01	$D = 5.04$ m	3.7
s.-c. limiting reactor	0.55	0	0.55	$X_D \approx Z_D$	3.8
Asynchronous machine	4.14	0.41	4.12	$S_{rM} = 1.53$ MVA	3.9
Rectifier	28.83	2.88	28.69	$R_M/X_M = 0.1$	3.10
System load	16.67	14.5	8.22	—	3.11

3.4.2 d.c. equipment

The impedance (resistance and reactance) of equipment in d.c. systems has to be calculated based on the data as below. Results are summarised in Table 3.18.

Conductor with joint:	Busbar arrangement, copper (120 × 10): $q_n = 1200$ mm^2; Distance $a = 50$ mm; Length of line loop 30 m
Capacitor:	MKP dry-type, self-healing; $C = 9000$ µF; $R_C = 0.5$ mΩ Connected to short-circuit location with conductor as above, $l = 20$ m Two bolted joints
Battery:	Sealed lead-acid-type; 108 cells, each: 150 Ah; $U_{nB} = 2.0$ V; $R_B = 0.83$ mΩ; $L_B = 0.21$ mH Connected to short-circuit location with conductor as above, $l = 15$ m Two bolted joints

Rectifier:	AC-system: $U_{nQ} = 600$ V; $S''_{kQ} = 40$ MVA; $R_Q/X_Q = 0.25$ Transformer: $t_{rT} = 600$ V/240 V; $S_{rT} = 400$ kVA; $u_{krT} = 3.5\%$; $P_{krT} = 4.2$ kW Rectifier: $I_{rD} = 1.2$ kA; commutating reactor: $L_S = 0.31$ μH; $R_S = 0.91$ mΩ Connected to short-circuit location with conductor as above, $l = 10$ m
d.c. motor (independent excitation):	$U_{rM} = 225$ V; $P_{rM} = 110$ kW; $I_{rM} = 500$ A; $R_M = 0.043$ Ω; $L_M = 0.41$ mH $R_F = 9.85$ M; $L_F = 9.97$ H Connected to short-circuit location with conductor as above, $l = 10$ m

Table 3.18 Results of calculation of impedance of equipment in d.c. installations (without common branch as per IEC 61660-1)

Equipment	R (mΩ)	L	Others	Remarks	Tab.
Conductor	926	0.653 μH	—	Loop length 60 m	3.12
	2.16	—	—	Resistance of bolted joint	
Capacitor	0.5	—	9000 μF		3.13
	309	218 nH	—	Conductor	
	4.32	—	—	Two joints	
	313.82	218 nH	9000 μF	Total	
Battery	$R_B = 89.6$ $R_{Bun} = 152.4$	$L_B = 21.6$ μH	$E_{Bge} = 226.8$ V $E_{Bun} = 194.4$ V	Voltage of discharged battery e.g., 1.8 V/cell	3.14
	231.5	163.2 nH	—	Conductor	
	4.32	—	—	Two joints	
	325.42 388.22	21.76 μH	—	Total	
Rectifier	0.367	4.667 μH	$Z_Q = 1.51$ mΩ	System feeder	3.15
	1.512	4.813 μH	$Z_T = 5.04$ mΩ	Transformer	
	1.879	9.48 μH	—	Total a.c. system	

Table 3.18 Continued

Equipment	R (mΩ)	L	Others	Remarks	Tab.
	0.91	0.31 μH	—	Commutating reactor	
	154.3	0.11 μH	—	Conductor	
	155.21	0.42 μH	—	Total d.c. system	
d.c. motor	43.3	0.41 mH	—	Motor	3.16
with	154.3	0.11 μH		Conductor	
independent	197.6	0.411 mH	$\tau_M = 2.08$ ms	Total	
excitation			$\tau_F = 1.01$ s		

Chapter 4

Calculation of short-circuit current in a.c. three-phase HV-systems

4.1 Types of short-circuits

In three-phase a.c. systems it has to be distinguished between different types of short-circuits (s.-c.), as outlined in Figure 4.1.

Short-circuit currents can be carried out with different methods and in different details, depending on the available data and the technical needs. IEC 60909-0 calculates characteristic parameters of the short-circuit current, which are necessary for the design of power system equipment. The course of time of short-circuit currents is outlined in Figure 4.2. Generally, one has to distinguish between near-to-generator and far-from-generator short-circuits. A near-to-generator short-circuit exists if the contribution to the short-circuit current of one synchronous generator is greater than twice its rated current, or if the contribution to the short-circuit current of synchronous or asynchronous motors is greater than 5 per cent of the short-circuit current without motors.

The analysis of the short-circuit current in the case of near-to-generator short-circuit as per Figure 4.2(a) indicates two components, besides the decaying d.c. component, a subtransient and a transient decaying a.c. component. The first or subtransient component is determined by the impedance between stator and damping winding, called subtransient reactance X''_{d}. The subtransient component decays with the subtransient time constant T'' which is normally in the range of some periods of the system frequency, i.e., $T'' < 70$ ms. The transient component is determined by the reactance between the stator and exciter winding, called transient reactance X'_{d}. The transient component decays with the transient time constant T', which can last up to 2.2 s for large generators. Finally, the short-circuit current is determined by the saturated synchronous reactance X_{d}.

In the case of a far-from-generator short-circuit as per Figure 4.2(b), the a.c. component remains constant throughout the total time duration of the short-circuit, as the influence of the changing reactance of generators can be neglected. The decaying

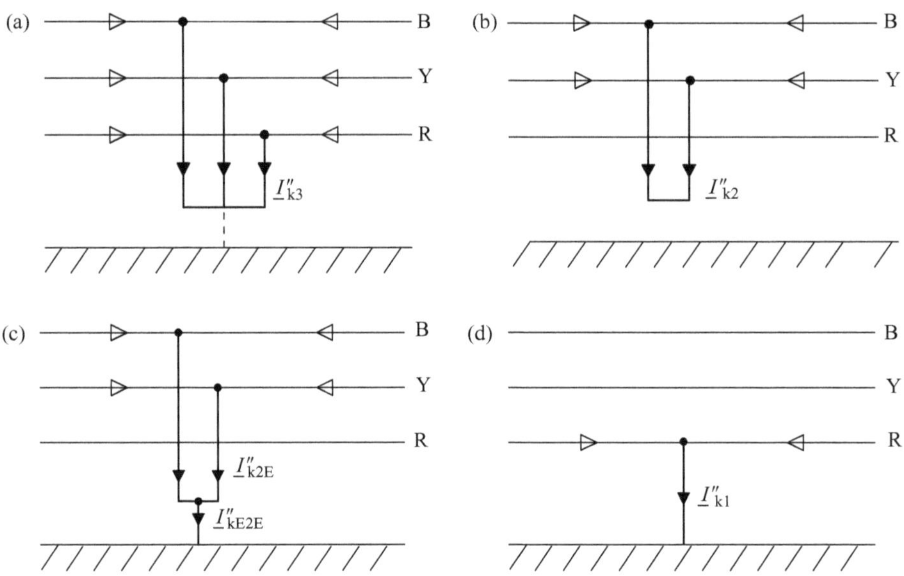

Figure 4.1 *Types of short-circuits and short-circuit currents. (a) Three-phase short-circuit, (b) double-phase short-circuit without earth/ground connection, (c) double-phase short-circuit with earth/ground connection and (d) line-to-earth (line-to-ground) short-circuit*

d.c. component is due to the ohmic-reactive short-circuit impedance and the instant of initiation of the short-circuit.

4.2 Methods of calculation

Short-circuit current calculation according to IEC 60909-0 is carried out based on the method of 'equivalent voltage source at the short-circuit location', which will be explained with the equivalent circuit diagram of a power system outlined in Figure 4.3. The method is based on the presuppositions as below:

- Symmetrical short-circuits are represented by the positive-sequence component; unsymmetrical (unbalanced) short-circuits are represented by connection of positive-, negative- and zero-sequence component as per Table 2.1.
- Capacitances and parallel admittances of non-rotating load of the positive- (and negative-) sequence component are neglected. Capacitances and parallel admittances of the zero-sequence component shall be neglected, except in systems with isolated neutral or with resonance earthing (systems with Petersen coil) as they have an influence on fault currents in power.
- Impedance of the arc at the short-circuit location is neglected.
- The type of short-circuit and the system topology remain unchanged during the duration of short-circuit.

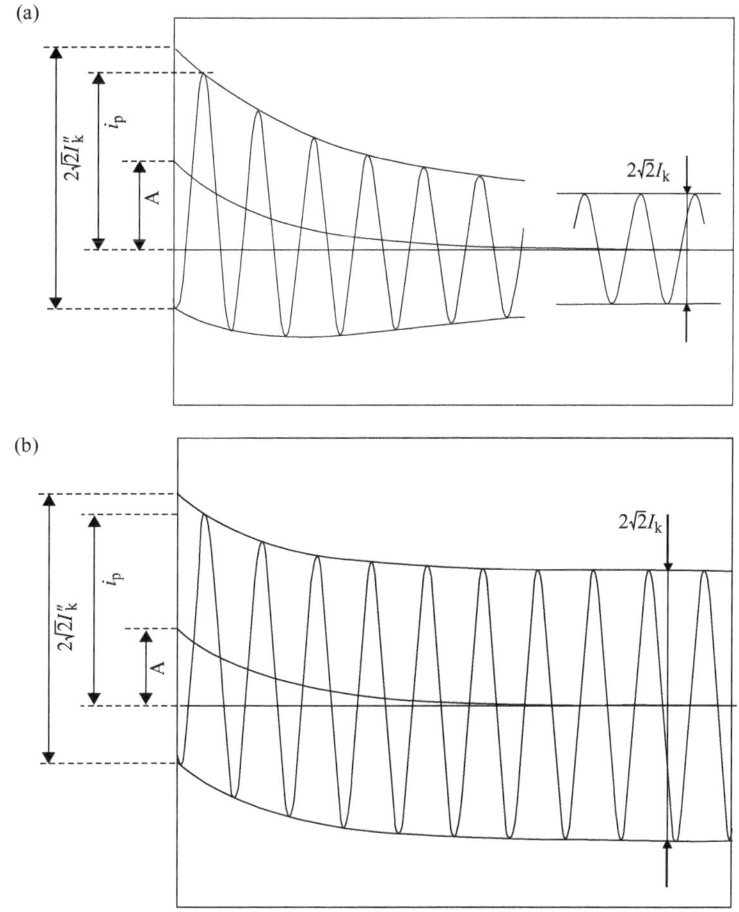

Figure 4.2 *Time-course of short-circuit currents. (a) Near-to-generator short-circuit (according to Figure 12 of IEC 60909:1988), (b) far-from-generator short-circuit (according to Figure 1 of IEC 60909:1988). I''_k – initial (symmetrical) short-circuit current, i_p – peak short-circuit current, I_k – steady-state short-circuit current and A – initial value of the aperiodic component i_{dc}*

- The tap-changers of all transformers are assumed to be in main-position (middle position).
- All internal voltages of system feeders, generators and motors are short-circuited and an equivalent voltage source with value $cU_n/\sqrt{3}$ is introduced at the short-circuit location. The voltage factor c shall be selected in accordance with Table 4.1.

The voltage factor c takes account of the differences between the voltage at the short-circuit location and the internal voltage of system feeders, motors and generators

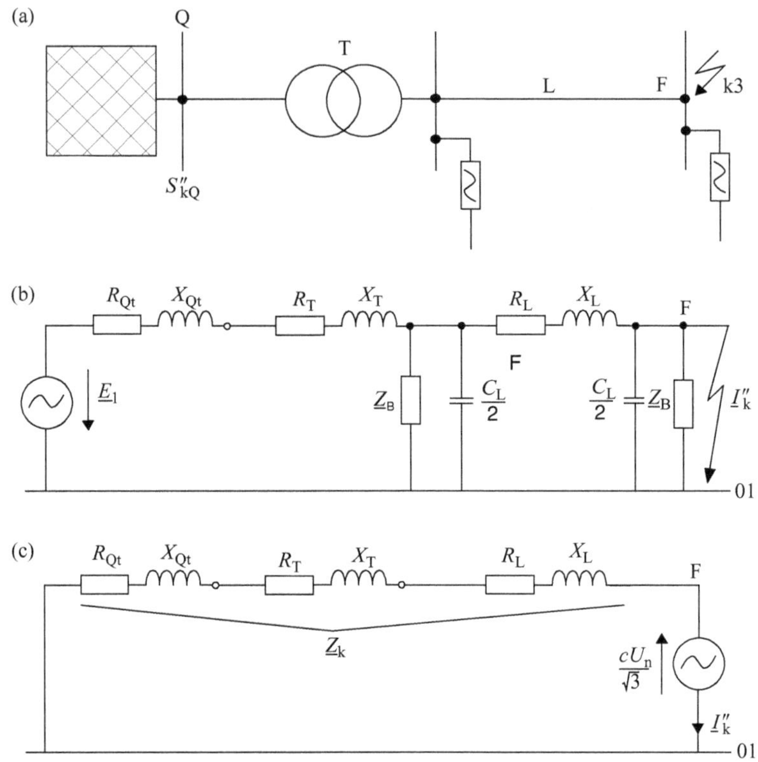

Figure 4.3 Example for short-circuit current calculation with an equivalent voltage source at s.-c. location. (a) Three-phase a.c. system with three-phase short-circuit, (b) equivalent circuit diagram in 012-system (positive-sequence system), (c) equivalent circuit diagram in 012-system with equivalent voltage source

due to voltage variations (time and place), operating of transformer tap-changer, etc. Assuming the voltage factor as per Table 4.1 will result in short-circuit currents on the safe side, that are higher than in the real power system, however, avoids an uneconomic high safety margin.

4.3 Calculation of parameters of short-circuit currents

4.3.1 General

The calculation of the impedances of power system equipment was explained in Chapter 3. It should be noted that the impedances shall be related to the voltage level of the short-circuit location and that all equipment belonging to the same voltage level shall have the same nominal voltage. The impedances, therefore, have to be calculated

Table 4.1 Voltage factor c according to IEC 60909-0

Nominal system voltage U_n	Voltage factor c for calculation of	
	Maximal s.-c. current c_{max}	Minimal s.-c. current c_{min}
LV: 100 V up to 1000 V (inclusive) (IEC 60038, Table I)		
Voltage tolerance +6%	1.05	0.95
Voltage tolerance +10%	1.10	0.95
MV: >1 kV up to 35 kV (inclusive) (IEC 60038, Table III)	1.10	1.00
HV: >35 kV (IEC 60038, Table IV)	1.10	1.00

Remark: $c_{max} U_n$ shall not exceed the highest voltage of equipment U_m as per IEC 60071.

in relation to the rated apparent power S_r of the equipment itself, respectively to the nominal system voltage U_n. In case the %/MVA-system or the p.u.-system is used, attention must be given to deviations of rated votages of the equipment from nominal system voltages, see Section 2.7.2.

The presentation within the following sections is closely related to IEC 60909-0. The IEC-document includes items for the calculation of impedances and for the calculation of the short-circuit parameters for balanced and unbalanced short-circuits, both near-to-generator and far-from-generator short-circuits. Calculation of short-circuit currents during two separate single-phase to earth short-circuits and the partial short-circuit currents flowing through earth are dealt with in IEC 60909-3, which is currently under review. These items are explained in Chapter 7.

Depending on the task, the maximal or minimal short-circuit current has to be calculated. The maximal short-circuit current is the main design criteria for the rating of equipment to withstand the effects of short-circuit currents. For the calculation of maximal short-circuit current, the items shall be considered as below:

- For the equivalent voltage source at the short-circuit location the voltage factor c_{max} as per Table 4.1 shall be used. National standards can define voltage factors different from those in Table 4.1.
- Short-circuit impedance of system feeders shall be minimal (Z_{Qmin}), so that the contribution to the short-circuit current is maximal.
- The contribution of motors has to be assessed and eventually be taken into account, see Section 4.4.
- Resistance of lines is to be calculated for a temperature of 20°C.
- Operation of power plants and system feeders shall be in such a way that the contribution to short-circuit currents will be maximal.
- System topology leading to the maximal short-circuit currents shall be taken into account.

The minimal short-circuit current is needed for the design of protection systems and the minimal setting of protection relays; details of the presuppositions for calculation are dealt with in Section 4.5.

4.3.2 Calculation of short-circuit current parameters according to IEC 60909-0

4.3.2.1 Initial symmetrical short-circuit current I_k''

The initial symmetrical short-circuit current I_k'' is calculated for balanced and unbalanced short-circuits based on the equivalent voltage source at the short-circuit location and the short-circuit impedance seen from the short-circuit location, which has to be determined with the system of symmetrical components. The results obtained for the short-circuit currents (and the voltages of the non-faulted phases, if required) in the 012-system have to be transferred back into the RYB-system. The results for the different types of short-circuits are outlined in Table 4.2.

Quantities as per Table 4.2:

c	Voltage factor according to Table 4.1
U_n	Nominal system voltage
$\underline{Z}_1; \underline{Z}_2; \underline{Z}_0$	Short-circuit impedance in the positive-, negative- and zero-sequence component

As can be obtained from Table 4.2, the value of the initial short-circuit current depends on the impedances in the positive-, negative- and zero-sequence component. Based on the impedance ratios Z_0/Z_1 and Z_2/Z_1, it can be estimated which type of short-circuit will cause the maximal initial short-circuit current. Figure 4.4 outlines the initial short-circuit currents for different types of short-circuits related to the short-circuit current of a three-phase short-circuit in variation of the impedance ratios mentioned above. As the phase angle of the impedances are different in the positive-, negative- and zero-sequence component, the diagram shall only be used for a preliminary estimate.

Figure 4.4 can be used as explained below. In the case of a far-from-generator short-circuit, Z_1 is equal to Z_2 ($Z_2/Z_1 = 1$). The maximal short-circuit current will occur in the case of a three-phase short-circuit if $Z_1/Z_0 \leq 1$. For ratios $Z_1/Z_0 > 1$, the single-phase short-circuit will result in the highest short-circuit currents. In the case of near-to-generator short-circuits the ratio of negative- to positive-sequence impedance Z_2/Z_1 mainly determines which type of short-circuit will cause the maximal short-circuit current. If $Z_1/Z_0 > 1$ the maximal short-circuit current will always occur in the case of a single-phase short-circuit.

4.3.2.2 Short-circuit currents inside power plant

When calculating short-circuit currents inside power plants, the short-circuit location itself and the installation of the unit transformer will result in a different

Table 4.2 Equations for the calculation of initial symmetrical short-circuit currents

Type of short-circuit	Equation	Section IEC 60909-0	Remarks
Three-phase	$I''_{k3} = \dfrac{cU_n}{\sqrt{3}\lvert\underline{Z}_1\rvert}$	4.2.1 4.3.1 4.6.1–4.6.3	Short-circuited phases R, Y and B
Double-phase short-circuit without earth connection	$I''_{k2} = \dfrac{cU_n}{\lvert 2\underline{Z}_1\rvert}$	4.2.2 4.3.2 4.6.4	Short-circuited phases Y and B
Double-phase short-circuit with earth connection	$I''_{kE2E} = \left\lvert \dfrac{-\sqrt{3}cU_n\underline{Z}_2}{\underline{Z}_1\underline{Z}_2 + \underline{Z}_1\underline{Z}_0 + \underline{Z}_2\underline{Z}_0} \right\rvert$	4.2.3 4.3.3 4.6.4	Current flowing through earth
General	$I''_{k2EY} = \left\lvert \dfrac{-jcU_n(\underline{Z}_0 - \underline{a}\,\underline{Z}_2)}{\underline{Z}_1\underline{Z}_2 + \underline{Z}_1\underline{Z}_0 + \underline{Z}_2\underline{Z}_0} \right\rvert$		Current of phase Y
	$I''_{k2EB} = \left\lvert \dfrac{jcU_n(\underline{Z}_0 - \underline{a}^2\underline{Z}_2)}{\underline{Z}_1\underline{Z}_2 + \underline{Z}_1\underline{Z}_0 + \underline{Z}_2\underline{Z}_0} \right\rvert$		Current of phase B Short-circuited phases Y and B
Far-from-generator ($\underline{Z}_1 = \underline{Z}_2$)	$I''_{kE2E} = \dfrac{\sqrt{3}cU_n}{\lvert\underline{Z}_1 + 2\underline{Z}_0\rvert}$	4.2.3 4.3.3 4.6.4	Current flowing through earth
	$I''_{k2EY} = \dfrac{cU_n\lvert\underline{Z}_0/\underline{Z}_1 - \underline{a}\rvert}{\lvert\underline{Z}_1 + 2\underline{Z}_0\rvert}$		Current of phase Y
	$I''_{k2EB} = \dfrac{cU_n\lvert\underline{Z}_0/\underline{Z}_1 - \underline{a}^2\rvert}{\lvert\underline{Z}_1 + 2\underline{Z}_0\rvert}$		Current of phase B Short-circuited phases Y and B
Line-to-earth single-phase short-circuit General	$I''_{k1} = \dfrac{\sqrt{3}cU_n}{\lvert\underline{Z}_1 + \underline{Z}_2 + \underline{Z}_0\rvert}$	4.2.4 4.3.4 4.6.4	Short-circuited phase R
Far-from-generator ($\underline{Z}_1 = \underline{Z}_2$)	$I''_{k1} = \dfrac{\sqrt{3}cU_n}{\lvert 2\underline{Z}_1 + \underline{Z}_0\rvert}$		

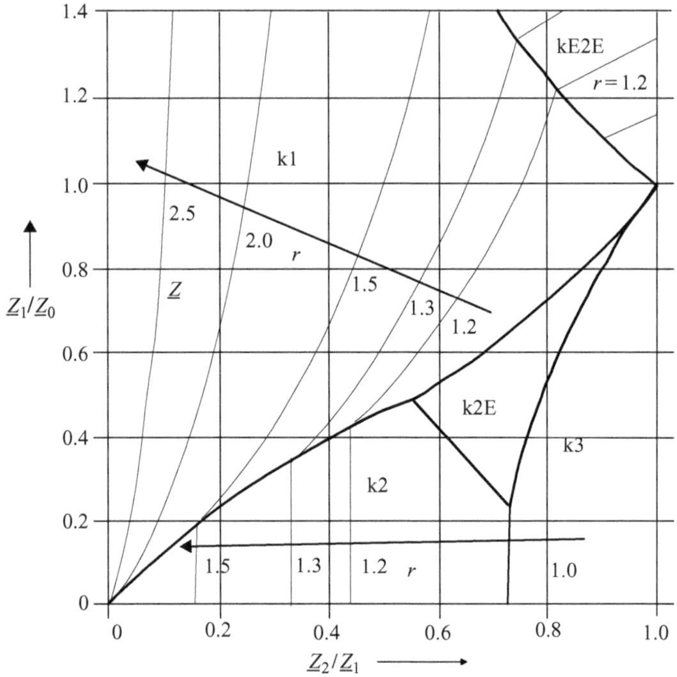

Figure 4.4 Estimate of maximal initial short-circuit current for different types of short-circuit and different impedance ratios Z_1/Z_0 and Z_2/Z_1. Phase angle of $\underline{Z}_0$, $\underline{Z}_1$ and $\underline{Z}_2$ assumed to be identical. Parameter r: ratio of asymmetrical short-circuit current to three-phase short-circuit current

approach, respectively considerations, concerning the impedances. According to Figure 4.5 different locations have to be considered, that is,

- Short-circuit between generator and unit transformer (F1)
- Short-circuit at HV-side of auxiliary transformer (F2)
- Short-circuit at MV-side of auxiliary transformer (F3)

Furthermore, the arrangement of the unit transformer, that is,

- Equipped with tap-changer
- Without tap-changer

has an important influence on the short-circuit currents.

The short-circuit current for location F1 between generator and unit transformer, if the unit transformer is equipped with tap-changer, is calculated by

$$I_{\mathrm{k}}'' = I_{\mathrm{kG}}'' + I_{\mathrm{kT}}'' = \frac{c * U_{\mathrm{rG}}}{\sqrt{3}} * \frac{1}{|K_{\mathrm{Gs}} * \underline{Z}_{\mathrm{G}}|} + \frac{1}{|\underline{Z}_{\mathrm{TLV}} + (1/t_{\mathrm{rT}}^2) * \underline{Z}_{\mathrm{Q\,min}}|} \tag{4.1}$$

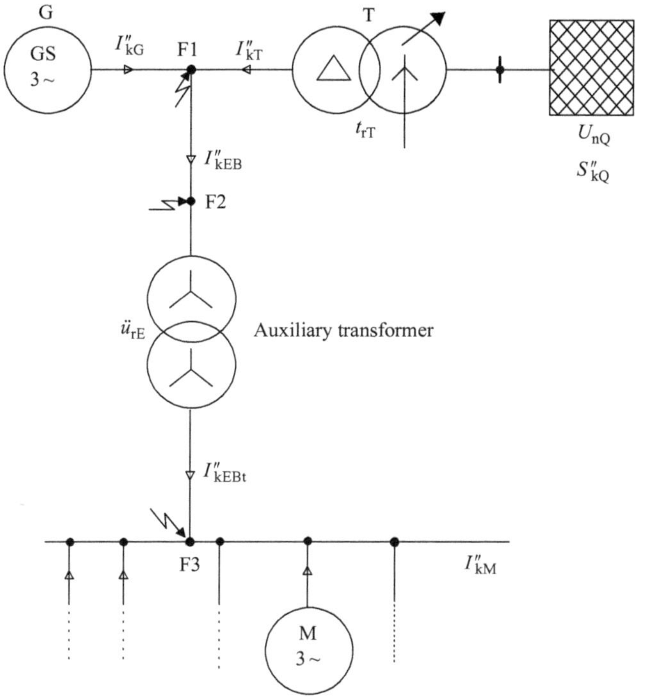

Figure 4.5 Equivalent circuit diagram for the calculation of short-circuit currents inside power plant

with the impedance correction factor K_{Gs}

$$K_{Gs} = \frac{c_{max}}{1 + x''_d * \sin \varphi_{rG}} \tag{4.2}$$

When the unit transformer is not equipped with tap-changer, the short-circuit current is given as

$$I''_k = I''_{kG} + I''_{kT} = \frac{c * U_{rG}}{\sqrt{3}} * \frac{1}{|K_{Go} * \underline{Z}_G|} + \frac{1}{|\underline{Z}_{TLV} + (1/t^2_{rT}) * \underline{Z}_{Q\,min}|} \tag{4.3}$$

with the impedance correction factor K_{Go}

$$K_{Go} = \frac{1}{1 + p_G} \frac{c_{max}}{1 + x''_d * \sin \varphi_{rG}} \tag{4.4}$$

The short-circuits at the HV-side of the auxiliary transformer at location F2 are given by

$$I''_{kEB} = \frac{c * U_{rG}}{\sqrt{3}} * \left(\frac{1}{|K_{Gs} * \underline{Z}_G|} + \frac{1}{|K_{Ts} * \underline{Z}_{TLV} + (1/t^2_{rT}) * \underline{Z}_{Q\,min}|} \right) \tag{4.5}$$

with impedance correction factor for the generator

$$K_{Gs} = \frac{c_{max}}{1 + x''_d * \sin\varphi_{rG}} \tag{4.6a}$$

and the unit transformer

$$K_{Ts} = \frac{c_{max}}{1 + x_T * \sin\varphi_{rG}} \tag{4.6b}$$

If the unit transformer is installed without tap-changer the impedance correction factors are given for the generator

$$K_{Go} = \frac{1}{1 + p_G} * \frac{c_{max}}{1 + x''_d * \sin\varphi_{rG}} \tag{4.7a}$$

and for the unit transformer

$$K_{To} = \frac{1}{1 + p_G} * \frac{c_{max}}{1 + x_T * \sin\varphi_{rG}} \tag{4.7b}$$

which shall be used instead of correction factors K_{Gs} and K_{Ts} as per Equations (4.6), respectively. The impedance Z_{rsl}

$$Z_{rsl} = \frac{1}{|K_{Gs} * \underline{Z}_G|} + \frac{1}{|K_{Ts} * \underline{Z}_{TLV} + (1/t_{rT}^2) * \underline{Z}_{Q\,min}|} \tag{4.8}$$

including the correction factors is called the coupling impedance.

Quantities of Equations (4.1)–(4.8) are

$\underline{Z}_G$	Impedance of the generator
$\underline{Z}_{TLV}$	Impedance of the unit transformer at LV-side (generator voltage)
$\underline{Z}_{Q\,min}$	Impedance of the system feeder for maximal short-circuit current
t_{rT}	Transformation ratio of the unit transformer ($t_{rT} \geq 1$)
U_{rG}	Rated voltage of generator
c; c_{max}	Voltage factor as per Table 4.1
x_T	Reactance of the transformer in p.u. (impedance voltage)
x''_d	Subtransient reactance of the generator in p.u.
p_G	Voltage control range of generator in p.u.
$\sin\varphi_{rG}$	Power factor of generator at rated operating conditions

Other quantities are explained previously.

Short-circuits at MV-side of the auxiliary transformer at location F3 or at the auxiliary busbar are a superposition of the partial short-circuit current I''_{kEBt} of the auxiliary transformer related to the voltage level of the short-circuit location and of the contribution of the motors in the auxiliary system to the short-circuit current. In the case of a unit transformer without tap-changer the impedance correction factors K_{Go} and K_{To} have to be considered accordingly.

4.3.2.3 Peak short-circuit current i_p

Depending on the feeding source of the short-circuit different considerations have to be taken to calculate the peak short-circuit current.

Figure 4.6 indicates an equivalent circuit diagram with **single-fed** short-circuit. The short-circuit impedance is represented by a series connection of the individual impedances.

The peak short-circuit current, which is a peak value, can be calculated for the different types of short-circuits based on the initial short-circuit current (r.m.s. value) by

$$i_{p3} = \kappa * \sqrt{2} I''_{k3} \tag{4.9a}$$

$$i_{p2} = \kappa * \sqrt{2} I''_{k2} \tag{4.9b}$$

$$i_{p1} = \kappa * \sqrt{2} I''_{k1} \tag{4.9c}$$

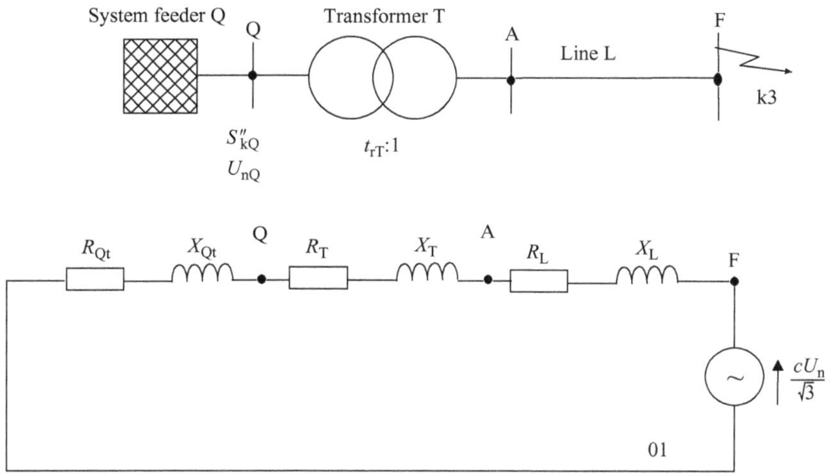

Figure 4.6 Equivalent circuit diagram for single-fed three-phase short-circuit

The peak short-circuit current i_{p2E} in the case of a double-phase short-circuit with earth connection is always smaller than either of a three-phase or single-phase short-circuit and need not be calculated separately. The factor κ can be obtained from Figure 4.7 or calculated by

$$\kappa = 1.02 + 0.98 * e^{-3(R/X)} \tag{4.10}$$

where I''_{k3}; I''_{k2}; I''_{k1} are the initial symmetrical short-circuit currents for three-phase, double-phase and line-to-earth short-circuit and R; X are the resistance and reactance of the short-circuit impedance.

Figure 4.8 indicates an equivalent circuit diagram in the case of a short-circuit *fed from non-meshed sources*. The peak short-circuit current is calculated by superposing the contributions of different sources.

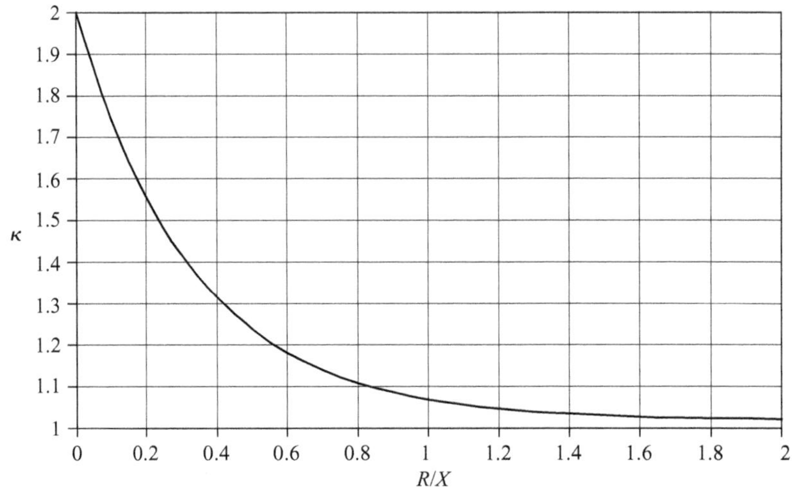

Figure 4.7 *Factor κ for the calculation of peak short-circuit current*

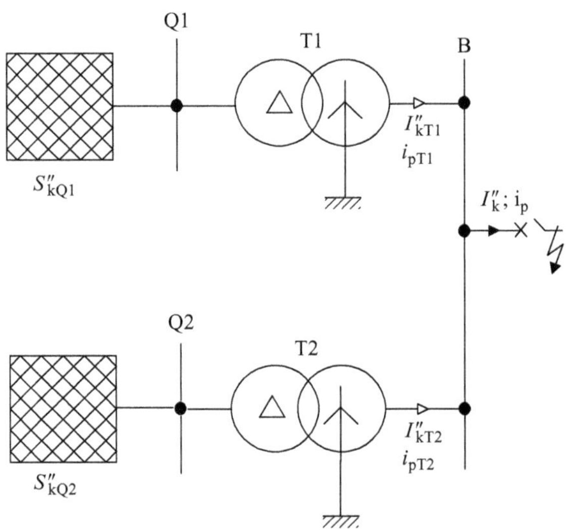

Figure 4.8 *Equivalent circuit diagram for three-phase short-circuit fed from non-meshed sources*

The peak short-circuit currents i_{p3T1} and i_{p3T2} of each branch, fed through the transformers T1 and T2, are calculated separately as well as the factors κ_{T1} and κ_{T2}.

$$i_{p3T1} = \kappa_1 * \sqrt{2} * I''_{k3T1} \tag{4.11a}$$

$$i_{p3T2} = \kappa_2 * \sqrt{2} * I''_{k3T2} \tag{4.11b}$$

The total peak short-circuit current i_{p3} is given by

$$i_{p3} = i_{p3T1} + i_{p3T2} \tag{4.11c}$$

Particular considerations are to be taken in the case of short-circuits in **meshed networks** according to Figure 4.9. The peak short-circuit current at the short-circuit location cannot be calculated by superposition as the R/X-ratios of the individual branches feeding the short-circuit are different and the direction of the branch short-circuit currents through the system is defined by the Ohmic law.

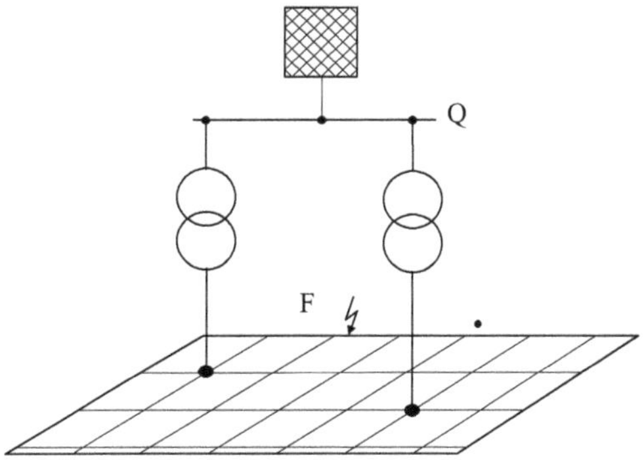

Figure 4.9 Equivalent circuit diagram of a three-phase short-circuit in a meshed system

In principle, the peak short-circuit current in a meshed system will be calculated by

$$i_{p3} = \kappa * \sqrt{2} I''_{k3} \tag{4.9a}$$

as given for three-phase short-circuits, and for other types of short-circuits accordingly. The factor κ, however, will be determined with different methods as below.

Uniform (smallest) ratio R/X. The factor κ is calculated based on the smallest ratio R/X of all branches of the network. Only those branches need to be taken into account which contribute to the short-circuit current in the power system corresponding to the short-circuit location, respectively those branches connected through transformers to the short-circuit location. The results are always on the safe side, however the accuracy is low.

Ratio R/X at short-circuit location. Based on the ratio R/X of the total system impedance at the short-circuit location, the factor κ is calculated taking account of a safety factor of 1.15 to allow for deviations due to the different ratios R/X in the different branches.

$$\kappa_b = 1.15 * \kappa \tag{4.12}$$

The factor $1.15 * \kappa$ should not exceed the value of 1.8 in LV-systems and shall not exceed 2.0 in HV-systems. The safety factor 1.15 is neglected when $R/X \leq 0.3$.

Equivalent frequency f_c. The factor κ is found from Figure 4.7 or can be calculated based on the ratio R/X:

$$\frac{R}{X} = \frac{R_c}{X_c} * \frac{f_c}{f} \tag{4.13}$$

R_c and X_c are the equivalent effective resistance and reactance at the short-circuit location at equivalent frequency f_c which is

$f_c = 20$ Hz (nominal power system frequency 50 Hz)
$f_c = 24$ Hz (nominal power system frequency 60 Hz)

The calculation of the equivalent impedance at equivalent frequency f_c is to be carried out additionally to the calculation of impedance at nominal power system frequency [34].

IEC 60909-1 mentions accuracy limits for the different methods of calculating short-circuit currents in meshed systems.

Results obtained by the method of **uniform (smallest) ratio R/X** are always on the safe side, if all branches contributing to the short-circuit current are taken into account. Errors can reach in rare cases up to 100 per cent. If only those branches are considered, which contribute up to 80 per cent to the short-circuit current and if the ratios R/X are in a wide range, the results can even be on the unsafe side. The method therefore shall be applied only if the ratios R/X are in a small bandwidth and if $R/X < 0.3$.

The method **ratio R/X at short-circuit location** (safety factor 1.15) will lead to results on the safe and unsafe side. Applying the method to ratios $0.005 \leq R/X \leq 1.0$ the error will be in the range +10 to −5 per cent.

The method of **equivalent frequency** has an accuracy of ±5 per cent, if the ratio R/X of each branch is in the range of $0.005 \leq R/X \leq 5.0$.

4.3.2.4 Decaying (aperiodic) component i_{dc}

The maximal decaying aperiodic component i_{dc} is calculated by

$$i_{dc} = \sqrt{2} * I_k'' * e^{-2\pi f t * (R/X)} \tag{4.14}$$

where I_k'' is the initial symmetrical short-circuit current, f is the power system frequency, t is the time parameter and R, X are the resistance and reactance of the short-circuit impedance.

The resistance of the short-circuit impedance shall be calculated with the real stator resistance R_G of generators instead of the fictitious resistance R_{Gf}, see Table 3.5. The ratio R/X shall be calculated in meshed systems with an equivalent frequency f_c which depends on the duration t of the short-circuit as outlined in Table 4.3.

4.3.2.5 Symmetrical short-circuit breaking current I_b

In the case of near-to-generator short-circuits, the a.c. component of the short-circuit current is decaying to a steady-state value, see Figure 4.2(a). The short-circuit current is interrupted by the switchgear at the instant of minimal time delay t_{min} of the

Table 4.3 Equivalent frequency for the calculation of decaying component

Factor	$f*t$	<1	<2.5	<5	<12.5
Ratio of equivalent frequency to power system frequency	f_c/f	0.27	0.15	0.092	0.055
Example: $f = 50$ Hz	t	<0.02 s	<0.05 s	<0.1 s	<0.25 s
	f_c	13.5 Hz	7.5 Hz	4.6 Hz	2.75 Hz
Example: $f = 60$ Hz	t	<0.02 s	<0.05 s	<0.1 s	<0.25 s
	f_c	16.2 Hz	9.0 Hz	5.52 Hz	3.3 Hz

protection. The calculation of the symmetrical short-circuit breaking current I_b is based on the initial short-circuit current and on the factor μ,

$$I_b = \mu * I''_k \tag{4.15}$$

The factor μ can be taken from Figure 4.10 or calculated by

$$\mu = 0.84 + 0.26 * e^{-0.26(I''_{kG}/I_{rG})} \quad \text{for } t_{min} = 0.02 \text{ s} \tag{4.16a}$$

$$\mu = 0.71 + 0.51 * e^{-0.30(I''_{kG}/I_{rG})} \quad \text{for } t_{min} = 0.05 \text{ s} \tag{4.16b}$$

$$\mu = 0.62 + 0.72 * e^{-0.32(I''_{kG}/I_{rG})} \quad \text{for } t_{min} = 0.10 \text{ s} \tag{4.16c}$$

$$\mu = 0.56 + 0.94 * e^{-0.38(I''_{kG}/I_{rG})} \quad \text{for } t_{min} \geq 0.25 \text{ s} \tag{4.16d}$$

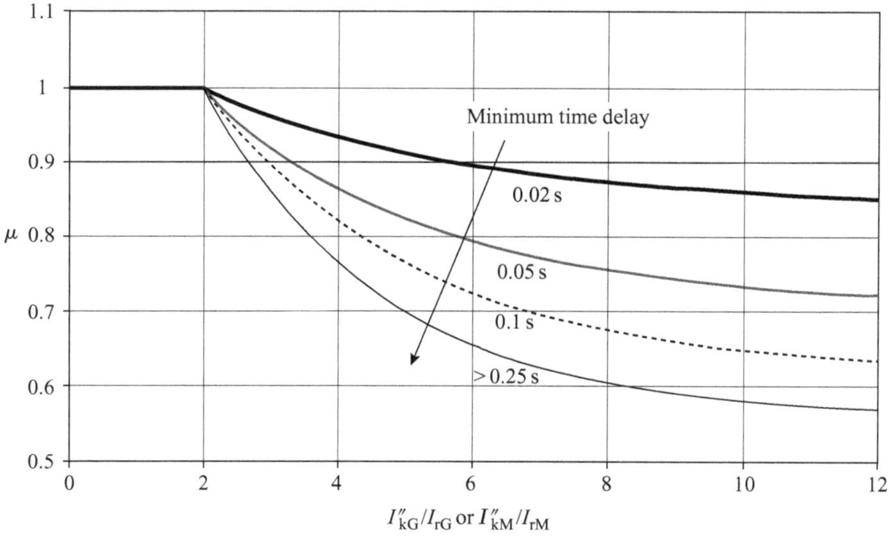

Figure 4.10 Factor μ for calculation of symmetrical short-circuit breaking current

where I''_{kG} is the initial symmetrical short-circuit current of the generator, I_{rG} is the rated current of the generator and t_{min} is the minimal time delay of the protection, switchgear and auxiliaries, that is, minimal time for switching the short-circuit current off.

The factor μ is valid for high-voltage synchronous generators, excited by rotating machines of rectifiers. If the excitation system is not known the factor shall be set to $\mu = 1$.

In the case of far-from-generator short-circuits, the symmetrical short-circuit breaking current I_b is equal to the initial short-circuit current I''_k as the a.c. component is not decaying.

4.3.2.6 Steady-state short-circuit current I_k

The steady-state short-circuit current I_k in the case of near-to-generator short-circuits depends on various factors such as saturation effects, power factor of generators, change of system topology due to operation of switching, etc. and can therefore only be determined with a certain inaccuracy. The method as per IEC 60909 determines lower and upper limits only when one synchronous machine is feeding the short-circuit. The calculation is based on the generator-rated current assuming a factor λ which depends on the ratio of initial symmetrical short-circuit current to rated current of the generator and on the saturated synchronous reactance.

Maximal excitation of the synchronous machine leads to the maximal symmetrical short-circuit breaking current

$$I_{k\,max} = \lambda_{max} * I_{rG} \tag{4.17a}$$

The factor λ_{max} is valid for turbine generators according to Figure 4.11 and for salient-pole generators as per Figure 4.12.

The minimal symmetrical short-circuit breaking current is calculated for constant no-load excitation of the generator with the factor λ_{min}.

$$I_{k\,min} = \lambda_{min} * I_{rG} \tag{4.17b}$$

The values for λ_{min} are included in Figures 4.11 and 4.12. Reference is made to the remarks in IEC 60909-0 on the factors.

Quantities as used above are

I_{rG}	Rated current of the generator
I''_{kG}	Initial synchronous short-circuit current of the generator
x_{dsat}	Saturated synchronous reactance of the generator, equal to the reciprocal of the short-circuit ratio of the generator

In the case of far-from-generator short-circuits, the symmetrical short-circuit breaking current I_b is equal to the initial short-circuit current I''_k as the a.c. component is not decaying.

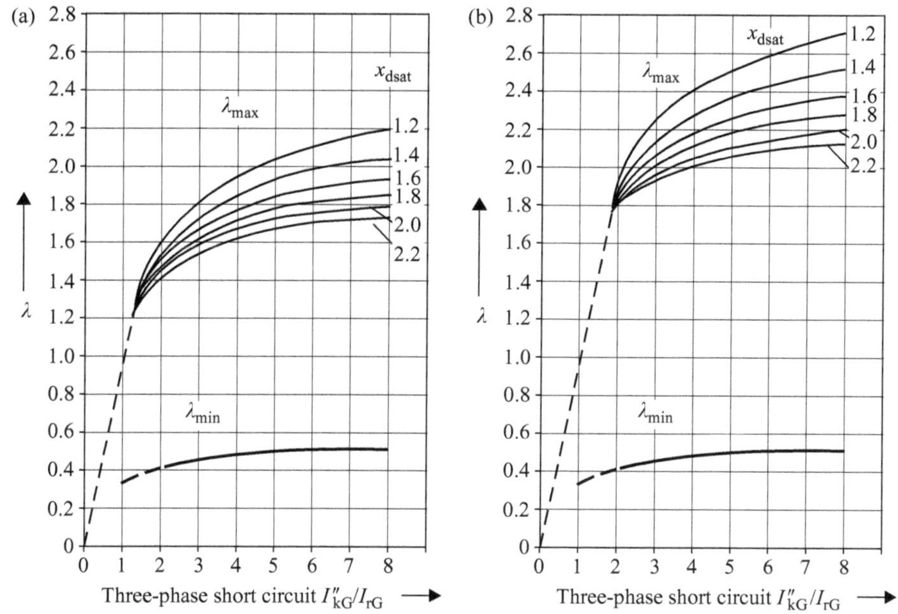

Figure 4.11 *Factors λ_{max} and λ_{min} for turbine generators (Figure 17 of DIN EN 60909-0 (VDE 0102)). (a) Series one and (b) series two*

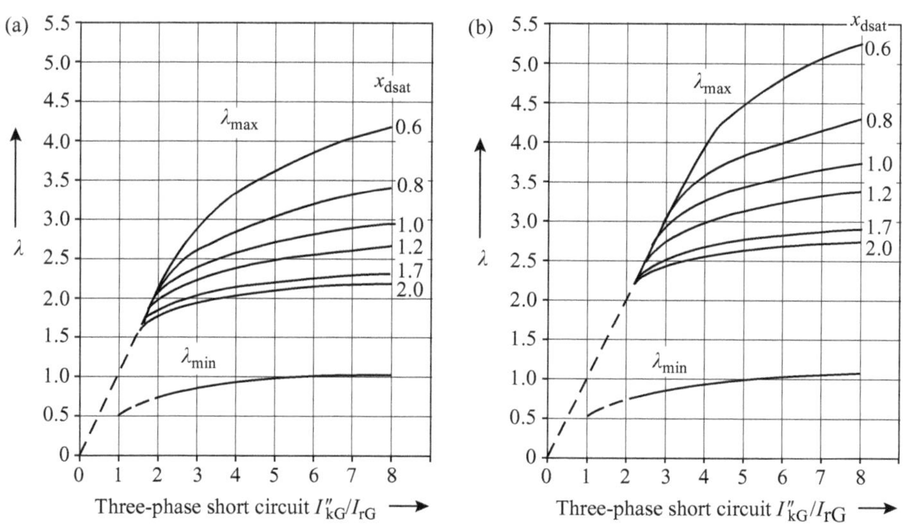

Figure 4.12 *Factors λ_{max} and λ_{min} for salient-pole generators (Figure 18 of DIN EN 60909-0 (VDE 0102) 1988). (a) Series one and (b) series two*

4.4 Influence of motors

Asynchronous motors and synchronous motors have to be taken into account in MV-systems and in auxiliary supply systems of power plants and industrial networks for the calculation of maximal short-circuit currents. They contribute to the initial symmetrical short-circuit current, to the peak short-circuit current, to the symmetrical short-circuit breaking current and in case of unbalanced short-circuits to the steady-state short-circuit current as well, see Table 4.4. Synchronous motors are modelled like generators and asynchronous generators are treated as asynchronous motors. Motors of any kind, which are not in operation at the same time, e.g., due to the process or due to any interlocking, can be neglected for the calculation of short-circuit currents. Motors fed by static-rectifiers need to be considered in the case of three-phase short-circuits only, if they are able to transfer energy for deceleration for the duration of the short-circuit, as they contribute to the initial symmetrical and to the peak short-circuit current.

Table 4.4 Calculation of short-circuit currents of asynchronous motors

Parameter	Type of short-circuit		
	Three-phase	Double-phase	Line-to-earth
Initial symmetrical short-circuit current	$I''_{\mathrm{k3M}} = \frac{cU_{\mathrm{n}}}{\sqrt{3}Z_{\mathrm{M}}}$	$I''_{\mathrm{k2M}} = \frac{\sqrt{3}}{2} I''_{\mathrm{k3M}}$	$I''_{\mathrm{k1M}} = \frac{c\sqrt{3}U_{\mathrm{n}}}{Z_{1\mathrm{M}}+Z_{2\mathrm{M}}+Z_{0\mathrm{M}}}$ in systems with earthed neutral only
Peak short-circuit current	$i_{\mathrm{p3M}} = \kappa_{\mathrm{M}}\sqrt{2}I''_{\mathrm{k3M}}$	$i_{\mathrm{p2M}} = \frac{\sqrt{3}}{2} i_{\mathrm{p3M}}$	$i_{\mathrm{p1M}} = \kappa_{\mathrm{M}}\sqrt{2}I''_{\mathrm{k1M}}$
	MV-motors: $\kappa_{\mathrm{M}} = 1.65$ $(R_{\mathrm{M}}/X_{\mathrm{M}} = 0.15)$ for active power per pole-pair <1 MW; $\kappa_{\mathrm{M}} = 1.75$ $(R_{\mathrm{M}}/X_{\mathrm{M}} = 0.10)$ for active power per pole-pair ≥ 1 MW; LV-motors including connection cables $\kappa_{\mathrm{M}} = 1.30$ $(R_{\mathrm{M}}/X_{\mathrm{M}} = 0.42)$		
Symmetrical short-circuit breaking current	$I_{\mathrm{b3M}} = \mu q I''_{\mathrm{k3M}}$	$I_{\mathrm{b2M}} \approx \frac{\sqrt{3}}{2} I''_{\mathrm{k3M}}$	$I_{\mathrm{b1M}} \approx I''_{\mathrm{k1M}}$
	μ as per Figure 4.10, q as per Figure 4.13		
Steady-state short-circuit current	$I_{\mathrm{k3M}} = 0$	$I_{\mathrm{k2M}} \approx \frac{\sqrt{3}}{2} I''_{\mathrm{k3M}}$	$I_{\mathrm{k1M}} \approx I''_{\mathrm{k1M}}$

Quantities used in the equations are explained in the text.

Asynchronous motors in public supply systems are considered when

- the sum of the rated currents is greater than 1 per cent of the initial symmetrical short-circuit current without motors;
- the contribution to the initial symmetrical short-circuit current without motors is greater or equal to 5 per cent of the initial symmetrical short-circuit current without motors.

Medium- and low-voltage asynchronous motors connected through two-winding transformers to the short-circuit are considered if

$$\frac{\sum P_{\text{rM}}}{\sum S_{\text{rT}}} > \frac{0.8}{|((c * 100 \sum S_{\text{rT}})/(\sqrt{3}U_{\text{nQ}}/I_{\text{k}}'')) - 0.3|} \tag{4.18}$$

where U_{nQ} is the nominal system voltage, $\sum S_{\text{rT}}$ is the sum of rated apparent power of all transformers, directly connected to motors feeding the short-circuit, I_{k}'' is the initial symmetrical short-circuit current without motors and $\sum P_{\text{rM}}$ is the sum of rated active power of all low- and medium-voltage motors.

In order to calculate the branch short-circuit current of asynchronous motors, Table 4.4 can be used. The factor q (three-phase short-circuit) depending on the minimal time delay of the protection t_{min} can be obtained from Figure 4.13 or by

$$q = 1.03 + 0.12 * \ln(m), \quad t_{\text{min}} = 0.02\ \text{s} \tag{4.19a}$$

$$q = 0.79 + 0.12 * \ln(m), \quad t_{\text{min}} = 0.05\ \text{s} \tag{4.19b}$$

$$q = 0.57 + 0.12 * \ln(m), \quad t_{\text{min}} = 0.10\ \text{s} \tag{4.19c}$$

$$q = 0.26 + 0.10 * \ln(m), \quad t_{\text{min}} \geq 0.25\ \text{s} \tag{4.19d}$$

where t_{min} is the minimal time delay of the protection, switchgear and auxiliaries, i.e., minimal time for switching the short-circuit current off and m is the active power of the motor per pole-pair in p.u. based on 1-MW-base. The factor q should not be greater than 1.

4.5 Minimal short-circuit currents

In order to calculate the minimal short-circuit current, the voltage factor c_{min} according to Table 4.1 for the equivalent voltage source at the short-circuit location has to be considered. Furthermore,

- System topology, generator dispatch and short-circuit power of feeding networks have to be defined in such a way that the minimal short-circuit current is expected. This normally applies for low-load conditions.
- Motors are to be neglected.
- Resistances of overhead lines and cables shall be calculated with the maximal permissible temperature at the end of the short-circuit, e.g., 80°C in low-voltage systems.

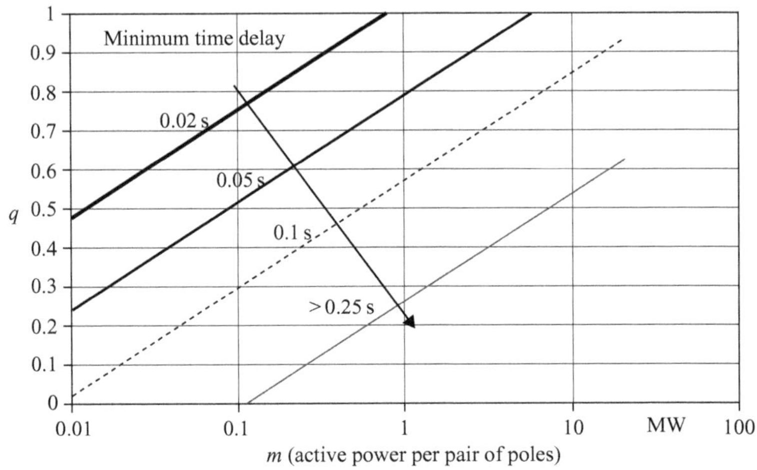

Figure 4.13 *Factor q for the calculation of symmetrical short-circuit breaking current*

These assumptions have to be taken in the case of balanced and unbalanced short-circuits, except where other presuppositions are mentioned.

4.6 Examples

Examples for the calculation of short-circuit currents are included in IEC 60909-4 besides those given below. Reference is made to the relevant chapters where the individual quantities are explained.

4.6.1 Three-phase near-to-generator short-circuit

Figure 4.14 outlines the equivalent circuit diagram of a 220-kV system. For three-phase short-circuit at busbar B (maximal s.-c. currents) the branch short-circuit currents of the generators and the system feeder as well as the contribution of the generators to the symmetrical short-circuit breaking current and to the steady-state short-circuit current ($t_{\min} = 0.1$ s; $x_{\mathrm{dsat}} = 140\%$; Turbine generator (Series 1)) are to be calculated.

Data of equipment taken from nameplates:

$$S_{\mathrm{rG1}} = 120\ \mathrm{MVA};\ U_{\mathrm{rG1}} = 10.5\ \mathrm{kV};\ \cos\varphi_{\mathrm{rG1}} = 0.8;\ x''_{\mathrm{dG1}} = 18\%$$
$$S_{\mathrm{rG2}} = 80\ \mathrm{MVA};\ U_{\mathrm{rG2}} = 10.5\ \mathrm{kV};\ \cos\varphi_{\mathrm{rG2}} = 0.85;\ x''_{\mathrm{dG2}} = 16\%$$
$$S_{\mathrm{rT1}} = 120\ \mathrm{MVA};\ U_{\mathrm{rT1HV}}/U_{\mathrm{rT1LV}} = 220\ \mathrm{kV}/10.5\ \mathrm{kV};\ u_{\mathrm{krT1}} = 14\%$$
$$S_{\mathrm{rT2}} = 80\ \mathrm{MVA};\ U_{\mathrm{rT2HV}}/U_{\mathrm{rT2LV}} = 220\ \mathrm{kV}/10.5\ \mathrm{kV};\ u_{\mathrm{krT2}} = 12\%$$
$$S_{\mathrm{rT3}} = 200\ \mathrm{MVA};\ U_{\mathrm{rT3HV}}/U_{\mathrm{rT3LV}} = 400\ \mathrm{kV}/220\ \mathrm{kV};\ u_{\mathrm{krT3}} = 12\%$$

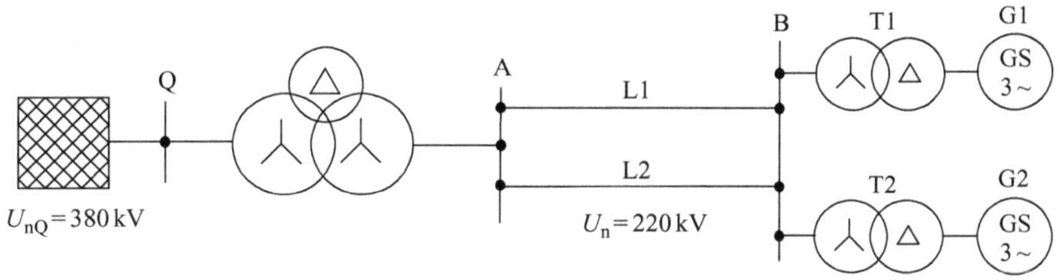

Figure 4.14 Equivalent circuit diagram of a 220-kV-system with short-circuit location

$S''_{kQ} = 5$ GVA; $U_{nQ} = 380$ kV
$X'_{L1} = X'_{L2} = 0.4\ \Omega/\text{km}; l = 50$ km

The impedances of equipment (positive-sequence component) including correction factors are

$X_{QK} = 9.632\ \Omega$
$X_{T3} = 28.31\ \Omega$
$X_{L1} = X_{L2} = 20.04\ \Omega$
$X_{KW1K} = 138.67\ \Omega$
$X_{KW2K} = 182.52\ \Omega$

The short-circuit impedance at the short-circuit location is $X_{kB} = 29.82\ \Omega$

The initial symmetrical short-circuit currents are:

$I''_{k3} = 4.69$ kA, at s.-c. location
$I''_{k3Q} = 1.69$ kA, branch s.-c. current of system feeder

The contribution of the generators to the short-circuit currents is outlined in the table below.

Parameter	Generator 1	Generator 2	Section
I''_{k3}	21.11 kA	16.04 kA	4.3.2.1
μ	0.879	0.844	
I_b	18.55 kA	13.54 kA	4.3.2.5
$\lambda_{max}/\lambda_{min}$	1.75/0.46	1.82/0.47	
$I_{k\,max}/I_{k\,min}$	11.55 kA/3.04 kA	8.01 kA/2.07 kA	

4.6.2 Line-to-earth (single-phase) short-circuit

The initial short-circuit current for a single-phase short-circuit at location F according to Figure 4.15 shall be calculated.

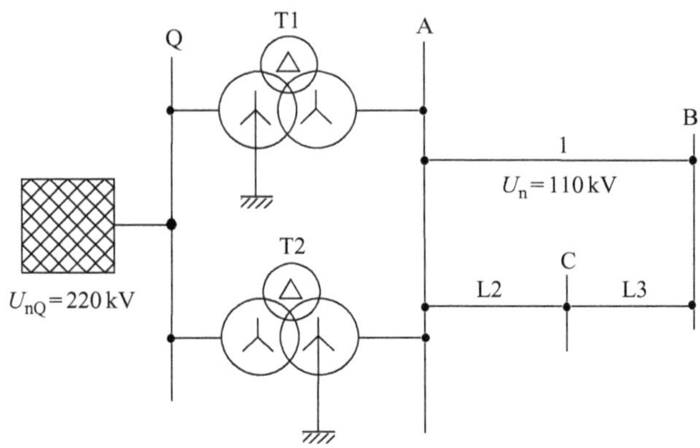

Figure 4.15 Equivalent circuit diagram of a 110-kV-system with 220-kV-feeder

The data of equipment are:

$S''_{kQ} = 5$ GVA; $X_0/X_1 = 4$
$S_{rT} = 100$ MVA; $u_{kr} = 12\%$; $t_{rT} = 220$ kV/115 kV; $X_0/X_1 = 3$
$X'_L = 0.13\ \Omega$/km; $l_L = 10$ km; $X_0/X_1 = 3.5$

The impedances in the positive- (negative-) and zero-sequence component including correction factors are:

$X_{1QK} = 2.904\ \Omega$; $X_{0QK} = 11.62\ \Omega$
$X_{1T1K} = X_{1T2K} = 15.49\ \Omega$; $X_{0T1K} = X_{0T2K} = 46.34\ \Omega$
$X_{1L1} = X_{1L2} = X_{1L3} = 1.295\ \Omega$; $X_{0L1} = X_{0L2} = X_{0L3} = 4.538\ \Omega$

The impedances at the short-circuit location are $X_{1k} = 11.507\ \Omega$; $X_{0k} = 49.368\ \Omega$ and the initial short-circuit current is $I''_{k1} = 2.895$ kA.

4.6.3 Calculation of peak short-circuit current

The peak short-circuit current for a three-phase short-circuit at location F according to Figure 4.16 shall be calculated with the different methods, i.e., 'superposition method', 'ratio R/X at s.-c. location' and 'equivalent frequency'. The accuracy of the results is to be assessed.

The data of equipment are

$S''_{kQ} = 1000$ MVA; $U_{nQ} = 110$ kV
$S_{rT} = 10$ MVA; $u_{krT} = 10\%$; $P_{kT} = 70$ kW; $t_{rT} = 125$ kV/12 kV
$S_{rG} = 20$ MVA; $U_{rG} = 10.5$ kV; $\cos\varphi_{rG} = 0.8$; $x''_{dG} = 10\%$
$X'_L = 0.09\ \Omega$/km; $R'_L = 0.123\ \Omega$/km; $l_L = 5$ km

The impedances of equipment including correction factors are

$X_{GK} = 0.5448\ \Omega$; $R_{GfK} = 0.5448\ \Omega$
$X_L = 0.45\ \Omega$; $R_L = 0.615\ \Omega$

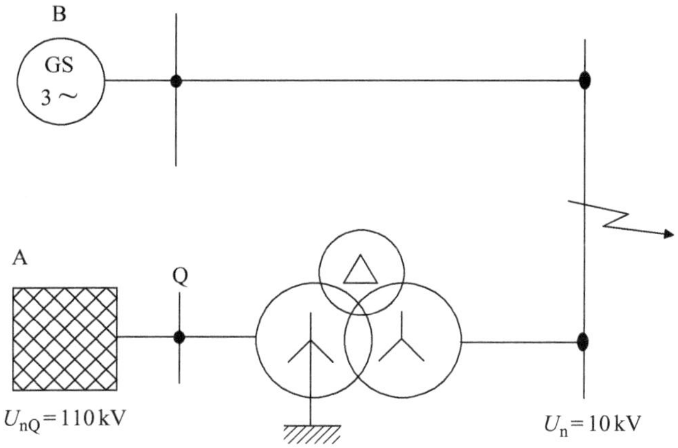

Figure 4.16 Equivalent circuit diagram of a 10-kV system, f = 50 Hz

$X_Q = 0.122\ \Omega$; $R_Q = 0.0122\ \Omega$
$X_T = 1.416\ \Omega$; $R_T = 0.0994\ \Omega$

The results of short-circuit calculation are outlined in the table below:

Method	Impedance	I''_{k3}	κ	i_p
Impedance at s.-c. location	$Z_k = (0.244 + 0.649)\ \Omega$	9.16 kA	—	—
Superposition of feeders A and B	$Z_{kA} = (0.1116 + 1.5381)\ \Omega$ $Z_{kB} = (0.653 + 0.9948)\ \Omega$	$I''_{k3A} = 4.118$ kA $I''_{k3B} = 5.337$ kA	1.837 1.157	$i_{pA} = 10.7$ kA $i_{pB} = 8.73$ kA $i_{pges} = 19.43$ kA
Ratio R/X at s.-c. location	$Z_k = (0.244 + 0.649)\ \Omega$	9.16 kA	1.337	19.92 kA
Equivalent frequency	$Z_{kAc} = (0.1116 + 0.6112)\ \Omega$ $Z_{kBc} = (0.653 + 0.3979)\ \Omega$ $Z_{kc} = (0.2509 + 0.4036)\ \Omega$	9.16 kA	1.485	19.24 kA

The results obtained with the superposition method are the correct results.

4.6.4 Short-circuit currents in a meshed 110-kV-system

The three-phase short-circuit current for the short-circuit location at busbar E in a meshed 110-kV-system shall be calculated. The system diagram is outlined in Figure 4.17.

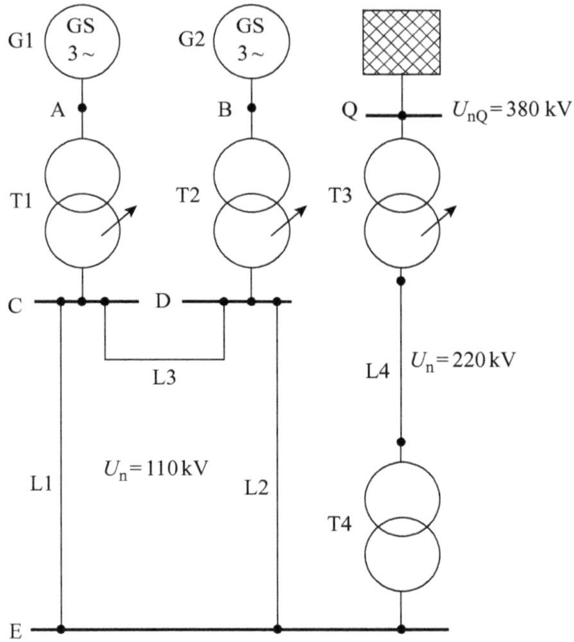

Figure 4.17 A 110-kV system with short-circuit location

The rated data of equipment (positive-sequence component) are given below:

System Q:	$U_{nQ} = 380$ kV; $S''_{kQ} = 4000$ MVA
G1:	$S_{rG1} = 200$ MVA; $U_{rG1} = 10.5$ kV; $x''_{d1} = 18\%$; $\cos\varphi_{rG1} = 0.85$; $p_{G1} = 10\%$
G2:	$S_{rG2} = 120$ MVA; $U_{rG2} = 10.5$ kV; $x''_{d2} = 15\%$; $\cos\varphi_{rG2} = 0.8$; $p_{G2} = 10\%$
T1:	$S_{rT1} = 200$ MVA; $u_{krT1} = 16\%$; $U_{rT1HV}/U_{rT1LV} = 110$ kV/10.5 kV; $p_{T1} = 12\%$
T2:	$S_{rT2} = 100$ MVA; $u_{krT2} = 14\%$; $U_{rT2HV}/U_{rT2LV} = 110$ kV/10.5 kV; $p_{T2} = 10\%$
T3:	$S_{rT3} = 300$ MVA; $u_{krT3} = 16\%$; $U_{rT3HV}/U_{rT3LV} = 400$ kV/220 kV; $p_{T3} = 15\%$
T4:	$S_{rT4} = 300$ MVA; $u_{krT4} = 16\%$; $U_{rT4HV}/U_{rT4LV} = 220$ kV/110 kV; $p_{T4} = 12\%$
Lines:	$X'_{L1} = 0.4$ Ω/km; $l_{L1} = 40$ km; $X'_{L2} = 0.36$ Ω/km; $l_{L2} = 30$ km $X'_{L3} = 0.4$ Ω/km; $l_{L1} = 20$ km; $X'_{L4} = 0.42$ Ω/km; $l_{L4} = 100$ km

The impedances of the equipment in %/MVA calculated according to the equations as per Section 3.2.2 are outlined in the table below:

No.	Equipment	x (%/MVA)	Correction as per Tables 3.2 and 3.6	Corrected impedance (%/MVA)	S''_{k3} (MVA)	I''_{k3} (kA)
1	Generator 1	0.09				
2	Transformer 1	0.08				
1 + 2	Power station 1	0.17	0.8996	0.153		
3	Generator 2	0.125				
4	Transformer 2	0.14				
3 + 4	Power station 2	0.265	0.8576	0.227		
5	Line 1	0.1322				
6	Line 2	0.0893				
7	Line 3	0.0661				
Y-Δ	L1L2			0.041		
5; 6; 7	L1L3			0.0304		
	L2L3			0.0205		
8	System Q	0.02736				
9	Transformer 3	0.0533	1.004	0.0535		
10	Line 4	0.0868				
11	Transformer 4	0.05	1.004	0.0502		
8 + 9 + 10 + 11				0.2212		
12	Total impedance at E	0.096			1146	6.01

4.6.5 *Influence of impedance correction factors on short-circuit currents*

The three-phase short-circuit current for the short-circuit location in the 110-kV-system in a single-fed system of different voltage levels as outlined in Figure 4.18 shall be calculated.

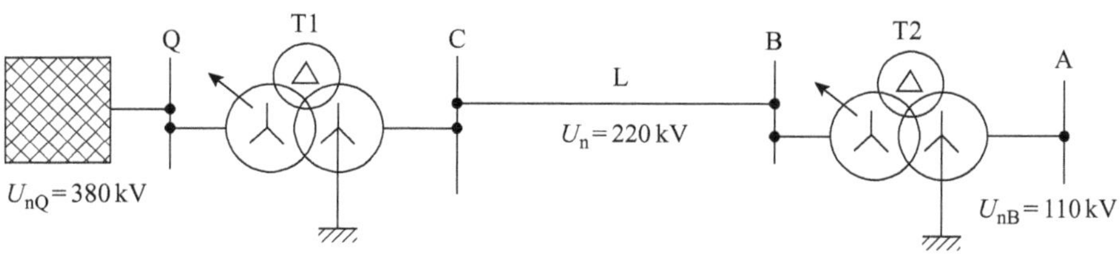

Figure 4.18 System with different voltage levels with short-circuit location

The rated data of equipment (positive-sequence component) are given below:

System Q: $U_{nQ} = 380$ kV; $S''_{kQ} = 5000$ MVA
T1: $S_{rT1} = 200$ MVA; $u_{krT1} = 16\%$; $U_{rT1HV}/U_{rT1LV} = 400$ kV/220 kV; $p_{T1} = 12\%$
T2: $S_{rT2} = 200$ MVA; $u_{krT2} = 16\%$; $U_{rT2HV}/U_{rT2LV} = 220$ kV/120 kV; $p_{T2} = 10\%$
Line: $X'_L = 0.4\ \Omega$/km; $l_L = 100$ km

The impedances of the equipment in %/MVA and the short-circuit current calculated according to the equations as per Section 3.2.2 are outlined in the table below:

No.	Equipment	x (%/MVA)	S''_{k3} (MVA)	I''_{k3} (kA)
1	System Q	0.0219		
2	Transformer 1	0.08		
3	Line	0.0826		
4	Transformer 2	0.08		
5	Total impedance at 110 kV	0.2645	415.9	2.18

As indicated by the rated data of equipment and as can be seen in Figure 4.18, the rated voltage of the transformers differ from the nominal voltages of the 380-kV- and the 110-kV-systems. Correction factors for those equipment connected to the short-circuit location through transformers must be taken into account. The results of the calculation with impedance correction factor as per Figure 2.16 and Equation (2.27) are given below.

No.	Equipment	x (%/MVA)	Correction factor as per Figure 2.16	Corrected impedance (%/MVA)	S''_{k3} (MVA)	I''_{k3} (kA)
1	System Q	0.0219	1.074	0.0235		
2	Transformer 1	0.08	1.19	0.0952		
3	Line	0.0826	1.19	0.0983		
4	Transformer 2	0.08	1.19	0.0952		
5	Total impedance at 110 kV			0.3122	352.3	1.85

The calculation of transformer impedances furthermore requires impedance correction factor as outlined in Table 3.3. The results of the analysis, taking account of both correction factors (Figure 2.16 and Table 3.3), are outlined in the table below:

No.	Equipment	x (%/MVA)	Correction factor as per Figure 2.16	Correction factor as per Table 3.3	Corrected impedance (%/MVA)	S''_{k3} (MVA)	I''_{k3} (kA)
1	System Q	0.0219	1.074		0.0235		
2	Transformer 1	0.08	1.19	1.004	0.0956		
3	Line	0.0826	1.19		0.0983		
4	Transformer 2	0.08	1.19	1.004	0.0956		
5	Total impedance at 110 kV	0.2645			0.313	351.4	1.84

Calculating the short-circuit current in the p.u.-system is similar to the calculation in the %/MVA-system. Correction factors as per Table 3.3 for the transformers and Figure 2.16 due to differences between the rated voltages of equipment and the nominal system voltages have to be taken into account. The reference voltage is equal to the nominal voltage at the short-circuit location $U_B = 110$ kV and the reference power is 100 MVA. The results are outlined in the table below:

No.	Equipment	x' (p.u.)	Correction factor as per Figure 2.16	Correction factor as per Table 3.3	Corrected impedance (p.u.)	s'_{k3} (p.u.)	i'_{k3} (p.u.)
1	System Q	0.0219	1.074		0.0235		
2	Transformer 1	0.08	1.19	1.004	0.0956		
3	Line	0.0826	1.19		0.0983		
4	Transformer 2	0.08	1.19	1.004	0.0956		
5	Total impedance at 110 kV	0.2645			0.313	3.514	2.024

If the short-circuit current is calculated using the Ω-system only the correction factor as per Table 3.3 for the transformers has to be taken into account. The results as per calculation in the Ω-system are outlined in the table below:

No.	Equipment	X (Ω)	Correction as per Table 3.3	Corrected impedance (Ω)	S''_{k3} (MVA)	I''_{k3} (kA)
1	System Q	2.844		2.844		
2	Transformer 1	11.519	1.004	11.565		
3	Line	11.894		11.894		
4	Transformer 2	11.519	1.004	11.565		
5	Total impedance at 110 kV	37.78		37.87	351.4	1.84

The greatest influence on the short-circuit current is given by the correction factors due to the difference of the rated voltage of the transformers and the nominal voltages of the 380-kV- and the 110-kV-systems as can be seen clearly from the results. The correction factor of the transformers as per Table 3.3, however, have only a negligible effect on the system under investigation. As the individual correction factors depend on the rated data of the equipment the influence on the short-circuit current may be different in other system configurations. Thus, it should be noted that the correction factors should be taken into account in general.

4.6.6 *Short-circuit currents in a.c. auxiliary supply of a power station*

Figure 4.19 indicates the high-voltage system configuration of the a.c. power supply of a power station. Auxiliary supply is connected to the 6-kV-busbar E. During start-up of the power station, i.e., prior to synchronization of the generator, the power supply is taken from the start-up supply through transformer T5 either from the 30-kV-system connected to busbar B or from the 110-kV-system (busbar A) or 220-kV-system (busbar Q). After synchronization the transformers T2 and T5 are both in operation for the auxiliary supply, transformer T5 is switched-off finally and the auxiliaries are supplied through transformer T2 only.

The rated data of equipment are given below:

System Q $U_{nQ} = 220$ kV; $S''_{kQ} = 10{,}000$ MVA
System A $U_{nA} = 110$ kV; $S''_{kA} = 3000$ MVA
System B $U_{nB} = 30$ kV; $S''_{kB} = 300$ MVA

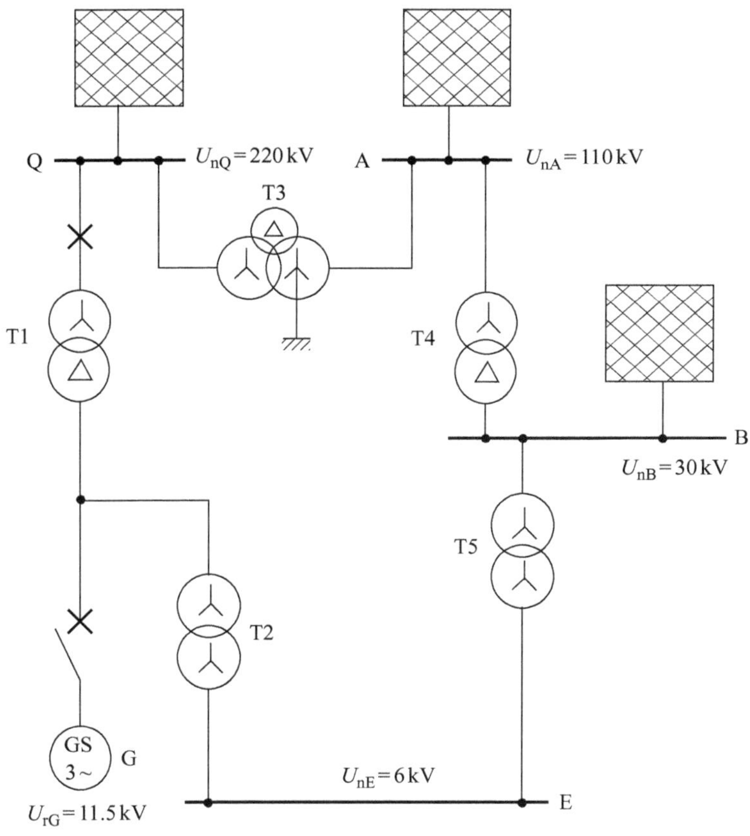

Figure 4.19 High-voltage system configuration for the auxiliary supply of a power station

G	$S_{rG} = 300$ MVA; $U_{rG} = 11.5$ kV; $x''_d = 18\%$; $\cos\varphi_{rG} = 0.85$
T1	$S_{rT1} = 300$ MVA; $u_{krT1} = 14\%$; $U_{rT1HV}/U_{rT1LV} = 220$ kV/11.5 kV
T2	$S_{rT2} = 25$ MVA; $u_{krT2} = 8\%$; $U_{rT2HV}/U_{rT2LV} = 11.5$ kV/6 kV
T3	$S_{rT3} = 150$ MVA; $u_{krT3} = 12\%$; $U_{rT3HV}/U_{rT3LV} = 220$ kV/110 kV
T4	$S_{rT4} = 40$ MVA; $u_{krT4} = 10\%$; $U_{rT4HV}/U_{rT4LV} = 110$ kV/30 kV
T5	$S_{rT5} = 25$ MVA; $u_{krT5} = 8\%$; $U_{rT5HV}/U_{rT5LV} = 11.5$ kV/6 kV

Data on the voltage control of the generator and the tap-changers of the transformers are not known. The short-circuit currents have to be calculated for three-phase short-circuit at the auxiliary busbar E for the three operating conditions as mentioned.

The impedances of the equipment are given in the table below.

No.	Equipment	x (%/MVA)	Correction factor as per Table 3.2 and Table 3.3	Corrected impedance (%/MVA)
1	System Q	0.0109		
2	System A	0.0365		
3	System B	0.3652		
4	Generator	0.06	1.005	0.0603
5	Transformer 1	0.0467	0.964	0.045
6	Transformer 2	0.32	0.997	0.31
7	Transformer 3	0.8	0.975	0.7799
8	Transformer 4	0.25	0.986	0.246
9	Transformer 5	0.32	0.997	0.31

For start-up operation of the power station the total impedance is $x_{kS} = 0.469\%/\text{MVA}$, resulting in a three-phase short-circuit current of $I''_{k3S} = 22.58$ kA. For the intermediate operation state (both transformers T2 and T5 are in operation) the total impedance is $x_{kI} = 0.197\%/\text{MVA}$, the three-phase short-circuit current is $I''_{k3I} = 53.82$ kA. For normal operation of the power station the total impedance is $x_{kN} = 0.339\%/\text{MVA}$, resulting in a three-phase short-circuit current of $I''_{k3N} = 31.23$ kA.

The highest short-circuit current appears in case the auxiliaries are supplied through the transformers T2 and T5. This condition is only present for a short-time while switching from one supply to the other. It is therefore not recommended to take this condition for the design rating of the switchgear and equipment, but to take the highest short-circuit current occurring under other operating conditions ($I''_{k3} = 31.23$ kA).

Calculation can also be done using p.u.-system, which gives identical numerical values for the impedances. The short-circuit currents in p.u.-system are calculated with reference voltage $U_B = 6$ kV (nominal system voltage at short-circuit location). Results of calculation in %/MVA-system and p.u.-system are outlined in the table below.

Operating condition	x (%/MVA)	I''_{k3} (kA)	x' (p.u.)	i'_{k3} (p.u.)
Start-up operation	0.469	22.58	0.469	1.35
Transformers T2 and T5 are in operation	0.197	53.82	0.197	3.23
Normal operation of the power station	0.339	31.23	0.339	1.87

Chapter 5

Influence of neutral earthing on single-phase short-circuit currents

5.1 General

The theoretical approach to calculate short-circuit (s.-c.) currents with symmetrical components in general and especially in the case of single-phase short-circuit was explained in detail in Chapter 2. Current and voltages in case of short-circuits with earth connection (e.g., single-phase short-circuits) depend on the positive- and zero-sequence impedances Z_1 and Z_0. If the ratio of zero-sequence to positive-sequence impedance is $k = Z_0/Z_1$ the voltages in the non-faulted phases (see Equation (2.25a)) and the single-phase short-circuit current (see Equation (2.25b)) are

$$|\underline{U}_{\mathrm{Y}}| = |\underline{U}_{\mathrm{B}}| = E_1 * \sqrt{3} * \frac{\sqrt{k_2 + k + 1}}{2 + k} \tag{5.1a}$$

$$I''_{\mathrm{k1}} = \frac{E_1}{Z_1} * \frac{3}{2 + k} \tag{5.1b}$$

If the voltage E_1 is set to $E_1 = U_{\mathrm{n}}/\sqrt{3}$, similar to the equivalent voltage at short-circuit location then

$$|\underline{U}_{\mathrm{Y}}| = |\underline{U}_{\mathrm{B}}| = U_{\mathrm{n}} * \frac{\sqrt{k_2 + k + 1}}{2 + k} \tag{5.1c}$$

$$I''_{\mathrm{k1}} = \frac{U_{\mathrm{n}}}{\sqrt{3} * Z_1} * \frac{3}{2 + k} \tag{5.1d}$$

The impedances in the positive-sequence (and negative-sequence) system are determined only by the network topology. The single-phase short-circuit current and the voltages of the non-faulted phases can be changed only by changing the ratio of positive-sequence to zero-sequence impedance, i.e., by changing the handling of transformer neutrals.

The type of neutral earthing determines the impedance Z_0 of the zero-sequence component and has a dominating influence on the short-circuit current through earth, i.e., I''_{k1} in case of single-phase short-circuits and I''_{kE2E} in case of two-phase short-circuit with earth connection. In order to change the zero-sequence impedance of the system, it is possible to earth any number of neutrals, i.e., none, a few or all transformer neutrals, leading to the highest zero-sequence impedance (no neutral earthed), respectively the lowest zero-sequence impedance (all neutrals earthed). The system is characterised less by the number of neutrals to be earthed, than by the value of the single-phase short-circuit current and by the voltages in the non-faulted phases.

The different types of neutral handling in power systems (high-voltage systems only) are outlined in Table 5.1.

Quantities as per Table 5.1

U_n	Nominal system voltage
U_{0max}	Maximal voltage in the zero-sequence system, i.e., at neutral of transformer
ω	Angular velocity of the power system
C_E	Line-to-earth capacitance of the power system
$\underline{Z}_0$; $\underline{Z}_1$	Zero-sequence, respectively positive-sequence, impedance of the system
δ_0	Damping of the power system (see Section 5.5)
v	Ratio indicating capacitance to reactance (see Section 5.5)

5.2 Power system with low-impedance earthing

Low-impedance earthing is applied in medium-voltage and high-voltage systems worldwide with nominal voltages above 10 kV. Power systems having nominal voltages $U_n \geq 132$ kV are generally operated with low-impedance earthing. In order to realise a power system with low-impedance earthing, it is not necessary that the neutrals of all transformers are earthed, but to fulfil the criteria, that the voltages of the non-faulted phases remain below 140 per cent of the nominal system voltage in the case of a single-phase short-circuit. The disadvantage while earthing all neutrals is seen in an increased single-phase short-circuit current, sometimes exceeding the three-phase short-circuit current. The neutral of unit transformers in power stations shall not be earthed at all, as the single-phase short-circuit current will then depend on the generation dispatch. As the contribution of one unit transformer is in the range of up to 8 kA, the influence on the single-phase short-circuit currents is significant.

Based on Figure 5.1 and assuming a far-from-generator short-circuit with positive-sequence impedance equal to negative-sequence impedance $\underline{Z}_1 = \underline{Z}_2$, the single-phase short-circuit current is calculated by

$$\underline{I}''_{k1} = \frac{c * \sqrt{3} * U_n}{2 * \underline{Z}_1 + \underline{Z}_0} \tag{5.1e}$$

Table 5.1 Characteristics of different types of neutral handling in power systems

	Isolated neutral	Low-impedance earthing	Earthing with current limitation	Resonance earthing
Single-phase fault current (short-circuit current)	Capacitive earth-faults current	Single-phase (earth-fault) short-circuit current	Single-phase (earth-fault) short-circuit current	Residual earth-fault current
	$\underline{I}_{CE} \approx j\sqrt{3}\omega C_E U_n$	$\underline{I}''_{k1} = c\sqrt{3}U_n/(2\underline{Z}_1 + \underline{Z}_0)$	$\underline{I}''_{k1} = c\sqrt{3}U_n/(2\underline{Z}_1 + \underline{Z}_0)$	$\underline{I}_{Rest} \approx j\sqrt{3}U_n\omega C_E(\delta_0 + jv)$
Increase of voltages at non-faulted phases	Present $U_{0max}/U_n \approx 0.6$	No increase $U_{0max}/U_n < 0.3$–0.45	No increase $U_{0max}/U_n \approx 0.45$–$0.6$	Present $U_{0max}/U_n \approx 0.6$
Earth-fault factor δ	$\approx\sqrt{3}$	<1.38	1.38–$\sqrt{3}$	$\approx\sqrt{3}$–$1.1 * \sqrt{3}$
Ratio of impedances Z_0/Z_1	Generally high	2–4	>4	→Infinite
Extinguishing of fault arc	Self-extinguishing (see Figure 5.11)	Not self-extinguishing	Self-extinguishing in rare cases	Self-extinguishing (see Figure 5.11)
Repetition of faults	Double earth-fault Reignition of earth-fault	None	None	Double earth-fault
Voltage at earthing electrode U_E	$U_E \leq 125$ V	$U_E > 125$ V permitted	$U_E > 125$ V permitted	$U_E \leq 125$ V
Touching voltage U_B	$U_B \leq 65$ V	See VDE 0141	See VDE 0141	$U_B \leq 65$ V

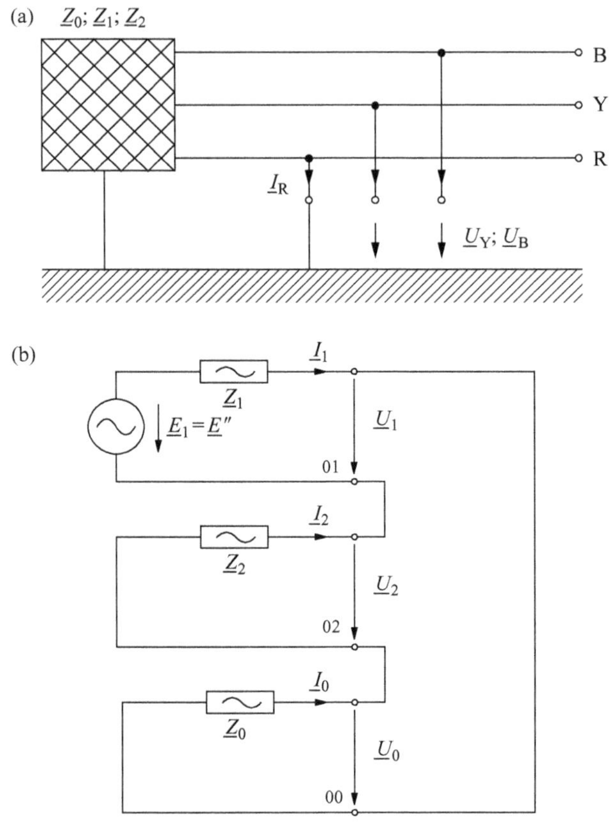

Figure 5.1 Equivalent circuit diagram of a single-phase short-circuit (system with low-impedance earthing). (a) Diagram in RYB-system, (b) equivalent circuit diagram in the system of symmetrical components

with voltage factor *c* according to Table 4.1. If the single-phase short-circuit current is related to the three-phase short-circuit current

$$\underline{I}''_{k3} = \frac{c * U_n}{\sqrt{3} * \underline{Z}_1} \tag{5.2}$$

it follows that

$$\frac{\underline{I}''_{k1}}{\underline{I}''_{k3}} = \frac{3 * \underline{Z}_1}{2 * \underline{Z}_1 + \underline{Z}_0} \tag{5.3}$$

The relation of single-phase to three-phase short-circuit current depending on the ratio of Z_1/Z_0 with the difference of phase angles $(\gamma_1 - \gamma_0)$ of the impedances as parameter is outlined in Figure 5.2. The phase angles γ_1 and γ_0 are defined by the arcustangens-function $\gamma_1 = \arctan(X_1/R_1)$ in the positive-sequence system respectively $\gamma_0 = \arctan(X_0/R_0)$ in the zero-sequence system.

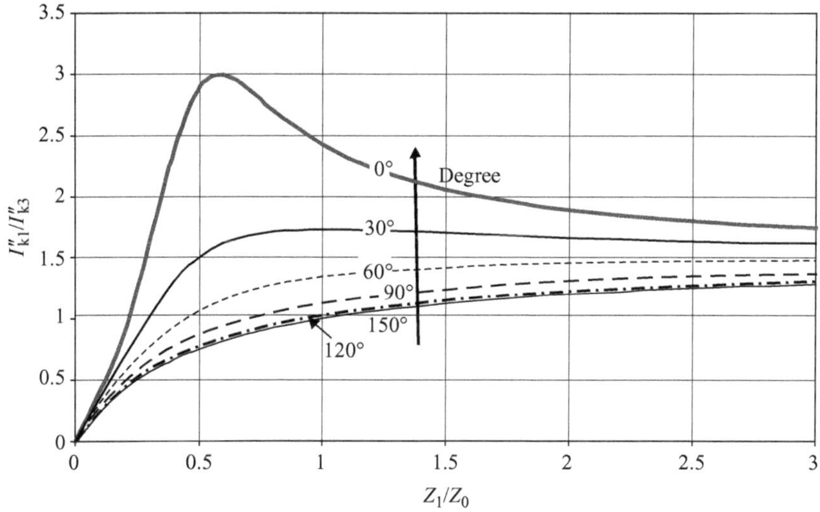

Figure 5.2 *Ratio of single-phase to three-phase short-circuit current depending on* Z_1/Z_0 *and* $(\gamma_1 - \gamma_0)$

The voltages (power-frequency voltage) of the non-faulted phases Y and B, as calculated in Chapter 2 in detail,

$$\underline{U}_{\text{Y}} = \underline{E}_1 * \frac{\underline{Z}_0 * (\underline{a}^2 - 1) + \underline{Z}_2 * (\underline{a}^2 - \underline{a})}{\underline{Z}_0 + \underline{Z}_1 + \underline{Z}_2} \tag{5.4a}$$

$$\underline{U}_{\text{B}} = \underline{E}_1 * \frac{\underline{Z}_0 * (\underline{a} - 1) + \underline{Z}_2 * (\underline{a} - \underline{a}^2)}{\underline{Z}_0 + \underline{Z}_1 + \underline{Z}_2} \tag{5.4b}$$

can be simplified if $\underline{Z}_1 = \underline{Z}_2$ is assumed and by taking account of the meaning of $\underline{a}$ and $\underline{a}^2$ as below:

$$\underline{U}_{\text{Y}} = -0.5\sqrt{3} * \underline{E}_1 * \frac{\sqrt{3}}{1 + (2\underline{Z}_1/\underline{Z}_0) + j} \tag{5.5a}$$

$$\underline{U}_{\text{B}} = -0.5\sqrt{3} * \underline{E}_1 * \frac{\sqrt{3}}{1 + (2\underline{Z}_1/\underline{Z}_0) - j} \tag{5.5b}$$

Relating the voltages to the voltage E_1 the earth-fault factors of the phases Y and B, δ_{Y} and δ_{B} are obtained.

$$\delta_{\text{Y}} = \left| -0.5 * \frac{3}{1 + (2\underline{Z}_1/\underline{Z}_0) + j} \right| \tag{5.6a}$$

$$\delta_{\text{B}} = \left| -0.5 * \frac{3}{1 + (2\underline{Z}_1/\underline{Z}_0) - j} \right| \tag{5.6b}$$

which are different from each other, depending on the impedances and the phase angle.

The effect of the earthing can be described by the earth-fault factor δ according to VDE 0141/07.89 and is defined to be the maximum of the earth-fault factors δ_Y and δ_B

$$\delta = \mathrm{MAX}\{\delta_Y; \delta_B\} = \frac{U_{\mathrm{LE\,max}}}{U/\sqrt{3}} \tag{5.7}$$

where U_{LEmax} is the highest value of the power-frequency voltage phase-to-earth of the non-faulted phases in the case of a short-circuit with earth connection and U is the voltage between phases prior to fault.

Power systems having an earth-fault factor $\delta < 1.4$ are defined as systems with low-impedance earthing. It should be noted that the single-phase short-circuit currents shall be below the permissible limits, which are defined by the breaking capability of circuit-breakers, the short-circuit withstand capability of switchgear, installations and equipment and by other criteria such as earthing voltage, induced voltages, etc.

Figure 5.3 indicates the earth-fault factors δ_Y and δ_B in dependence of the ratio $\underline{Z}_1/\underline{Z}_0$ and the difference of impedance angles $(\gamma_1 - \gamma_0)$. An impedance angle above 90° is only possible in the case of a capacitive impedance of the zero-sequence component but not in systems with low-impedance earthing.

Figure 5.4 presents the earth-fault factor δ in relation to X_0/X_1 with the parameter R_0/X_0, whereas the impedance angle in the positive-sequence component remains constant. The earth-fault factor δ remains below 1.4 if $X_0/X_1 \leq 5$ can be achieved and if R_0/X_0 is kept below 0.2 (alternatively $X_0/X_1 \leq 4$ and $R_0/X_0 < 0.3$).

An impedance ratio $X_0/X_1 = 2$–4 can easily be achieved in power systems as the relation of zero-sequence to positive-sequence impedances of equipment is

$X_0/X_1 \approx 4$	Parallel double-circuit overhead lines
$X_0/X_1 \approx 3$	Single-circuit overhead lines and HV-transformers Yy(d)
$X_0/X_1 \approx 0.3$	Unit transformers Yd in power stations (normally not to be earthed)

5.3 Power system having earthing with current limitation

Earthing with current limitation can be seen in some cases as a special case of the low-impedance earthing, provided the earth-fault factor is below 1.4. Earthing with current limitation is applied in urban power systems having rated voltage $U_n \leq 20$ kV. Some applications are known in systems with nominal voltage up to 132 kV.

The criterion for the design of the earthing conditions is the value of the single-phase short-circuit current, which can be limited to some kA (1 kA or 2 kA) in medium-voltage systems or to some 10 kA in high-voltage systems (e.g., below the three-phase short-circuit current). To realise the scheme of earthing with current limitation, the neutrals of some or all transformers are earthed through reactances or resistances

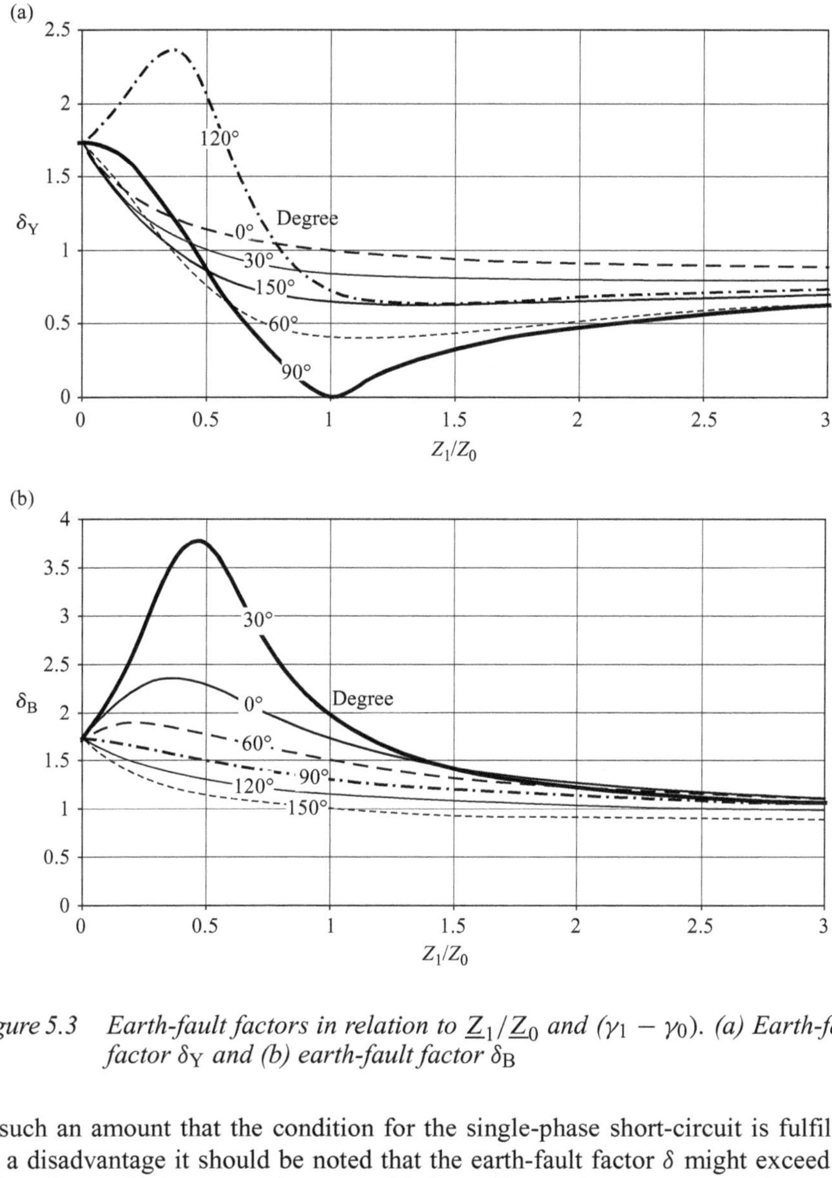

Figure 5.3 Earth-fault factors in relation to $\underline{Z}_1/\underline{Z}_0$ and $(\gamma_1 - \gamma_0)$. (a) Earth-fault factor δ_Y and (b) earth-fault factor δ_B

to such an amount that the condition for the single-phase short-circuit is fulfilled. As a disadvantage it should be noted that the earth-fault factor δ might exceed the value of 1.4, which seems to be acceptable in medium-voltage systems with nominal voltages $U_n = 10$–20 kV. In high-voltage systems with $U_n = 110$–132 kV the advantages and disadvantages have to be analysed in more detail.

In order to estimate the required value of the earthing impedance, the zero-sequence impedance is considered based on Figure 5.5, indicating the ratio of I''_{k1}/I''_{k3} as well as the earth-fault factor δ in relation to X_0/X_1. As an example, a medium-voltage system with $U_n = 10$ kV having an initial three-phase short-circuit power $S''_{k3} = 100$–250 MVA ($I''_{k3} = 5.8$–14.4 kA) is regarded. In this case for the limitation

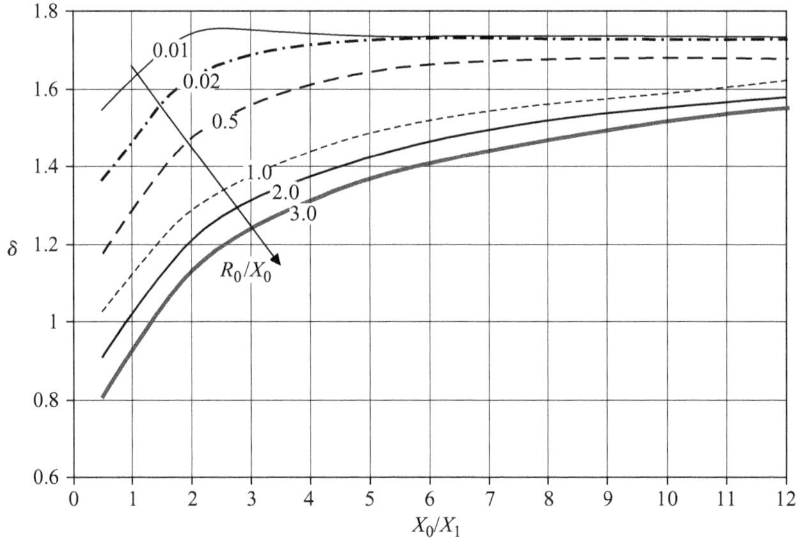

Figure 5.4 *Earth-fault factor δ depending on X_0/X_1 for different ratios R_0/X_0 and $R_1/X_1 = 0.01$*

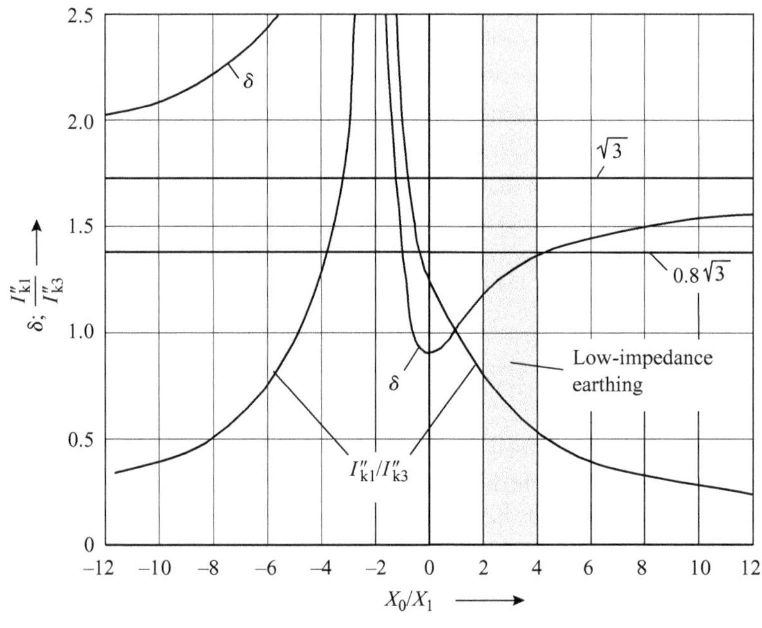

Figure 5.5 *Earth-fault factor δ and ratio I''_{k1}/I''_{k3} depending on X_0/X_1*

of the single-phase short-circuit current to $I''_{k1} = 2$ kA the ratio $X_0/X_1 = 6.7$–19.6 is required. The earth-fault factor in this case will be $\delta = 1.44$–1.61. By this, the system is no longer a system with low-impedance earthing.

5.4 Power system with isolated neutral

The operation of power systems with isolated neutrals is applicable to systems with nominal voltages up to 60 kV, however the main application is seen in power station auxiliary installations and industrial power systems with voltages up to 10 kV. In public supply systems, isolated neutrals are not very common.

The analysis of a single-phase earth-fault is based on Figure 5.6.

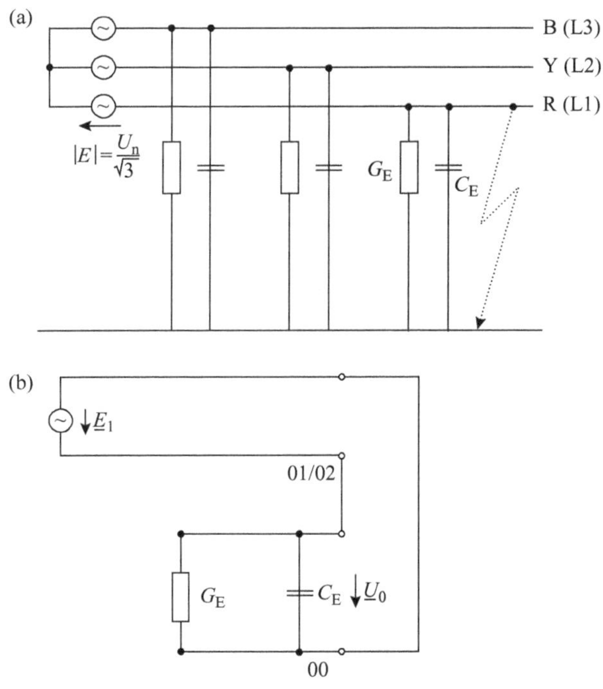

Figure 5.6 Power system with isolated neutral with single-phase earth-fault. (a) Equivalent circuit diagram in RYB-system and (b) equivalent circuit diagram in the system of symmetrical components

Contrary to power systems with low-impedance earthing or earthing with current limitation the capacitances phase-to-earth capacitances in the zero-sequence component cannot be neglected in power systems with isolated neutral as can be seen from Figure 5.6. To determine the respective parameters of the equipment, no-load measurements are necessary. The single-phase earth-fault current, in general, is calculated by

$$I_1'' = \frac{c * \sqrt{3} * U_n}{|2 * \underline{Z}_1 + \underline{Z}_0|} \tag{5.8}$$

where U_n is the nominal system voltage, c is the voltage factor as per Table 4.1 and $\underline{Z}_1$; $\underline{Z}_0$ are the positive- and zero-sequence impedances, respectively.

The zero-sequence impedance $\underline{Z}_0$ is determined by the capacitance phase-to-earth C_E, and is significantly higher than the positive-sequence impedance $\underline{Z}_1$. The single-phase earth-fault current is determined through the capacitive component by

$$\underline{I}_R = \underline{I}_{CE} = j\omega * C_E * \sqrt{3} * U_n \tag{5.9}$$

and is called capacitive earth-fault current $\underline{I}_{CE}$. As the capacitive earth-fault current is significantly lower than a typical short-circuit current, in most of the cases even lower than the normal operating current, the single-phase fault in a system with isolated neutral is called earth-fault instead of short-circuit. The earth-fault current increases with increasing capacitance phase-to-earth and by this with increasing line length as can be seen from Equation (5.9). Small capacitive currents in the case of faults through air can be extinguished by themselves if they remain below some 10 A depending on the voltage level. Figure 5.7 indicates the limits for self-extinguishing of capacitive currents I_{CE} according to VDE 0228 part 2/12.87.

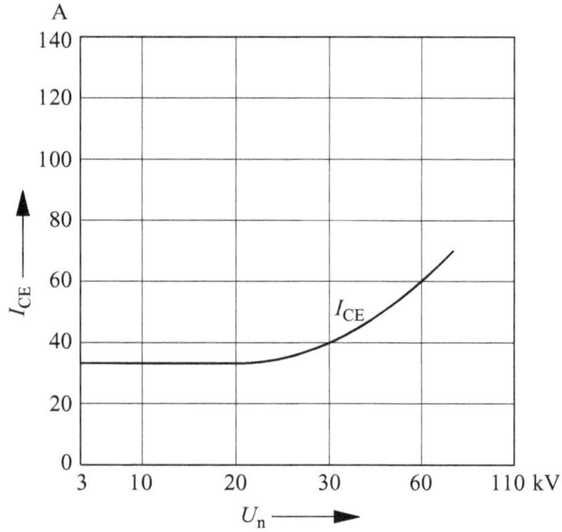

Figure 5.7 Limit for self-extinguishing of capacitive currents in air according to VDE 0228 part 2

The voltages (phase-to-earth) of the non-faulted phases in the case of an earth-fault are increasing to the amount of the phase-to-phase voltage, as can be seen from Figure 5.8. Prior to fault the voltage potential of earth (E) and neutral (N) are identical, the phase-to-earth voltages are symmetrical as well as the line-to-line voltages. During the earth-fault, the voltage of the faulted phase (R) is identical to

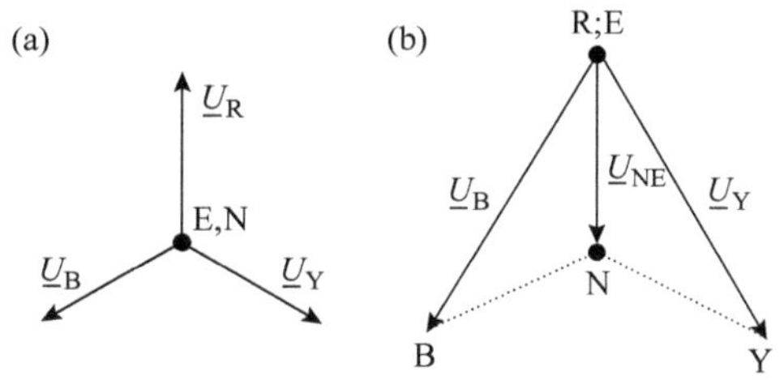

Figure 5.8 Vector diagram of voltages, power system with isolated neutral. (a) Prior to fault and (b) during earth-fault

the voltage of the earth (E). The voltage potential of the neutral (N) is given, by definition, as the mean value of the three phases R, Y and B which is not changed by the earth-fault. A voltage displacement $\underline{U}_{NE}$ between neutral and earth equal to the line-to-earth voltage is originating from the earth-fault. The voltage displacement is equal to the voltage $\underline{U}_0$ of the zero-sequence component. As the impedance of the zero-sequence component is significantly higher than the impedances of the positive- and negative-sequences system, the displacement voltage is identical with the voltage at the transformer neutral. The voltages of the non-faulted phases are increased, but the three voltages phase-to-phase remain symmetrical as outlined in Figure 5.8(b).

The capacitive earth-fault current and the recovery voltage at the fault location have a phase displacement of nearly 90°. At the instant of the maximum of the recovery voltage or shortly after it, a reignition of the fault arc is possible and probable. The time courses of the phase-to-earth voltages u_R, u_Y and u_B and of the displacement voltage u_{NE} as well as the earth-fault current i_{CE} are outlined in Figure 5.9 indicating the time prior, during and after the occurrence of the earth-fault.

The earth-fault occurs at time instant t_1, phase R having the maximal voltage. The phase-to-earth voltage of the non-faulted phases Y and B are increasing to the value of the phase-to-phase voltage. The displacement voltage u_{NE} increases from a very low value, ideally zero, to the phase-to-earth voltage. The transient frequency can be calculated by

$$f \approx \frac{1}{2 * \pi * \sqrt{3 * L_1 * C_0}} \tag{5.10}$$

where L_1 is the inductance of the positive-sequence system and C_0 the capacitance of the zero-sequence system.

The earth-fault arc is extinguished at time t_2 approximately 10 ms after ignition of the earth-fault; the current i_{CE} has its zero-crossing, whereas the displacement voltage has nearly reached its peak value. The three phase-to-earth voltages u_R, u_Y and u_B are symmetrical to each other, however with a displacement determined by the displacement voltage at the time of arc extinguishing, i.e., the displacement voltage is equal to the peak value of the phase-to-earth voltage. Approximately 10 ms after the extinguishing of the arc the phase-to-earth voltage of phase R reaches the new peak value $2 * \sqrt{2} * U_n$.

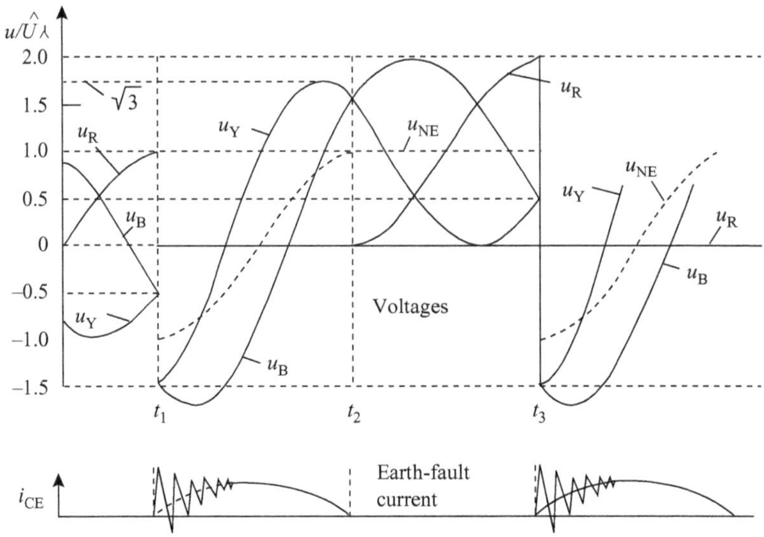

Figure 5.9 Time courses of phase-to-earth voltages, displacement voltage and earth-fault current. System with isolated neutral, earth-fault in phase R

This voltage may cause a reignition of the earth-fault due to the very high-voltage stress. This reignition takes place at time instant t_3 with the phase-to-earth voltage of phase R having its peak value. The voltages of the non-faulted phases again are increasing, this time starting from a higher value and reaching the peak value nearly to $\sqrt{3} * \sqrt{2} * U_n$.

Besides the power-frequency overvoltage in the case of an earth-fault, the transient overvoltage with frequency according to Equation (5.10) has to be considered. The overvoltage factor k_{LE}, taking account of both types of overvoltages, is given by the maximal peak voltage related to the peak value of phase-to-earth voltage

$$k_{LE} = \frac{u_{ü}}{\sqrt{2} * U/\sqrt{3}} \tag{5.11}$$

where $u_{ü}$ is the maximal peak voltage during the earth fault and U the phase-to-earth voltage (power-frequency).

In theory, the overvoltage factor after multiple reignition of the earth-fault can reach $k_{LE} = 3.5$. Due to the system damping, the overvoltage factor will be below $k_{LE} < 3$ in most of the cases.

5.5 Power system with resonance earthing (Petersen-coil)

5.5.1 General

Power systems with resonance earthing are widely in operation in Central European countries. The German power system statistic [3] indicates that 87 per cent of the

MV-systems having nominal voltages U_n = 10–30 kV and nearly 80 per cent of 110-kV-systems are operated with resonance earthing (Criteria: Total line lengths). Some MV-systems are operated with a combined scheme of resonance earthing under normal operating conditions and low-impedance earthing in case of earth-fault. Resonance earthing, therefore, is the dominating type of system earthing in Germany for power systems with voltage 10 kV up to 110 kV. In other countries such as India, South Africa and China, power systems with resonance earthing have gained an increasing importance during the last decades, however are still not so common as systems with low-voltage earthing.

Resonance earthing is realised by earthing of one or several neutrals of transformers through reactances (Petersen-coils), normally adjustable, which will be set in resonance to the phase-to-earth capacitances of the system. The principal arrangement of a power system with resonance earthing is outlined in Figure 5.10.

The impedances of transformers and lines of the positive-sequence component can be neglected compared with those of the zero-sequence component due to the order of magnitude of the impedances. The admittance of the zero-sequence component is

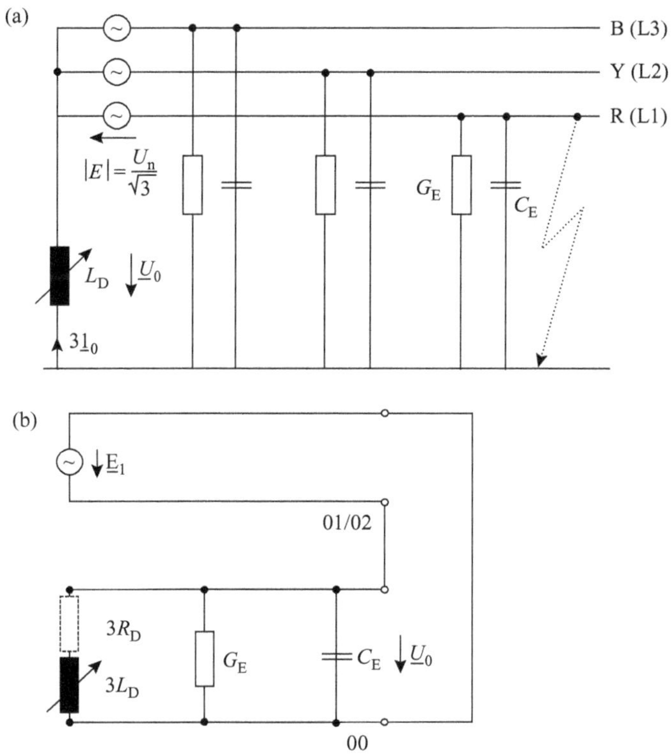

Figure 5.10 System with resonance earthing, earth-fault in phase R. (a) Equivalent diagram in RYB-system and (b) equivalent diagram in the system of symmetrical components

given by

$$\underline{Y}_0 = j\omega * C_E + \frac{1}{3 * R_D + j3 * X_D} + G_E \tag{5.12}$$

where C_E is the phase-to-earth capacitance of the system, ω is the angular frequency of the system, R_D is the resistance of the Petersen-coil, X_D is the reactance of the Petersen-coil $X_D = \omega L$ and G_E is the admittance representing the phase-to-earth line losses.

After some conversions it follows that

$$\underline{Y}_0 = j\omega * C_E * \left(1 - \frac{1}{3 * \omega^2 * L_D * C_E * (1 - j(R_D/X_D))}\right) + G_E \tag{5.13a}$$

The impedance of the Petersen-coil appears with its threefold value in the zero-sequence component [1]. It is assumed that $R_D \ll X_D$ and that the losses of the Petersen-coil are summed up with the phase-to-earth losses and are represented as admittance G_E of the line. The admittance in the zero-sequence component is then

$$\underline{Y}_0 = j\omega * C_E * \left(1 - \frac{1}{3 * \omega^2 * L_D * C_E}\right) + G_E \tag{5.13b}$$

The maximal impedance is obtained if the imaginary part as per Equation (5.13b) is equal to zero; the current from the Petersen-coil I_D is equal to the capacitive current I_{CE} of the system. As indicated in Figure 5.10, the capacitance phase-to-earth C_E, the reactance $3L_D$ and the ohmic losses $R_0 = 1/G_E$ are forming a parallel resonance circuit with the resonance frequency

$$\omega = \frac{1}{\sqrt{3 * L_D * C_E}} \tag{5.14}$$

The resonance frequency in the case of resonance earthing shall be the nominal frequency of $f = 50$ Hz or $f = 60$ Hz, respectively. Defining the detuning factor v

$$v = \frac{I_D - I_{CE}}{I_{CE}} = 1 - \frac{1}{3 * \omega^2 * L_D * C_E} \tag{5.15a}$$

and the damping d

$$d = \frac{G_E}{\omega * C_E} \tag{5.15b}$$

the admittance of the zero-sequence component is given by

$$\underline{Y}_0 = \omega * C_E * (jv + d) \tag{5.16}$$

The admittance will be minimal and the impedance will be maximal in the case of resonance tuning ($v = 0$). The earth-fault current $\underline{I}_{Res}$, in general, is obtained by

$$\underline{I}_{Res} \approx \sqrt{3} * U_n * \omega * C_E * (jv + d) \tag{5.17a}$$

In case of resonance tuning ($v = 0$) the earth-fault current is a pure ohmic current

$$\underline{I}_{Res} \approx \sqrt{3} * U_n * \omega * C_E * d \tag{5.17b}$$

The phase-to-earth voltages of the non-faulted phases increase to the value of the phase-to-phase voltage in the case of a single-phase earth-fault, which is furthermore increased due to asymmetrical system voltages resulting in a higher displacement voltage between neutral and earth. In order to avoid the high voltages in the case of exact resonance tuning a small detuning of 8–12 per cent is chosen in practice.

The task of resonance earthing is to reduce the earth-fault current at the fault location to the minimum or nearly to the minimum by adjusting the Petersen-coil to resonance or nearly to resonance with the phase-to-earth capacitances. The ohmic part of the residual current I_{Res} cannot be compensated by this. If the residual current is small enough, a self-extinguishing of the arc at the fault location is possible. VDE 0228 part 2:12.87 defines the limits for self-extinguishing of residual currents I_{Res} (and capacitive earth-fault currents I_{CE}) for different voltage levels as outlined in Figure 5.11. It can be seen from Figure 5.11 that the limit for ohmic currents, e.g., in 30-kV-systems, is twice the limit for capacitive currents.

The Petersen-coil can only be tuned for one frequency (nominal frequency) in resonance; harmonics present in the system voltage are increasing the residual current at the fault location.

As the phase-to-earth capacitances are changing during system operation, e.g., due to switching of lines, the Petersen-coil has to be changed also to keep system operation with resonance tuning. Reliable criteria have to be established to tune the Petersen-coil in resonance with the phase-to-earth capacitances.

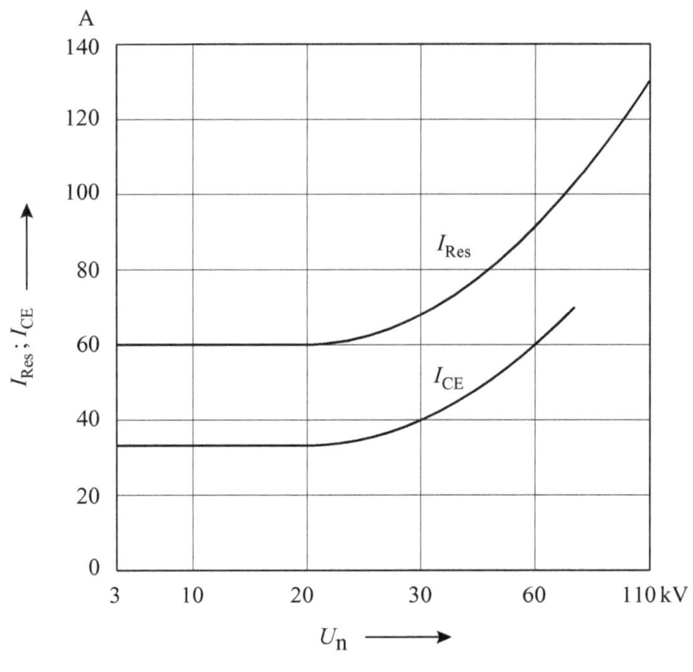

Figure 5.11 Current limits according to VDE 0228 part 2:12.87 of ohmic currents I_{Res} and capacitive currents I_{CE}

5.5.2 Calculation of displacement voltage

In real power systems, the phase-to-earth capacitances are unequal, e.g., in the case of a transmission line due to different clearance of the phase-wires above ground or the case of cables due to manufacturing tolerances. Under normal operating conditions, a displacement voltage between transformer neutral and earth $\underline{U}_{NE}$ can be measured. As mentioned in previous sections, this voltage is equal to the voltage $\underline{U}_0$ in the zero-sequence component. The calculation of the displacement voltage can be carried out in the RYB-system (Figure 5.12(a)) as well as with the system of symmetrical components (Figure 5.12(b)).

Based on Figure 5.12(a) the displacement voltage is calculated as

$$\underline{U}_{NE} = \frac{U_n}{\sqrt{3}} * \frac{j\omega * (C_{RE} + \underline{a}^2 * C_{YE} + \underline{a} * C_{BE})}{j\omega * (C_{RE} + C_{YE} + C_{BE}) - j(1/(\omega * L_D)) + 3 * G_E} \quad (5.18)$$

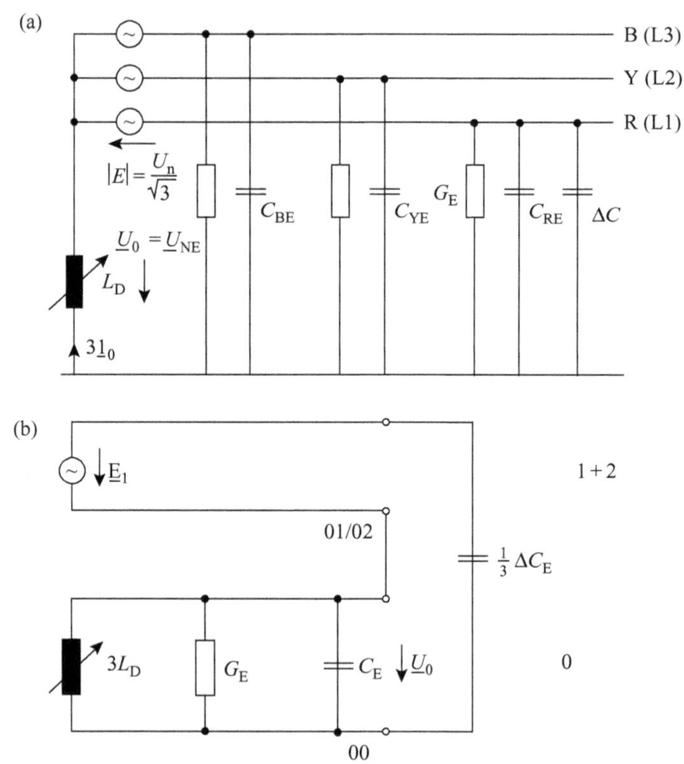

Figure 5.12 Equivalent circuit diagram of a power system with asymmetrical phase-to-earth capacitances. (a) Equivalent circuit diagram in the RYB-system and (b) equivalent circuit diagram in the system of symmetrical components

where U_n is the nominal system voltage, ω is the angular frequency of the system, C_{RE}; C_{YE}; C_{BE} are the line-to-earth capacitances as per Figure 5.12(a), L_D is the inductance of the Petersen-coil and G_E is the admittance representing the phase-to-earth line losses.

If the phase-to-earth capacitances are different and if the asymmetry is assumed to be placed in phases R and Y, the capacitances are

$$C_{RE} = C_E + \Delta C_{RE} \tag{5.19a}$$

$$C_{YE} = C_E + \Delta C_{YE} \tag{5.19b}$$

$$C_{BE} = C_E \tag{5.19c}$$

where ΔC_{RE}; ΔC_{YE} are the asymmetry of the line-to-earth capacitances.

The displacement voltage is given by

$$\underline{U}_{NE} = \frac{U_n}{\sqrt{3}} * \frac{\Delta C_{RE} + \underline{a}^2 * \Delta C_{YE}}{(3 * C_E + \Delta C_{RE} + \Delta C_{YE}) - j(1/(\omega * L_D)) + 3 * G_E} \tag{5.20}$$

Defining the asymmetry factor $\underline{k}$

$$\begin{aligned} \underline{k} &= \frac{C_{RE} + \underline{a}^2 * C_{YE} + \underline{a} * C_{BE}}{C_{RE} + C_{YE} + C_{BE}} \\ &= \frac{\Delta C_{RE} + \underline{a}^2 * \Delta C_{YE}}{3 * C_E + \Delta C_{RE} + \Delta C_{YE}} \end{aligned} \tag{5.21a}$$

the system damping d

$$\begin{aligned} d &= \frac{3 * G_E}{\omega * (C_{RE} + C_{YE} + C_{BE})} \\ &= \frac{3 * G_E}{\omega * (3 * C_E + \Delta C_{RE} + \Delta C_{YE})} \end{aligned} \tag{5.21b}$$

and the detuning factor v

$$\begin{aligned} v &= \frac{1/(\omega * L_D) - \omega * (C_{RE} + C_{YE} + C_{BE})}{\omega * (C_{RE} + C_{YE} + C_{BE})} \\ &= \frac{1/(\omega * L_D) - \omega * (3 * C_E + \Delta C_{RE} + \Delta C_{YE})}{\omega * (3 * C_{RE} + \Delta C_{RE} + \Delta C_{YE})} \end{aligned} \tag{5.21c}$$

the displacement voltage $\underline{U}_{NE}$ is calculated by

$$\underline{U}_{NE} = \frac{U_n}{\sqrt{3}} * \frac{\underline{k}}{v + jd} \tag{5.22}$$

Assuming the asymmetrical capacitance ΔC_E concentrated in phase R ($\Delta C_E \gg \Delta C_{RE}$ and $\Delta C_E \gg \Delta C_{YE}$) the displacement voltage $\underline{U}_{NE}$, equal to the voltage in the zero-sequence component $\underline{U}_0$, is calculated with the system of symmetrical components based on Figure 5.12(b)

$$\underline{U}_0 = \frac{U_n}{\sqrt{3}} * \frac{j\omega * \Delta C_E}{j\omega * 3 * C_E} * \frac{1}{1 - (1/(3 * \omega^2 * L_D * C_E)) - j(G_E(\omega * C_E))} \tag{5.23}$$

The asymmetry factor k, the system damping d and the detuning factor v can be calculated based on these assumptions:

$$k = \frac{\Delta C_E}{3 * C_E} \tag{5.24a}$$

$$d = \frac{G_E}{\omega * C_E} \tag{5.24b}$$

$$v = 1 - \frac{1}{3 * \omega^2 * L_D * C_E} \tag{5.24c}$$

The displacement voltage $\underline{U}_{NE}$, equal to the voltage in the zero-sequence component $\underline{U}_0$, is calculated by

$$\underline{U}_{NE} = \underline{U}_0 = \frac{U_n}{\sqrt{3}} * \frac{k}{v + jd} \tag{5.25}$$

The polar plot of the displacement voltage $\underline{U}_{NE}$ as per Equations (5.22) and (5.25) and outlined in Figure 5.13 indicates a circular plot through the zero point. The phase angle of the diameter location at $v = 0$ is determined by the phase angle of the capacitive asymmetry. The diameter of the polar plot is defined as per Equation (5.25) as the ratio of capacitive asymmetry k and damping d.

The capacitive asymmetry is comparatively high in power systems with overhead transmission lines, resulting in a sufficient high-displacement voltage. Cable systems have a comparative small asymmetry, resulting for most of the cable systems in

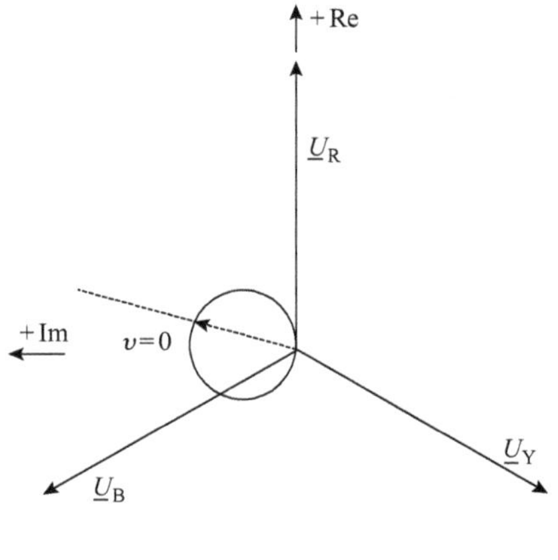

Figure 5.13 Polar plot of the displacement voltage in a power system with resonance earthing

an insufficient low-displacement voltage and problems while tuning the Petersen-coil into resonance. Capacitors between two phases or between one phase and earth will increase the displacement voltage to the required value.

5.5.3 Tuning of the Petersen-coil

The Petersen-coil can be constructed as a plunger-coil (tuning-coil) with continuous adjustment of the reactance, which can be tuned into resonance by successive operation. The displacement voltage measured at the Petersen-coil is maximal in the case of resonance tuning; the value depends on the capacitive asymmetry and on the losses of the reactor. The earth-fault current will be minimal in this case and the power frequency component of the capacitive earth-fault current is compensated by the reactive current of the Petersen-coil. Figure 5.14 indicates the displacement voltage and the residual current for different tuning of the reactor.

The displacement voltage shall be limited to $U_{NE} < 10$ kV. It is obvious that the residual current is increased as can be seen from Figure 5.14. Residual currents above 130 A in 110-kV-systems, respectively 60 A in 10-kV-systems, are not self-extinguishing; both parameters define the tuning limits of the Petersen-coil as indicated in Figure 5.14. Tuning of the Petersen-coil can be done in such a way that the resonance circuit is either capacitive (undertuning; $v < 0$), resulting in an

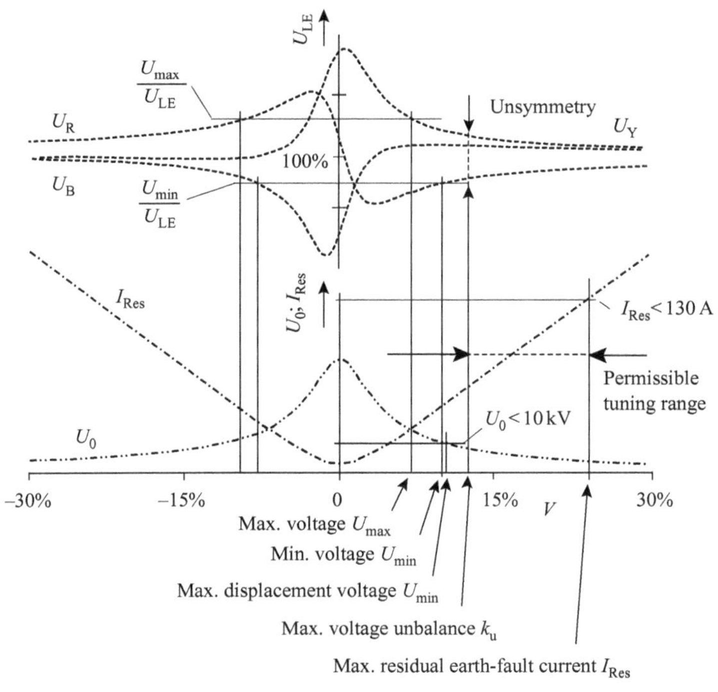

Figure 5.14 Voltages and residual current in the case of an earth-fault; displacement voltage without earth-fault

ohmic-capacitive residual current or inductive (overtuning; $v > 0$), resulting in an ohmic-inductive residual current at the earth-fault location. A small overtuning (overcompensation) up to $v = 10\%$ is often recommended as the displacement voltage will not increase in the case of switching of lines, because the capacitances will be reduced by this and the resonance circuit will be detuned without any further adjustment. The limits for the displacement voltage and the residual current as indicated in Figure 5.14 have to be guaranteed even under outage conditions.

Figure 5.14 also indicates the phase-to-earth voltages for different tuning factors (system parameters are: $U_n = 110$ kV; $I_{CE} = 520$ A; $d = 3\%$; $k = 1.2\%$) which also limit the range of detuning of the Petersen-coil. Assuming a minimal permissible voltage of $U_{min} = 0.9 * (U_n/\sqrt{3})$ according to IEC 60038, a maximal permissible voltage according to IEC 60071-1 of $U_{max} = 123\ \text{kV}/\sqrt{3}$ and a permissible asymmetry of the three voltages according to DIN EN 50160 of $p = 2\%$ it can be seen that the permissible tuning range of the Petersen-coil is $v = 12$–22%.

All considerations carried out so far are based on a linear current-voltage-characteristic of the Petersen-coil. Figure 5.15 indicates the non-linear characteristic of a Petersen-coil ($U_r = 20\ \text{kV}/\sqrt{3}$; $I_r = 640$ A) for minimal and maximal adjustment.

Due to the non-linear characteristic, the minimum of the residual current is not achieved at the maximal displacement voltage (adjustment criteria of the Petersen-coil). The difference is typically in the range of 3–15% of the rated current as outlined in Figure 5.16.

5.6 Handling of neutrals on HV-side and LV-side of transformers

Special attention must be placed while selecting the type of neutral handling on HV-side and LV-side of transformers. The neutral earthing on one side of the transformer has an influence on the system performance on the other side, in case of earth-faults or single-phase short-circuits as the voltages in the zero-sequence component are transferred from one side of the transformer to the other. The neutral earthing of a 110/10-kV-transformer (vector group Yyd) according to Figure 5.17 is taken as an example. It is assumed [15] that $X_{C0}/R_0 = 0.1$–0.05, first value is applied for systems with overhead lines, second value with cables.

The impedances $\underline{Z}_{E1}$ and $\underline{Z}_{E2}$ as per Figure 5.17 representing the earthing are different depending on the type of neutral earthing. In case of a single-phase fault in the high-voltage system (110 kV), the voltage U_0 in the zero-sequence component is transferred to the medium-voltage system (10 kV) with the same amount. Similar consideration indicates that the voltage in the zero-sequence component is transferred to the HV-side in case of a single-phase fault in the LV-system. In both cases, a fault current is measured in the system, having no fault. Table 5.2 indicates the results of a fault-analysis [16] with the voltages transferred through the transformer in case of faults.

The 110/10-kV-transformer can be operated with low-impedance earthing on both sides if a third winding (compensation winding, vector group d) is available, as can be seen from Table 5.2. If the transformer is not equipped with compensation winding,

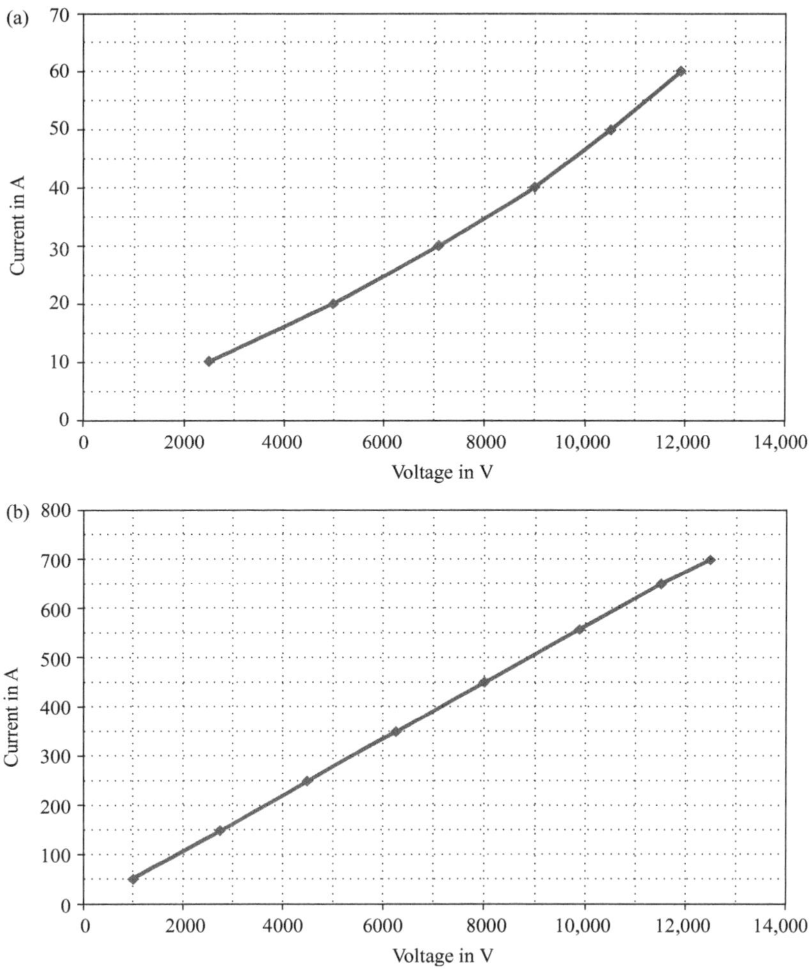

Figure 5.15 Current-voltage characteristic of a Petersen-coil; $U_r = 20$ kV/$\sqrt{3}$; $I_r = 640$ A. (a) Minimal adjustment (50 A) and (b) maximal adjustment (640 A)

the voltages in the zero-sequence component may reach values up to 70 per cent of the phase-to-earth voltage.

Low-impedance earthing on the 110-kV-side and resonance earthing on the 10-kV-side should be avoided due to high voltages in the zero-sequence component, which furthermore depend on the tuning of the Petersen-coil. The maximal voltage in this case is not reached for resonance tuning but depends on the ratio X_{C0}/R_0. The strategy to limit the displacement voltage under normal operation conditions as per Section 5.5.2 may result in an increased displacement voltage in the 10-kV-system with resonance earthing in the case of an earth-fault in the 110-kV-system with low-impedance earthing.

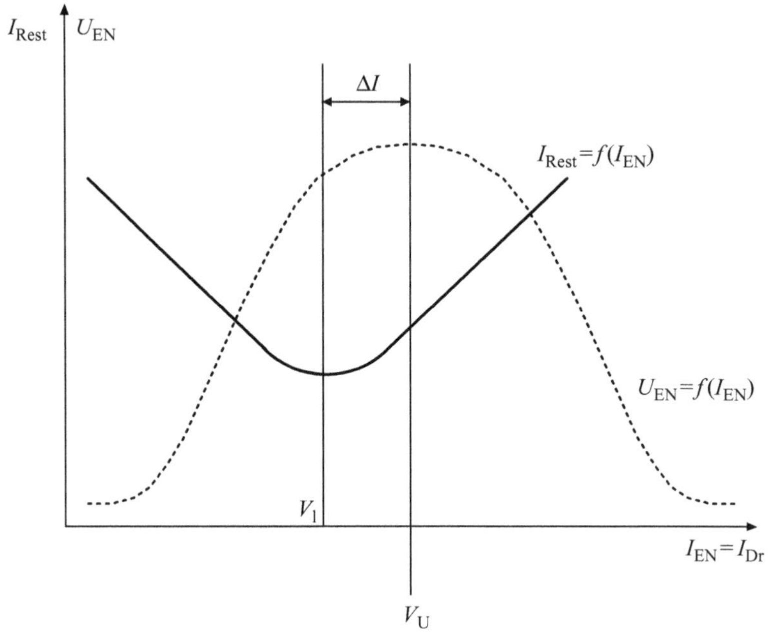

Figure 5.16 Displacement voltage in non-faulted operation and residual current under earth-fault conditions; non-linear characteristic of the Petersen-coil

Resonance earthing in the 110-kV-system can be combined with all types of neutral earthing in the 10-kV-system if the transformer is equipped with a compensation winding. The connection of Petersen-coils to both neutrals (110- and 10-kV) has to be investigated for special cases and is not generally recommended. The voltage transfer by stray capacitances in the case of isolated neutral in the 10-kV-system can be reduced by installing capacitances in the 10-kV-system. If the earthing of both neutrals of transformers by Petersen-coils cannot be avoided in the same substation, the earthing should be alternate in the case of two parallel transformers as indicated in Figure 5.18(a). If only one transformer is installed, the connection of one Petersen-coil X_{D1} can be carried out directly to the transformer, the second one X_{D2} should be connected at an artificial neutral as per Figure 5.18(b).

If the feeding system (e.g., 110 kV) is operated with low-impedance earthing and the medium-voltage system (e.g., 20 kV) is earthed through Petersen-coils or by fault limiting impedance, fault currents will occur in the medium-voltage system in the case of a single-phase short-circuit in the high-voltage system, as outlined in Figure 5.19, the value of which depends on the impedance of the earthing in the medium-voltage system. In some cases, this current may exceed the rated current of the transformer, thus causing operation of power system protection on MV-side [5,16].

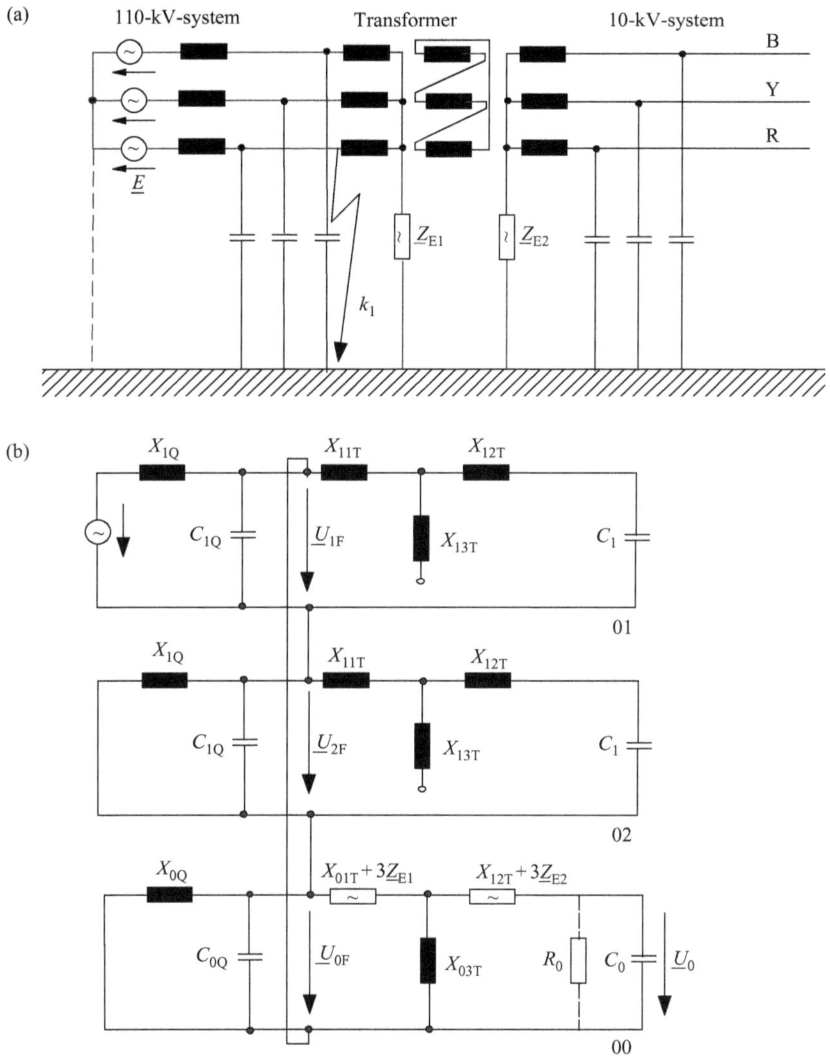

Figure 5.17 Transformation of voltage in the zero-sequence component of transformers in the case of single-phase faults. (a) Equivalent circuit diagram in RYB-system and (b) equivalent circuit diagram in the system of symmetrical components

5.7 Examples

5.7.1 Increase of displacement voltage for systems with resonance earthing

The capacitive asymmetry k of cable systems normally is below $k < 0.1\%$; the system damping is in the range of $d \approx 2$–4% resulting in a displacement voltage

Table 5.2 Voltages in the zero-sequence component U_0 transferred through 110/10-kV-transformer in the case of single-phase fault in the 110-kV-system according to Figure 5.17; U_n: nominal system voltage

10-kV-system		110-kV-system and compensation winding of the transformer		
Z_{E2}	Limitation I_{k1}	Low-impedance earthing $u_0 = U_0/(U_n/\sqrt{3})$		Resonance earthing
		With compensation winding	Without compensation winding	With compensation winding
Low-resistance earthing $Z_{E2} = 0\ \Omega$		0.2	0.6	0.03
Current limitation	2000 A	0.2	0.6	0.03
Z_{E2} inductive	500 A	0.25	0.7	0.04
Current limitation	2000 A	0.2	0.6	0.03
Z_{E2} ohmic	500 A	0.2	0.6	0.03
Resonance earthing		<7	>10	<0.3
Isolated neutral		Voltage transfer through stray capacitances		

which is too low for the operation of the resonance controller. For a given capacitive asymmetry of $k = 0.1\%$ and system damping $d = 2\%$ the displacement voltage will be $U_{NE} = 0.005 * U_n/\sqrt{3}$ for resonance tuning ($v = 0\%$); if the detuning shall be $v = 5\%$ the displacement voltage will be $U_{NE} = 0.002 * U_n/\sqrt{3}$ only [17]. Resonance controller normally operates sufficiently if the displacement voltage is above $U_{NE} > 0.03 * U_n/\sqrt{3}$. The displacement voltage, therefore, had to be increased by installing an additional capacitor in one phase [18]. The required capacitance ΔC_E for different parameters is given in Table 5.3 for a 20-kV-system earthed through Petersen-coil.

Documentation of system data normally do not indicate exact values of the line-to-earth capacitances C_E in each phase, furthermore the system damping d, the exact length of the cables and the non-linear characteristic of the Petersen-coil are also unknown or only to an insufficient extent and are not suitable to determine the displacement voltage U_{NE} without measurement. It should be noted that the system damping changes with the system load as well as in the case of a power system with a significant number of overhead lines and with external climatic conditions. It can be deducted from this, that:

- the asymmetrical capacitance (absolute value)
- the angle of the asymmetrical capacitance (with respect to the three phases)
- the value of the displacement voltage, and
- the resonance curve for different detuning factors

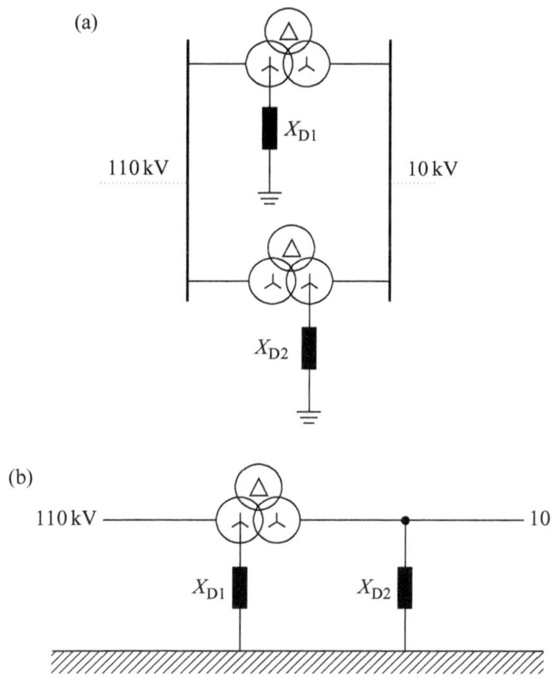

Figure 5.18 Alternate earthing of transformer neutrals by Petersen-coils. (a) Two parallel transformers and (b) earthing at artificial neutral with reactor X_{D2}

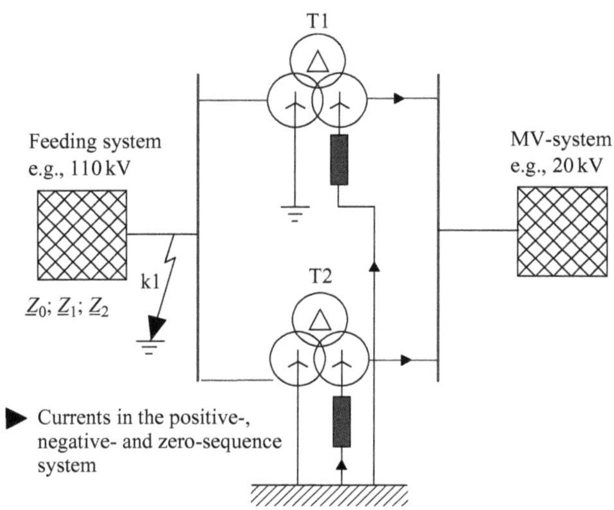

Figure 5.19 Fault current in the MV-system in the case of a short-circuit in the HV-system

Table 5.3 Capacitive asymmetry ΔC_E for different parameters in a 20-kV-system

$U_{NE} = U_0$ (V)	Damping d (%)	Detuning v (%)	Asymmetry k (%)	ΔC_E (nF)
346.4	2	4	0.134	177
		6	0.19	251
	4	4	0.17	225
		6	0.216	285
577.4	2	4	0.224	296
		6	0.316	417
	4	4	0.283	374
		6	0.361	477

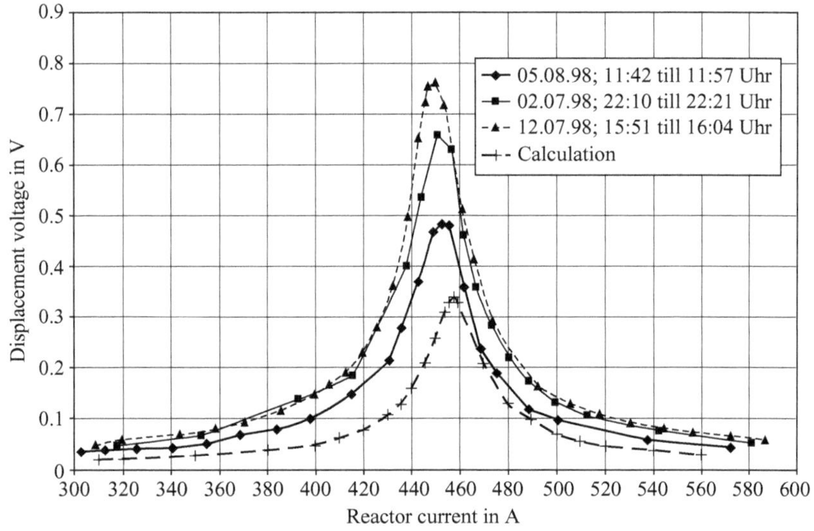

Figure 5.20 Resonance curve (displacement voltage) for different detuning factors in a 20-kV-system for different conditions

cannot be determined from system studies, but require measurement for different load conditions.

The resonance curves of a 20-kV-system with total cable length of 176 km as measured for different load conditions are outlined in Figure 5.20. Furthermore, the resonance curve calculated from documented system data is also given in Figure 5.20. Voltages refer to the secondary side of a voltage transformer in the neutral (ratio $20\ \text{kV}/\sqrt{3} : 100\ \text{V}$). The maximal value of each of the resonance curves differ by

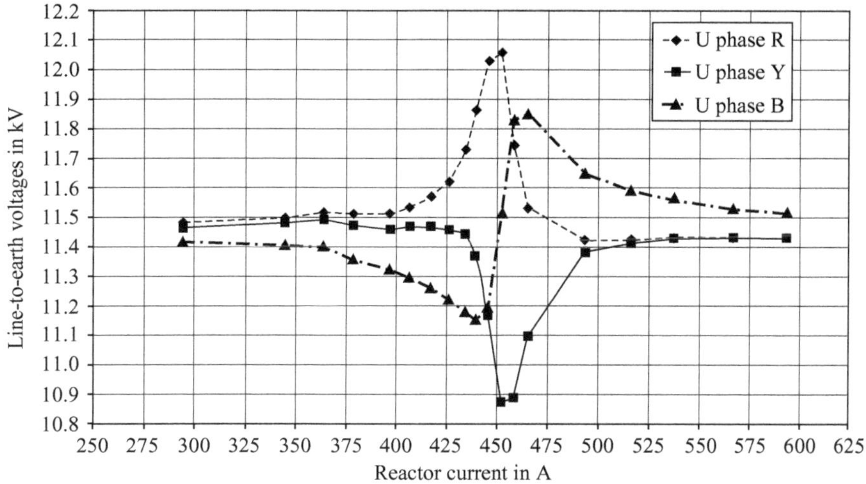

Figure 5.21 *Voltages in a 20-kV-system with resonance earthing for different tuning factors. (a) Phase-to-earth voltages and (b) displacement voltage (resonance curve)*

more than 60 per cent for different load conditions and is higher than the calculated value. The resonance tuning ($v = 0$) of calculated and measured resonance curves differs also by 6 per cent as can be seen from Figure 5.20.

A capacitance of $\Delta C_E = 173.3$ nF was installed in phase B in order to increase the displacement voltage up to 3–5 per cent (350–580 V) of the nominal system voltage for resonance tuning. The total phase-to-earth capacitance of the 20-kV-system was determined to be $C_E = 44\ \mu$F. The results of measurement of the phase-to-earth voltages and the displacement voltage for different detuning of the Petersen-coil are outlined in Figure 5.21 (voltage transformer ratio 20 kV/$\sqrt{3}$: 100 V).

It can be seen from Figure 5.21 that the maximal displacement voltage is in the required range between 3 and 5 per cent of the phase-to-earth voltage. For resonance tuning, the maximal phase-to-earth voltage appears in phase R with $U_R \approx 12.1$ kV and the minimal voltage in phase Y with $U_Y \approx 10.9$ kV, which is only 94 per cent of the nominal phase-to-earth voltage. Exact resonance tuning shall be avoided in this system due to the low voltage in one phase and the resulting high asymmetry factor.

Other measures to increase the displacement voltage such as installation of a reactor in one phase to earth and voltage additions are not considered in detail here [19].

5.7.2 Limitation of single-phase short-circuit current by earthing through impedance

An urban 11.5-kV-system is fed from a 132-kV-system; both systems are pure cable systems. The substations are equipped with four transformers each, having rated values of $S_r = 40$ MVA, $u_{kr1} = 14\%$ and $u_{kr0} = 16\%$. Both systems are earthed by low impedance in such a way that all transformer neutrals are solidly connected to earth. The short-circuit current level in the 132-kV-system is high, whereas the single-phase short-circuit currents is higher than the three-phase short-circuit currents ($I''_{k3} \approx 29.3$ kA; $I''_{k1} \approx 37.3$ kA). The maximal permissible short-circuit current in the 132-kV-system is $I''_{kmax} = 40$ kA; in the 11.5-kV-system $I''_{kmax} = 25$ kA. The 132/11.5-kV-transformers cannot be operated in parallel due to the high short-circuit currents. In the case of low-impedance earthing the neutrals of each transformer, the single-phase short-circuit current of one transformer is $I''_{k1} = 15.04$ kA in case of a short-circuit at the 11.5-kV-busbar. In the case of two transformers in parallel, the short-circuit current is increased to $I''_{k1} = 29.27$ kA, which is above the maximal permissible short-circuit current of the 11.5-kV-system.

The limitation of single-phase short-circuit currents to 25 kA in case two transformers shall be operated in parallel is possible by earthing of the 11.5-kV-transformer neutrals through a resistor having $R_E = 0.31\ \Omega$ or through a reactor having $X_E = 0.1\ \Omega$ (see also Figure 11.13).

5.7.3 *Design of an earthing resistor connected to an artificial neutral*

The short-circuit limitation in a 20-kV-system having initial short-circuit power $S''_{kQ} = 700$ MVA and an impedance ratio $Z_0/Z_1 = 4$ is explained below. The single-phase short-circuit current in case of low-impedance earthing is $I''_{k1} = 12.1$ kA. The single-phase short-circuit current shall be limited to $I''_{k1max} = 1.5$ kA ($I''_{k1} = 3I_0$) realised by earthing through a resistor to be connected to an artificial neutral. As the current through the artificial neutral is only one-third of the single-phase short-circuit current, the rated power shall be

$$S_{rS\,min} = 0.33 * \sqrt{3} * I''_{k1\,max} * U_n = 17.55 \text{ MVA}$$

A transformer with vector group Zz is selected for the artificial neutral with $S_{rS} = 20$ MVA; $P_{Cu} = 1.3$ kW and $u_{k0} = 45\%$ resulting in a required zero-sequence impedance of $Z_{0S} = 9\ \Omega$ ($R_{0S} = 3.75\ \Omega$; $X_{0S} = 8.18\ \Omega$). The required rating of the earthing resistor resulting from the calculation is $R_E = 4.21\ \Omega$. The value of the initial short-circuit power of the system has only a marginal effect on the rating of the earthing.

A typical design rating of the resistor is given below:

$R_n = 4.21\ \Omega$ at 20°C (Tolerance ±5 per cent or ±10 per cent); CrNi-alloyed steel
$I_r = 1.5$ kA, rated short-time duration $t = 5$ s or 10 s
$U_r = 13.3$ kV
Isolation according to IEC 60071-1, Table 5.2: $U_m = 17.5$ kV IP 00

5.7.4 *Resonance earthing in a 20-kV-system*

Figure 5.22 shows a 20-kV-system fed from two sides by one transformer in each of the substations S/S A and S/S B. The system is split at S/S C into two subsystems A and B during summer months. During winter months, a part of system A is connected to system B, but both systems are still operated separately.

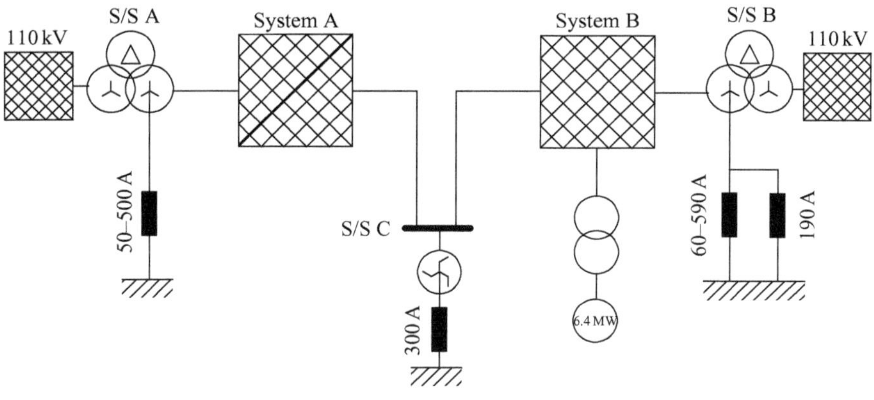

Figure 5.22 Equivalent circuit diagram of a 20-kV-system with resonance earthing

The 20-kV-system A is a pure cable-system having a total system length of 490 km of the XLPE-type. The capacitive earth-fault current is calculated to be $I_{res} = 453$ A. The transformer in substation A has a rating of 12.5 MVA; current carrying capacity of the neutral is 361 A. The system shall be operated with resonance earthing, i.e., a reactor has to be installed in the neutral having a rated current of more than 450 A to compensate the capacitive earth-fault current. It is obvious that the reactor cannot be connected to the neutral of the transformer in substation S/S A as the required current for resonance earthing exceeds the permissible current of the transformer neutral. An additional fixed reactor with rated current 300 A is installed in substation S/S C and connected at an artificial neutral of a Zz-transformer.

System B is a pure cable-system as well, having a total system length of 386 km of different type (XLPE and mass-impregnated cables); the earth-fault current is calculated to be $I_{res} = 632$ A. The feeding transformer in substation B has a rated power $S_r = 25$ MVA and a current carrying capacity of the neutral of 721 A. Resonance earthing is done with a fixed reactor (190 A) in parallel with a tuning-reactor (60–590 A) capable of compensating the total capacitive earth-fault current of system B.

Operation in wintertime differs from the described scheme as additional generation in a combined-cycle plant is in operation with a power of 6.4 MW. In order to avoid back-feeding into the 110-kV-system additional load from system A is supplied from system B. The capacitive earth-fault current of system A is reduced and the current in system B is increased accordingly (see Table 5.4 for details). As the earthing reactor in S/S C is connected in wintertime as well to system B, the compensation scheme with different reactors at different substations are capable of realising the resonance earthing for both systems under different operating conditions.

5.7.5 Calculation of capacitive earth-fault current and residual current

A 10-kV-system with isolated neutral (overhead lines, system length 170 km, $C'_B = 9.5$ nF/km; $C_E/C_L = 5$) shall be extended by cables. The existing earthing

Table 5.4 Characteristics of a 20-kV-system with respect to resonance earthing

Condition	System A			System B			
	I_{res} (A)	Available reactors		I_{res}	Available reactors		
Summertime	453	30–300 A S/S A	300 A S/S C	721 A	60–590 A S/S B	190 A S/S B	—
Wintertime	261	30–300 A S/S A	—	913 A	60–590 A S/S B	190 A S/S B	300 A S/S C

concept shall be checked with respect to its suitability. The phase-to-earth capacity is calculated as $C_E = 1009\ \mu F$ and the earth-fault current is given according to Equation (5.9) to be $I_E = 6.05$ A which is self-extinguishing as can be seen in Figure 5.7. The capacitive earth-fault current is increased by 1.42 A/km if cables of the NAKBA-type will be installed ($C_{E\ cable}/C_{E\ OHL} = 40$) reaching the permissible limit for self-extinguishing $I_{CE} = 35$ A for a cable length of 23 km.

If the system earthing shall be changed to resonance earthing, the Petersen-coil shall have a reactance for resonance tuning ($v = 0$) of $X_D = 165\ \Omega$, the inductance will be $L_D = 0.525$ H for 50 Hz. The resistance of the reactor is assumed to be $R_D = 6\ k\Omega$ (parallel equivalent diagram). The residual current is given by $I_{Res} = 2.9$ A. The ratio of residual current to capacitive current is $I_{Res}/I_{CE} = 8.3$ per cent.

5.7.6 Voltages at neutral of a unit transformer

A power station with rated power 400 MVA is connected to a 220-kV-system. The neutral of the unit transformer can be earthed either through an impedance (reactor), directly without impedance or kept isolated without earthing as outlined in Figure 5.23. The rating of the reactor shall be determined in such a way to guarantee a ratio $X_0/X_1 = 2$ in case of a short-circuit at location F. The single-phase short-circuit

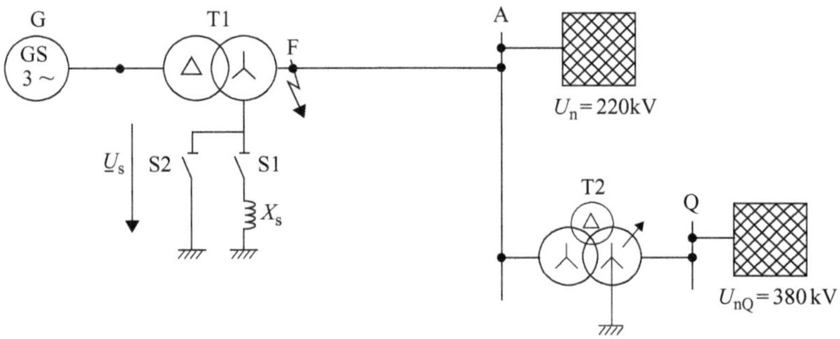

Figure 5.23 Connection of a power station to a 220-kV-system with short-circuit location

current and the voltage at the transformer neutral shall be calculated for the different operating conditions.

The rated data of the equipment are given below:

System Q	$U_{nQ} = 380$ kV; $S''_{kQ} = 15{,}000$ MVA; $X_{0Q}/X_{1Q} = 3$
System A	$U_{nQ} = 220$ kV; $S''_{kQ} = 5{,}000$ MVA; $X_{0A}/X_{1A} = 3$
G	$S_{rG} = 400$ MVA; $U_{rG} = 21$ kV; $x''_{d1} = 15\%$; $\cos\varphi_{rG1} = 0.8$; $p_{G1} = 10\%$
T1	$S_{rT1} = 400$ MVA; $u_{krT1} = 14\%$; $U_{rT1HV}/U_{rT1LV} = 220\ \text{kV}/21\ \text{kV}$; $p_{T1} = 12\%$
T2	$S_{rHVT2} = 660$ MVA; $S_{rMVT2} = 660$ MVA; $S_{rLVT2} = 198$ MVA
	$u_{krHVMVT2} = 10.2\%$; $u_{krHVLVT2} = 13.5\%$; $u_{krMVLVT2} = 10.5\%$
	$U_{rHVT2} = 380$ kV; $U_{rMVT2} = 220$ kV; $U_{rLVT2} = 30$ kV; $p_{T2} = 10\%$
	$X_{0T2}/X_{1T2} = 1$ (three single-phase transformers)
Line L	$X'_{1L} = 0.3\ \Omega/\text{km}$; $X'_{0L} = 1.0\ \Omega/\text{km}$; $l_L = 10$ km

The impedances of the equipment calculated in the Ω-system are given below.

No.	Equipment	X_1 (Ω)	X_0 (Ω)	Correction as per Table 3.3	X_{1K} (Ω)	X_{0K} (Ω)	I''_{k3} (kA)	I''_{k1} (kA)
1	System Q	10.595	42.38		10.595	42.38		
2	System A	3.532	10.596		3.532	10.596		
3	Generator G	18.15						
4	Transformer 1	16.94	16.94					
3 + 4	Power station	35.09	16.94 $+3X_S$	1.093	38.35	From right 18.681 From left 16.94+ $3X_s$		
5	Line	3.0	10.0		3.0	10.0		

No.	Equipment	X_1 (Ω)	X_1 (Ω)	Correction as per Table 3.2	X_{1K} (Ω)	X_{0K} (Ω)	I''_{k3} (kA)	I''_{k1} (kA)
6	Transformer 2 HV	7.422	7.422	HV-MV 0.985	7.023	7.023		
	MV	0.3436	0.3436	MV-LV 0.983	0.399	0.399		
	LV	25.59	25.59	HV-LV 0.967	25.586	25.586		
	Total impedance at 220 kV ($X_0/X_1 = 2$)				6.945	13.89 !!	18.29	15.09

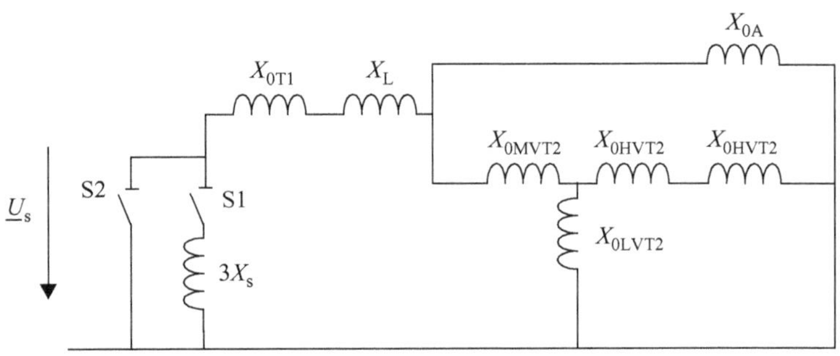

Figure 5.24 Equivalent diagram in the zero-sequence component for fault location F

It should be noted that the impedance of the earthing reactor appears threefold ($X_{0S} = 3X_S$) in the zero-sequence component. As the rating of the reactor shall be determined to guarantee the ratio $X_0/X_1 = 2$ for short-circuit at location F the single-phase short-circuit current is $I''_{k1} = 15.09$ kA. By this the impedance of the reactor will be $X_S = 12.4\ \Omega$.

In order to calculate the voltage $\underline{U}_S$ at the transformer neutral while earthed with the reactor, the zero-sequence current through the reactor $\underline{I}_{0S}$ needs to be calculated as outlined in Figure 5.24 indicating the equivalent diagram in the zero-sequence component.

The zero-sequence current through the reactor is determined by the ratio of the impedances

$$\underline{I}_{0S} = \frac{X_{0k}}{X_{0\text{left}}} * \underline{I}_0 = \frac{X_{0k}}{X_{0\text{left}}} * \frac{\underline{I}''_{k1}}{3} = \frac{13.89\ \Omega}{54.16\ \Omega} * \frac{15.09\ \text{kA}}{3} = 1.29\ \text{kA}$$

The voltage across the reactor U_{Se} at the transformer neutral is calculated by

$$\underline{U}_{Se} = \underline{I}_{0S} * 3 * X_S = 1.29 \text{ kA} * 3 * 12.41 \ \Omega = 48.03 \text{ kV}$$

In this case, the transformer neutral is kept isolated (S1 and S2 open); the zero-sequence current through the reactor will be Zero. The voltage is determined by the zero-sequence current determined by the remaining equipment (see Figure 5.24). The zero-sequence impedance is $X_0 = 18.681 \ \Omega$. The single-phase short-circuit current without earthing is calculated by

$$\underline{I}''_{k1} = \frac{1.1 * \sqrt{3} * U_n}{X_0 + 2 * X_1} = \frac{1.1 * \sqrt{3} * 220 \text{ kV}}{18.681 \ \Omega + 2 * 6.945 \ \Omega} = 12.87 \text{ kA}$$

The voltage across the reactor $\underline{U}_{Si}$ is calculated by

$$\underline{U}_{Si} = \frac{\underline{I}''_{k1}}{3} * X_0 = \frac{12.87 \text{ kA}}{3} * 18.681 \ \Omega = 80.14 \text{ kV}$$

The insulation of the transformer neutral must be designed for the maximal voltage, i.e., 80.14 kV.

Chapter 6
Calculation of short-circuit currents in low-voltage systems

6.1 General

IEC 60781 presents an application guide for the calculation of short-circuit currents in low-voltage radial systems. The methods described there are identical to those as per IEC 60909-0 and as outlined in Chapter 3. The short-circuits are treated as far-from-generator short-circuits. This assumption is valid in the future as well, even with an increasing number of distributed generation units in low-voltage systems. New generation will be connected to the system in the case of

- Photovoltaic installations by rectifier (six-pulse bridge or PWM-rectifier).
- Small wind turbines, low-rated combined heat and power units and small hydro power plants by asynchronous generators (rotor fed).

Installations with synchronous generators are comparatively rare.

The following chapters describe the approach and special conditions for the calculation of short-circuit currents in low-voltage systems.

6.2 Types of faults

Depending on the type of protection against electrical shock in low-voltage systems all types of short-circuit, i.e., three-phase, double-phase with and without earth connection and single-phase-to-earth short-circuits can occur. The maximal short-circuit current depends on the impedances of the positive- and zero-sequence component. Reference is made to Figure 4.4. The ratio Z_2/Z_1 can be set to 1, as low-voltage systems for the most part have no generation by synchronous generators. The three-phase short-circuit will lead in those cases to the maximal short-circuit current. Special attention must be given to the currents flowing through earth, as the phase angles of the impedances in the positive- and zero-sequence components differ a lot from each other in low-voltage systems.

6.3 Method of calculation

The method of the equivalent voltage source at the short-circuit location is applied for the calculation of short-circuit currents in low-voltage systems.

- Symmetrical short-circuits are represented by the positive-sequence component, asymmetrical (unbalanced) short-circuits are represented by connection of positive-, negative- and zero-sequence components (see also Table 2.1).
- Capacitances and parallel admittances of non-rotating load of the positive-sequence component are neglected. Capacitances and parallel admittances of the zero-sequence component have only an influence on fault currents in power systems with isolated neutral or with resonance earthing.
- Impedance of the arc at the short-circuit location is neglected.
- The type of short-circuit and the system topology remain unchanged during the duration of the short-circuit.
- The tap-changer of any transformer is assumed to be in main-position.
- All internal voltages are short-circuited and an equivalent voltage source with value $cU_n/\sqrt{3}$ is introduced at the short-circuit location. The voltage factor c shall be selected in accordance with Table 6.1.

6.4 Calculation of short-circuit parameters

6.4.1 Impedances

Calculation of impedances of equipment and the analysis of the short-circuit impedance at short-circuit location was explained in Sections 2.5 and 3.2. It should be noted that the impedances of the equipment must be related to the voltage level of the short-circuit location.

Table 6.1 Voltage factor c according to IEC 60909-0. (Voltage factors as per IEC 60781 are of different values. The standard is under review)

Nominal system voltage U_n	Voltage factor c for calculation of	
	Maximal short-circuit current, c_{max}	Minimal short-circuit current, c_{min}
LV: 100 V up to 1000 V (inclusive) (IEC 60038, Table 1)		
Voltage tolerance +6%	1.05	0.95
Voltage tolerance +10%	1.10	0.95

Remark: $c_{max}U_n$ shall not exceed the highest voltage of equipment U_m according to IEC 60071

When data for generation in LV-systems are not known, approximation as below shall be used for the branch short-circuit currents [12]:

Synchronous generator	Branch short-circuit current equal to eight times rated current
Asynchronous generator	Branch short-circuit current equal to six times rated current
Generator with rectifier	Branch short-circuit current equal to rated current

This approximation does not include the effect of conductors between the generation unit and the short-circuit location.

6.4.1.1 Initial symmetrical short-circuit current I''_{k}

The initial symmetrical short-circuit current I''_{k} is calculated for balanced and unbalanced short-circuits based on the equivalent voltage source at the short-circuit location. The short-circuit impedance seen from the short-circuit location has to be determined with the system of symmetrical components. The results obtained for the short-circuit currents (and the voltages of the non-faulted phases, if required) in the 012-system have to be transferred back into the RYB-system. The results for the different types of short-circuits are outlined in Table 4.2.

6.4.1.2 Peak short-circuit current i_{p}

Depending on the feeding source of the short-circuit, different considerations have to be taken to calculate the peak short-circuit current. Short-circuits in low-voltage systems normally are single-fed short-circuits. The short-circuit impedance is represented by a series connection of the individual impedances. The peak short-circuit current can be calculated for the different types of short-circuit by

$$i_{\mathrm{p3}} = \kappa * \sqrt{2} I''_{\mathrm{k3}} \tag{6.1a}$$

$$i_{\mathrm{p2}} = \kappa * \sqrt{2} I''_{\mathrm{k2}} \tag{6.1b}$$

$$i_{\mathrm{p1}} = \kappa * \sqrt{2} I''_{\mathrm{k1}} \tag{6.1c}$$

The factor κ

$$\kappa = 1.02 + 0.98 * \mathrm{e}^{-3R/X} \tag{6.2}$$

obtained from the ratio R/X for three-phase short-circuit shall be taken for all types of short-circuits as well. All other assumptions as per Section 4.3.2.3 are valid in low-voltage systems as well.

6.4.2 *Symmetrical short-circuit breaking current I_{b}*

Short-circuits in low-voltage systems normally are far-from-generator short-circuits. The symmetrical short-circuit breaking current is identical to the initial symmetrical short-circuit current.

6.4.3 *Steady-state short-circuit current* I_k

Short-circuits in low-voltage systems normally are far-from-generator short-circuits. The steady-state short-circuit current is identical to the initial symmetrical short-circuit current.

6.4.3.1 Influence of motors

Asynchronous motors contribute to the initial symmetrical short-circuit current, to the peak short-circuit current, to the symmetrical short-circuit breaking current and in the case of unbalanced short-circuits to the steady-state short-circuit current as well. Synchronous motors are modelled like generators and asynchronous generators are treated as asynchronous motors. Motors of any kind, which are not in operation at the same time, e.g., due to the process or due to any interlocking, can be neglected for the calculation of short-circuit current to such an extent that only those motors are to be taken into account which lead to the highest contribution of the short-circuit current under realistic operating conditions. Motors fed by static-rectifiers need to be considered in the case of three-phase short-circuits only if they are able to transfer energy for deceleration at the time of short-circuit, as they contribute to the initial symmetrical and to the peak short-circuit current.

Asynchronous motors in public supply systems are considered when

- the sum of the rated currents is greater than 1 per cent of the initial symmetrical short-circuit current without motors;
- the contribution to the initial symmetrical short-circuit current is greater or equal to 5 per cent without motors.

Medium-and low-voltage motors connected through two-winding transformers to the short-circuit are considered if

$$\frac{\sum P_{\mathrm{rM}}}{\sum S_{\mathrm{rT}}} > \frac{0.8}{|(c100\sum S_{\mathrm{rT}}/(\sqrt{3}U_{\mathrm{nQ}}/I_{\mathrm{k}}'')) - 0.3|} \tag{6.3}$$

The influence of asynchronous motors in low-voltage systems can be neglected if

$$\sum I_{\mathrm{rM}} \leq 0.01 * I_{\mathrm{k}}'' \tag{6.4}$$

where U_{nQ} is the nominal system voltage at short-circuit location Q, $\sum S_{\mathrm{rT}}$ is the sum of rated apparent power of all transformers, directly connected to motors feeding the short-circuit, I_{k}'' is the initial symmetrical short-circuit current without motors, $\sum P_{\mathrm{rM}}$ is the sum of rated active power of all motors and $\sum I_{\mathrm{rM}}$ is the sum of rated currents of all motors.

6.5 Minimal short-circuit currents

In order to calculate the minimal short-circuit current the voltage factor c_{min} according to Table 6.1 for the equivalent voltage source at the short-circuit location has to be considered. Furthermore,

- System topology, generator dispatch and short-circuit power of feeding networks have to be defined in such a way that the minimal short-circuit current is expected. This normally applies for low-load conditions.
- Motors are to be neglected.
- Resistances of overhead lines and cables shall be calculated with the maximal permissible temperature at the end of the short-circuit, e.g., 80°C in low-voltage systems.

These assumptions have to be made in the case of balanced and unbalanced short-circuits, except when other presuppositions are mentioned.

6.6 Examples

Examples for the calculation of short-circuit currents in LV-systems are included in IEC 60781. The calculation is carried out with form-sheets, which are used for the calculation of impedances of equipment as well as for short-circuit current calculation itself. The voltage factor $c_{max} = 1.0$ as given in the examples shall be $c_{max} = 1.05$ or $c_{max} = 1.1$ according to IEC 60909-0. Reference is made to Table 6.1.

Short-circuit current calculation can easily be carried out with spreadsheet analysis using, e.g., EXCEL [14]. Figure 6.1 outlines an example for a low-voltage installation. The initial symmetrical and the peak short-circuit current both for minimal and maximal conditions shall be calculated for short-circuit either at location A or at location B. The data of the equipment are given below:

System feeder	$U_n = 10$ kV; $S''_{kQmax} = 240$ MVA; $S''_{kQmin} = 190$ MVA
Transformer	$S_{rT} = 630$ kVA; $u_{krT} = 6\%$; $u_{rRT} = 1.1\%$; $t_{rT} = 10/0.4$ kV; $R_0/R_1 = 1$; $X_0/X_1 = 1$
Cable	Each 4×240 mm^2; $R'_K = 77.4$ mΩ/km; $X'_K = 78$ mΩ/km; $R_0/R_1 = 4$; $X_0/X_1 = 4.1$; $l = 35$ m
Motor	$P_{rM} = 50$ kW; $U_{rM} = 0.41$ kV; $\cos\varphi_{rM} = 0.84$; $\eta_{rM} = 0.94$; $I_{anM}/I_{rM} = 6$

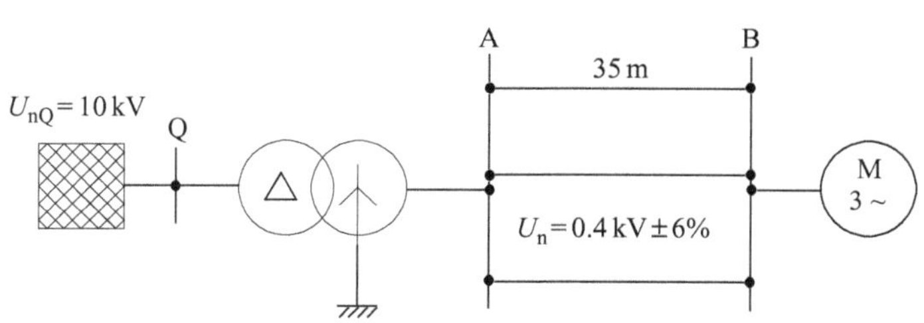

Figure 6.1 Equivalent circuit diagram of a LV-installation

Table 6.2 Example for the calculation of maximal short-circuit currents in LV-system

	Data Input	R mΩ	X mΩ	Z_1 mΩ	Z_0 mΩ
Low voltage					
Nominal voltage in kV	0.4				
c_{max}/c_{min}	1.1				
System feeder					
Pos. seq. system					
S''_{kQmax} in MVA	240	0.0421	0.4203	0.4225	
S''_{kQmin} in MVA	190	0.0333	0.3328	0.3345	
Transformer					
S_{rT} in kVA	630				
HHV-side	10				
LV-side	0.4				
Pos. seq. system					
u_{1kT} in %	6		15.2381	15.493	
u_{1rRT} in %	1.1	2.7937			
Zero seq. system					
u_{0kT} in %	6		15.2381		15.493
u_{0rRT} in %	1.1	2.7937			
Cable					
Length in m	35				
Cables in parallel	3				
Max. temp. °C	80				
Pos. seq. system					
X'_1 in mΩ/km	78		0.91	1.282	
R'_1 in mΩ/km	77.4	0.903			
Zero seq. system					
X'_0 in mΩ/km	319.8		3.731		5.193
R'_0 in mΩ/km	309.6	3.612			
Motor					
Active power in kW	50	171.34	407.93	442.46	
U_r in V	410				
cos	0.84				
η	0.94				
I_a/I_r	6				
Zero seq. Imp. in mΩ	99999999			99999999	

Rated current	Ampere	89.17
Assessment of motors	1% of s.-c. current in Ampere	147.8
Assessment of motors	5% of s.-c. current in Ampere	739

s.-c. at location A

I''_{k3min} kA	R_k/X_k	i_{p3} kA	I''_{k2min} kA	i_{p2} kA	I''_{kE2E} kA	i_{pE2E} kA	I''_{k1max} kA	i_{p1} kA
15.97	0.182	35.86	13.83	31.83	16.25	36.49	16.11	36.18

s.-c. at location B

I''_{k3min} kA	R_k/X_k	i_{p3} kA	I''_{k2min} kA	i_{p2} kA	I''_{kE2E} kA	i_{pE2E} kA	I''_{k1max} kA	i_{p1} kA
14.78	0.226	31.72	12.8	27.47	13.02	27.95	13.84	29.7

Branch s.-c. current of motor

I''_{k3max} kA	R_k/X_k	i_{p3} kA	I''_{k2max} kA	i_{p2} kA	I''_{kE2E} kA	i_{pE2E} kA	I''_{k1max} kA	i_{p1} kA
0.58	0.42	1.07	0.51	0.93	0.01	0.02	0.01	0.02

Branch s.-c. current of motor to be neglected

Table 6.3 Example for the calculation of minimal short-circuit currents in LV-system

	Input data	R mΩ	X mΩ	Z_1 mΩ	Z_0 mΩ
Low voltage					
Nominal voltage in kV	0.4				
c_{max}/c_{min}	0.95				
System feeder					
Pos. seq. system					
S''_{kQmax} in MVA2	240	0.0421	0.4203	0.4225	
S''_{kQmin} in MVA1	190	0.0333	0.3328	0.3345	
Transformer					
S_{rT} in kVA	630				
HV-side	10				
LV-side	0.4				
Pos. seq. system					
u_{1kT} in %	6		15.2381	15.493	
u_{1rRT} in %	1.1	2.7937			
Zero seq. system					
u0kT in %	6		15.2381		15.493
u0rRT in %	1.1	2.7937			
Cable					
Length in m	35				
Cables in parallel	3				
Max. temp. °C	80				
Pos. seq. system					
X'_1 in mΩ/km	78		0.91	1.443	
R'_1 in mΩ/km	77.4	1.1198			
Zero seq. system					
X'_0 in mΩ/km	319.8		3.731		5.83
R'_0 in mΩ/km	309.6	4.4789			

s.-c. at location A

I''_{k3min} kA	R_k/X_k	i_{p3} kA	I''_{k2min} kA	i_{p2} kA	I''_{kE2E} kA	i_{pE2E} kA	I''_{kmin} kA	i_{p1} kA
13.78	0.182	30.87	11.84	26.81	14.04	31.53	13.81	31.24

s.-c. at location B

I''_{k3min} kA	R_k/X_k	i_{p3} kA	I''_{k2min} kA	i_{p2} kA	I''_{kE2E} kA	i_{pE2E} kA	I''_{k1min} kA	i_{p1} kA
12.64	0.239	26.79	10.95	23.21	10.97	23.25	11.75	24.9

The calculation is carried out with EXCEL; the spreadsheet is shown in Tables 6.2 and 6.3. Data for the calculation of minimal short-circuit currents are automatically transferred; results are highlighted in the tinted boxes. All fields, except the input fields, are blocked against unintentional modification.

Chapter 7

Double earth-fault and short-circuit currents through earth

7.1 General

IEC 60909-3 describes methods and procedures for the calculation of currents during two separate simultaneous single-phase line-to-earth short-circuits (s.-c.) at different locations of the system, which are called 'double earth-fault' in the context of this section. The double earth-fault is not identical to a double-phase short-circuit, where two phases have a short-circuit at the same location. Furthermore, the branch short-circuit currents flowing through earth are dealt with.

7.2 Short-circuit currents during double earth-faults

7.2.1 Impedances and initial symmetrical short-circuit current I''_{k}

In order to calculate the short-circuit currents I''_{kEE} in the case of a double earth-fault, the mutual impedance in the positive- and zero-sequence component between the two short-circuit locations is needed. As the mutual impedance comparatively difficult to determine, the double earth-fault can only be analysed in a simple manner for special system configurations. The following cases are considered:

- Both short-circuit locations are on the same line.
- The short-circuit locations are on different lines.
- Single-fed line.
- Double-fed line.

In the case of a single-fed (radial) line, both short-circuit locations are on the same line and the double earth-fault is identical to a double-phase short-circuit without earth connection, as seen from the feeding point of the line. In the case of a double-fed single-circuit line, the voltage line-to-earth is changing significantly between the two short-circuit locations, whereas the line-to-line voltages are remaining almost unchanged. Details are outlined in [3].

In general, the short-circuit current in the case of a double earth-fault with short-circuit locations A and B is calculated according to [4] by

$$I''_{\mathrm{kEE}} = \frac{3 * c * U_{\mathrm{n}}}{|\underline{Z}_{1\mathrm{A}} + \underline{Z}_{2\mathrm{A}} + \underline{Z}_{1\mathrm{B}} + \underline{Z}_{2\mathrm{B}} + \underline{M}_1 + \underline{M}_2 + \underline{Z}_0|} \tag{7.1}$$

where $\underline{Z}_{1\mathrm{A}}$; $\underline{Z}_{1\mathrm{B}}$ are the short-circuit impedances in the positive-sequence system at location A and B, respectively, $\underline{Z}_{2\mathrm{A}}$; $\underline{Z}_{2\mathrm{B}}$ the short-circuit impedances in the negative-sequence system at location A and B, respectively, $\underline{M}_1$; $\underline{M}_2$ the mutual impedances in the positive-sequence system between the short-circuit locations A and B, respectively, and $\underline{Z}_0$ the short-circuit impedance in the zero-sequence system between locations A and B.

The impedance $\underline{M}_1$ and $\underline{M}_2$, representing the mutual impedances of the positive- and negative-sequence components between the two short-circuit locations, can be measured by short-circuiting all voltages in the system and feeding the voltage $\underline{U}_{1\mathrm{A}}$ (positive-sequence component), respectively the voltage $\underline{U}_{2\mathrm{A}}$ (negative-sequence component), at short-circuit location A. The mutual impedances $\underline{M}_1$ and $\underline{M}_2$ are calculated using the voltage $\underline{U}_{1\mathrm{B}}$ at short-circuit location B and the current $\underline{I}_{1\mathrm{A}}$ at short-circuit location A by

$$\underline{M}_1 = \frac{\underline{U}_{1\mathrm{B}}}{\underline{I}_{1\mathrm{A}}} \tag{7.2a}$$

$$\underline{M}_2 = \frac{\underline{U}_{2\mathrm{B}}}{\underline{I}_{2\mathrm{A}}} \tag{7.2b}$$

In case feeding shall be carried out at location B, the mutual impedances given are obtained by the voltage $\underline{U}_{1\mathrm{A}}$ at short-circuit location A and the current $\underline{I}_{1\mathrm{B}}$ at short-circuit location B as below:

$$\underline{M}_1 = \frac{\underline{U}_{1\mathrm{A}}}{\underline{I}_{1\mathrm{B}}} \tag{7.3a}$$

$$\underline{M}_2 = \frac{\underline{U}_{2\mathrm{A}}}{\underline{I}_{2\mathrm{B}}} \tag{7.3b}$$

7.2.2 Power system configurations

In the case of far-from-generator short-circuits the impedances in the positive- and negative-sequence components $\underline{Z}_1 = \underline{Z}_2$ and $\underline{M}_1 = \underline{M}_2$ are equal. Table 7.1 outlines simple systems configurations in accordance with IEC 60909-3.

Table 7.1 System configurations and equations for the calculation of short-circuit currents in the case of double earth-faults

Configuration	Equation
Single-fed radial line Both short-circuit locations on the same line (circuit) 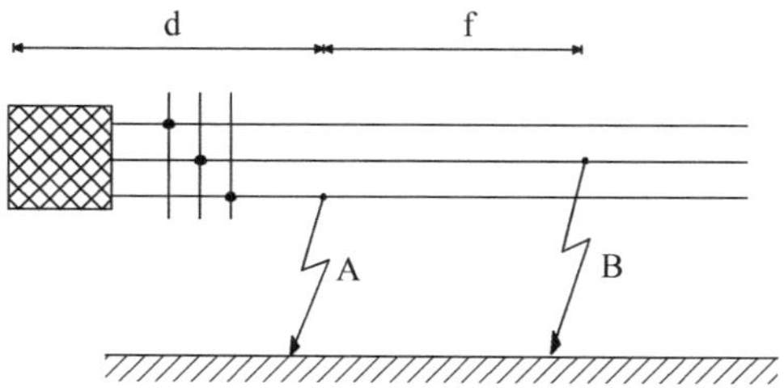 	$I''_{\mathrm{kEE}} = \dfrac{3 * c * U_{\mathrm{n}}}{\lvert 6 * \underline{Z}_{1\mathrm{d}} + 2 * \underline{Z}_{1\mathrm{f}} + \underline{Z}_{0\mathrm{f}} \rvert}$ d: Impedance to short-circuit location A f: Impedance between short-circuit locations A and B
Two single-fed radial lines Short-circuit locations on different lines (circuits) 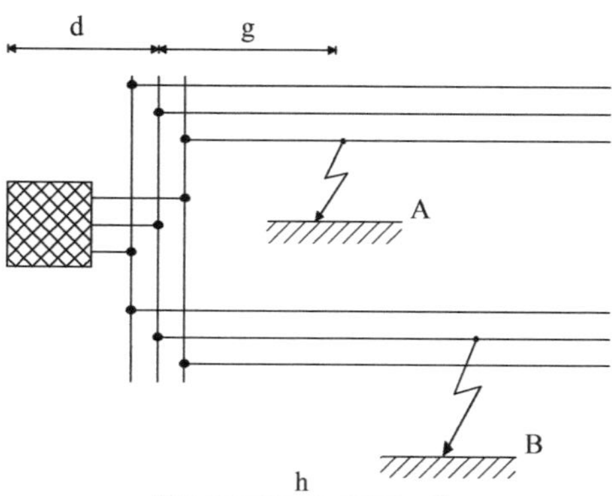 	$I''_{\mathrm{kEE}} = \dfrac{3 * c * U_{\mathrm{n}}}{\lvert 6 * \underline{Z}_{1\mathrm{d}} + 2 * \underline{Z}_{1\mathrm{g}} + \underline{Z}_{1\mathrm{h}} + \underline{Z}_{0\mathrm{g}} + \underline{Z}_{0\mathrm{h}} \rvert}$ d: Impedance of system feeder including busbar connection g: Impedance between busbar and short-circuit location A h: Impedance between busbar and short-circuit location B

Table 7.1 Continued

Double-fed line
Both short-circuit locations on the same line (circuit)

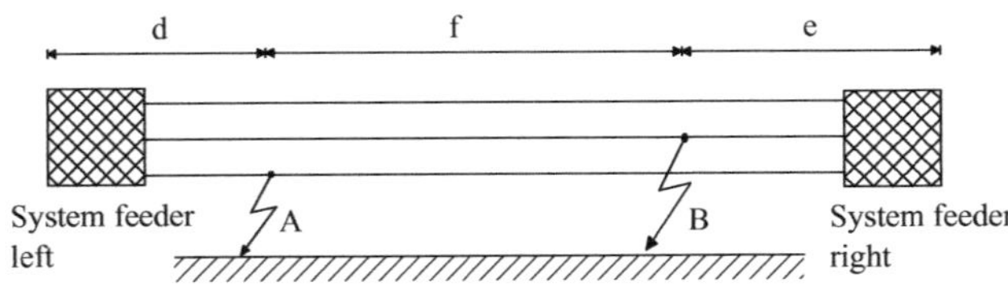

$$I''_{\text{kEE}} = \frac{3 * c * U_{\text{n}}}{|(6 * \underline{Z}_{1\text{d}} * \underline{Z}_{1\text{e}} + 2 * \underline{Z}_{1\text{f}}(\underline{Z}_{1\text{d}} + \underline{Z}_{1\text{e}}))/(\underline{Z}_{1\text{d}} + \underline{Z}_{1\text{f}} + \underline{Z}_{1\text{e}}) + \underline{Z}_{0\text{f}}|}$$

d: Impedance from left till short-circuit location A
e: Impedance from right till short-circuit location B
f: Impedance between short-circuit locations A and B

Earthing impedances in the switchyard and at the overhead tower can be neglected.

The voltage factor c shall be selected in accordance to Table 4.1.

Indices: 0, 1 and 2: zero-, positive- and negative-sequence components. d, e, f, g, h: length of the line as indicated in the drawing of the first column.

7.2.3 Peak short-circuit current i_p

The calculation of peak short-circuit current is carried out in the same manner as described in Section 4.3.2.3. The peak short-circuit current is calculated by

$$i_{pEE} = \kappa * \sqrt{2} * I''_{kEE} \tag{7.4}$$

The factor κ shall be the maximum of the factors obtained for three-phase short-circuit at location A or B.

$$\kappa = \mathrm{MAX}\{\kappa_A; \kappa_B\} \tag{7.5}$$

For explanations and calculation method of quantities reference is made to Chapter 4.

7.2.4 Symmetrical short-circuit breaking current I_b and steady-state short-circuit current I_k

In the case of far-from-generator short-circuits, the symmetrical short-circuit breaking current and the steady-state short-circuit current are identical to the initial symmetrical short-circuit current. In other cases, reference is made to Chapter 4.

7.3 Short-circuit currents through earth

7.3.1 Introduction

Branch short-circuit currents can flow through earth in the case of unbalanced short-circuits with earth connection, i.e., line-to-earth (single-phase) short-circuit and double-phase short-circuit with earth connection. Single-phase short-circuits are the dominating fault type in power systems with earthed neutrals and are leading to the maximal branch short-circuit currents flowing through earth. The current I''_{k1} flowing through earth is equal to three-times the current in the zero-sequence component flowing towards the short-circuit location. Type, number and arrangement of earth conductors of overhead lines, the installation of counterpoise, shielding, armouring and sheaths of cables, and their connection to the earthing grid of the switchyard determine the part of the short-circuit current that will flow through the earthing installations. This part is described by means of the reduction factor $\underline{p}_E$, sometimes represented as $\underline{r}$. The reduction factor is defined as the ratio of branch short-circuit current $\underline{I}_{Etot}$ flowing through earth to the total short-circuit current $3\underline{I}_0$, which can be described by the ratio of impedances

$$\underline{p}_E = \frac{\underline{I}_{Etot}}{3\underline{I}_0} = 1 - \frac{\underline{Z}_{LE}}{\underline{Z}_E} \tag{7.6}$$

where $\underline{Z}_E$ is the impedance of earth conductor with earth return, $\underline{Z}_{LE}$ is the mutual impedance between earth and line conductor with common earth return, $\underline{I}_{Etot}$ is the branch short-circuit current through earth and $\underline{I}_0$ is the total zero-sequence short-circuit current.

The loop-impedance $\underline{Z}_E$ of the earth conductor and earth return is given by

$$\underline{Z}'_E = R' + \frac{\mu_0}{8} * \omega + j\left(\frac{\mu_0}{2\pi} * \omega * \left(\ln\left(\frac{\delta}{r}\right) + \frac{\mu_r}{4}\right)\right) \tag{7.7}$$

The mutual impedance $\underline{Z}_{LE}$ of the loop earth conductor and conductor with earth return is given by

$$\underline{Z}'_{LE} = \frac{\mu_0}{8} * \omega + j\left(\frac{\mu_0}{2\pi} * \omega * \left(\ln\frac{\delta}{d_{LE}}\right)\right) \tag{7.8}$$

where r is the radius of earth conductor, μ_0 is the absolute permeability, μ_r is the relative permeability, ω is the angular frequency, R' is the resistance of earth wire per unit length, δ is the depth of the earth return path $\delta = 1.85/\sqrt{\omega * \mu_0/\rho_E}$, ρ_E is the resistivity of soil depending on soil conditions and d_{LE} is the distance between the earth conductor and phase conductor. Equations for the calculation of the impedances $\underline{Z}_E$ and $\underline{Z}_{LE}$ are also given in [1,6].

7.3.2 Short-circuit inside a switchyard

Figure 7.1 outlines the equivalent circuit diagram of a power system (HV) with short-circuit inside the switchyard B. All quantities are defined according to Figure 7.1.

The single-phase short-circuit current is calculated using

$$\underline{I}''_{k1} = 3 * \underline{I}_{0A} + 3 * \underline{I}_{0B} + 3 * \underline{I}_{0C} \tag{7.9}$$

The total current through the earth grid at location B is

$$\underline{I}_{ZB} = \underline{p}_A * 3 * \underline{I}_{0A} + \underline{p}_C * 3 * \underline{I}_{0C} \tag{7.10}$$

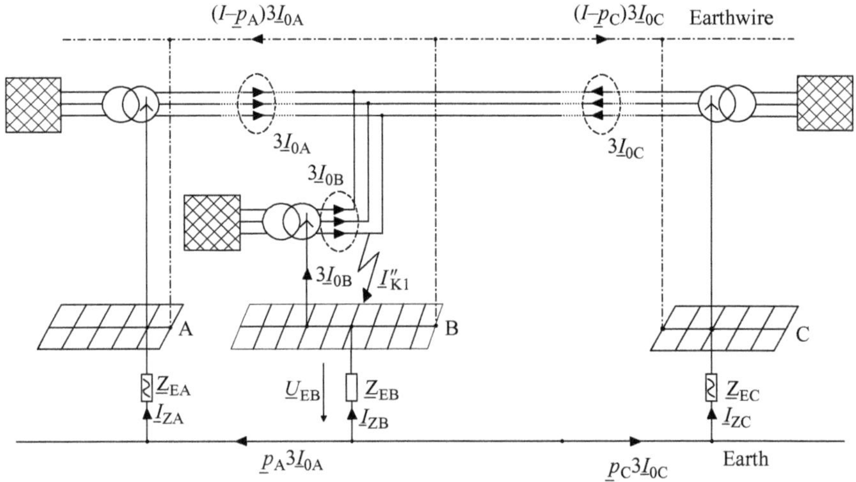

Figure 7.1 Equivalent circuit diagram with short-circuit inside switchyard B

The potential at the earth grid is then

$$\underline{U}_{EB} = \underline{Z}_{EB} * \underline{I}_{ZB} \tag{7.11}$$

with the earthing impedance of the switchyard

$$\underline{Z}_{EB} = \frac{1}{1/R_E + \sum 1/\underline{Z}_P} \tag{7.12}$$

where $\underline{R}_E$ is the resistance of the earth grid and $\underline{Z}_P$ is the driving point impedance (impedance of earth conductor with earth return and earthing impedance of overhead towers, respectively, input impedance of shielding, sheaths and armouring of cables).

The considerations mentioned above are only valid when the short-circuit location (switchyard B) is far away from other switchyards (A and C in Figure 7.1). The current through earth is lower when the distance to the short-circuit is lower than the far-from-station distance d_F defined as

$$d_F = 3 * \sqrt{R_{To}} * \frac{d_{To}}{\mathrm{Re}\{\sqrt{\underline{Z}_E}\}} \tag{7.13}$$

where R_{To} is the footing resistance of tower, d_{To} is the distance between two towers and $\underline{Z}_E$ is the impedance of earth conductor with earth return.

7.3.3 *Short-circuit at overhead-line tower*

Figure 7.2 outlines the equivalent circuit diagram of a power system with short-circuit outside the switchyard (distance $d > d_F$, see Equation (7.13)). The single-phase short-circuit current is calculated by

$$\underline{I}''_{k1} = 3 * \underline{I}_{0A} + 3 * \underline{I}_{0B} + 3 * \underline{I}_{0C} \tag{7.14}$$

The branch current through the earthing of the tower at short-circuit location is

$$\underline{I}_{ZF} = \underline{p}_C * (3 * \underline{I}_{0A} + 3 * \underline{I}_{0C}) + \underline{p}_C * 3 * \underline{I}_{0C} = \underline{p}_C * \underline{I}''_{k1} \tag{7.15}$$

and the potential at the earth grid

$$\underline{U}_{EF} = \underline{Z}_{EF} * \underline{I}_{ZF} \tag{7.16}$$

with the earthing impedance of the installation

$$\underline{Z}_{EF} = \frac{1}{1/R_{To} + 2/\underline{Z}_P} \tag{7.17}$$

where R_{To} is the footing resistance of tower and $\underline{Z}_P$ is the driving point impedance (impedance of earth conductor with earth return and earthing impedance of overhead-line towers, respectively, input impedance of shielding, sheaths and armouring of cables).

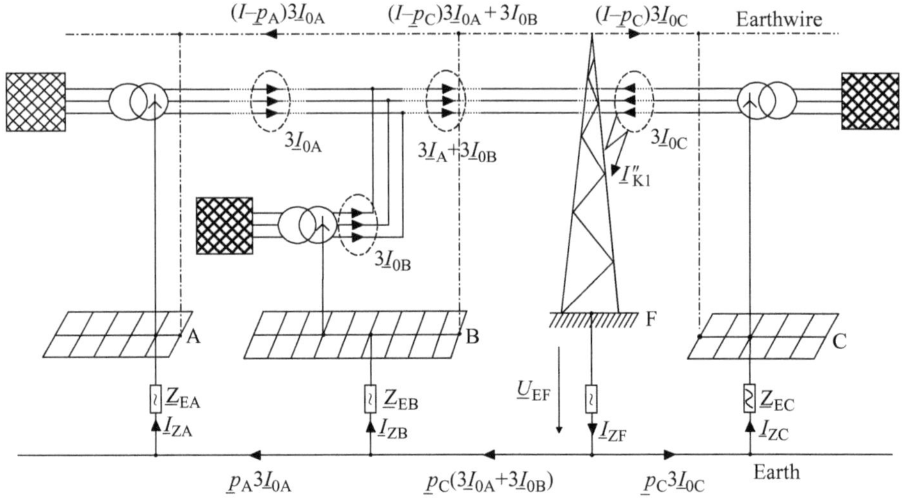

Figure 7.2 Equivalent circuit diagram with short-circuit at overhead-line tower

If the short-circuit occurs in a short distance from the switchyard, the branch short-circuit current through the earth conductor and back to the switchyard can be comparatively high. The branch short-circuit current through earth is reduced accordingly. The branch short-circuit current through the earthing grid at switchyard B in the case of a short-circuit at location F is

$$\underline{I}_{ZB} = \underline{p}_C * (3 * \underline{I}_{0A} + 3 * \underline{I}_{0C}) - \underline{p}_A * 3 * \underline{I}_{0A} \tag{7.18}$$

The branch short-circuit current through the earthing grid can be higher or lower for the short-circuit location inside the switchyard or at any overhead-line tower outside depending on the actual earthing conditions.

IEC 60909-3 presents the method to calculate the reduction factor on overhead lines. Further reference is made to [2,3,5,6].

7.4 Examples

Examples for the calculation of short-circuit currents in the case of double earth-fault and for the calculation of branch short-circuit currents flowing through earth are included in IEC 60909-3 Annexes A and B. Configurations with nominal system voltage $\underline{U}_n = 132$ kV similar to those as per Figures 7.1 and 7.2 are presented.

7.4.1 Double earth-fault in a 20-kV-system

Figure 7.3 represents the equivalent circuit diagram of a 20-kV-system with overhead lines to calculate the short-circuit current in the case of a double earth-fault. Data of

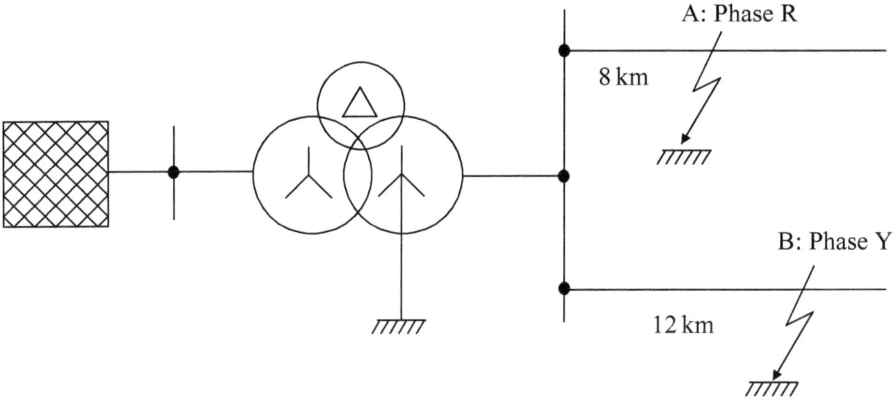

Figure 7.3 Equivalent circuit diagram of a 20-kV-system

equipment are

System feeder	$S''_{kQ} = 1$ GVA; $U_n = 110$ kV
Transformer	$S_{rT} = 40$ MVA; $u_{krT} = 14\%$; $t_{rT} = 110$ kV/20 kV
Overhead lines	ACSR 95/15; $R'_1 = 0.384$ Ω/km; $X'_1 = 0.35$ Ω/km; $R'_0 = 1.35$ Ω/km; $X'_0 = 0.6$ Ω/km.

The impedances of equipment are calculated in accordance with Section 7.3.

Equipment	Positive-sequence component	Zero-sequence component
System feeder	$(0.044 + j0.438)$ Ω	—
Transformer	$(0.0 + j1.35)$ Ω	—
Line to short-circuit location A	$(3.07 + j2.8)$ Ω	$(10.08 + j3.07)$ Ω
Line to short-circuit location B	$(4.61 + j4.2)$ Ω	$(16.2 + j4.61)$ Ω
Total impedance (Table 7.1)	$(42.624 + j32.408)$ Ω	

The short-circuit current in the case of a double earth-fault is calculated as $I''_{kEE} = 1.23$ kA. When both short-circuits occur at a tower, a part of the short-circuit current flows through the tower and the footing resistance. Depending on the resistivity of the earth, the footing resistance and the surge impedance of the line, this part is approximately 10–25 per cent smaller than the total short-circuit current.

7.4.2 Single-phase short-circuit in a 110-kV-system

Figure 7.4 indicates the equivalent circuit diagram of a 110-kV-system with short-circuit location F. The single-phase short-circuit current, the branch short-circuit currents flowing through the earth and the potential of the earth grid shall be calculated. The data of equipment are

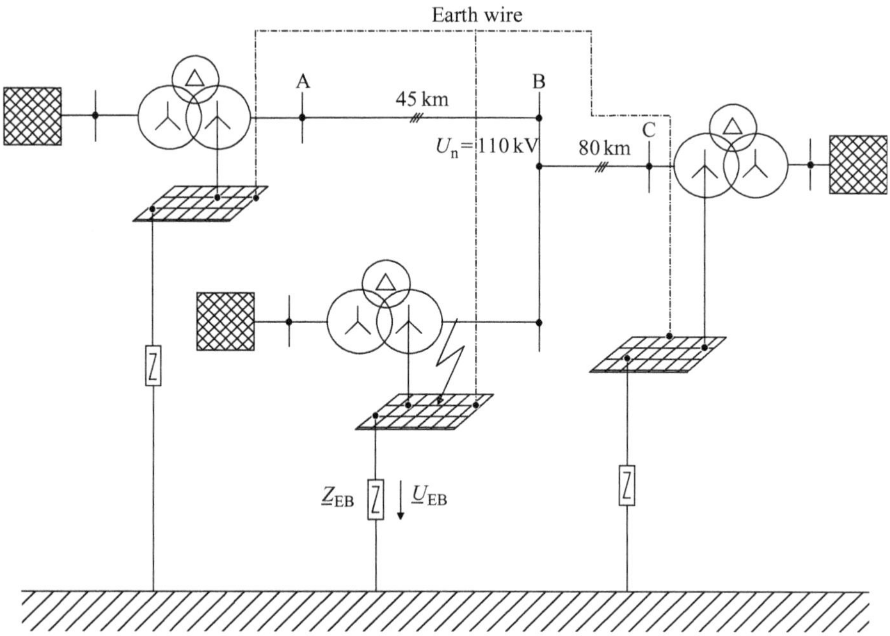

Figure 7.4 Equivalent circuit diagram of a 110-kV-system with short-circuit location

Feeder A $S''_{kQA} = 1.5$ GVA; $S_{rT} = 250$ MVA; $u_{1kT} = 16\%$; $u_{0kT} = 20\%$; $u_{0rR} = 0.2\%$

Feeder B $S''_{kQB} = 2.0$ GVA; $S_{rT} = 350$ MVA; $u_{1kT} = 14\%$; $u_{0kT} = 18\%$; $u_{0rR} = 0.2\%$

Feeder C $S''_{kQC} = 1.1$ GVA; $S_{rT} = 200$ MVA; $u_{1kT} = 15\%$; $u_{0kT} = 19\%$; $u_{0rR} = 0.2\%$

Line ACSR $2 \times 240/40$; earth wire 240/40; reduction factor $p \approx 0.6$

$\underline{Z}'_1 = (0.059 + j0.302)\ \Omega/\text{km}$; $\underline{Z}'_0 = (0.27 + j1.51)\ \Omega/\text{km}$

The impedances of equipment are calculated in accordance to Section 7.3 and Section 3.2.

Equipment	Positive-sequence component	Zero-sequence component
Feeder A including transformer	$(0.0 + j8.87)\ \Omega$	$(0.1 + j9.68)\ \Omega$
Feeder B including transformer	$(0.0 + j6.66)\ \Omega$	$(0.06 + j6.22)\ \Omega$
Feeder C including transformer	$(0.0 + j12.1)\ \Omega$	$(0.12 + j11.5)\ \Omega$
Line A–B	$(2.655 + j13.59)\ \Omega$	$(12.15 + j21.6)\ \Omega$
Line B–C	$(4.72 + j24.16)\ \Omega$	$(67.95 + j120.8)\ \Omega$
Impedance at short-circuit location	$(0.178 + j4.512)\ \Omega$	$(0.407 + j5.11)\ \Omega$

The single-phase short-circuit current is calculated as $\underline{I}''_{k1} = (0.798 - j14.785)$ kA and $|\underline{I}''_{k1}| = 14.8$ kA, respectively. The current in the zero-sequence component is $\underline{I}_0 = \frac{1}{3}\,\underline{I}''_{k1} = (0.266 - j4.928)$ kA and the branch short-circuit currents of the three feeders A, B and C are

$$\underline{I}_{0A} = (0.179 - j0.732)\ \text{kA}$$

$$\underline{I}_{0B} = (0.502 - j4.036)\ \text{kA}$$

$$\underline{I}_{0C} = (0.057 - j0.16)\ \text{kA}$$

The total current through the earth grid is $\underline{I}_{ZB} = (0.425 - j1.426)$ kA. The impedance of the earth grid depends on the resistivity of the soil, the footing resistance of the tower and the surge impedance of the line. Assuming (in accordance with IEC 60909-3) the impedance of the earthing grid to be $\underline{Z}_{EB} = (0.68 + j0.49)\ \Omega$ the potential of the earth grid is $\underline{U}_{EB} = (0.987 - j0.761)$ kV and $|\underline{U}_{EB}| = 1.246$ kV, respectively.

Chapter 8

Factors for the calculation of short-circuit currents

8.1 General

Several factors for the calculation of short-circuit (s.-c.) currents have been introduced in previous sections, the origin of which will be explained within this section.

- Voltage factor c_{max} and c_{min} for different voltage levels as per Table 4.1.
- Correction factor using the %/MVA- or the p.u.-system as mentioned in Chapter 2.
- Impedance correction factors for synchronous machines, power station units and transformers as per Tables 3.2, 3.3, 3.5 and 3.6.
- Factors for the calculation of different parameters of the short-circuit current based on the initial short-circuit current as per Chapter 4.

The factors are necessary as the method of the equivalent voltage source at the short-circuit location is used for the calculation of short-circuit currents which is based on some simplifications such as neglecting the load current prior to fault, assuming the tap-changer of transformers in middle-position, calculating the impedance of equipment based on the name-plate data or on data for rated operating conditions and neglecting voltage control gear for generators and transformers. The main task of short-circuit analysis is to determine the maximal short-circuit current which is one of the main criteria for the rating of equipment in electrical power systems. It is obvious that the parameters of the short-circuit current as calculated with the equivalent voltage source at the short-circuit location will differ from those currents, which may be measured during short-circuit tests or may be calculated with transient network analysing programmes. In order to obtain results on the safe side without uneconomic safety margin the correction factors will be applied. Detailed deductions of the various correction factors are given in IEC 60909-1:1991-10.

8.2 Correction using %/MVA- or p.u.-system

The need to use special correction factors for the impedances using the %/MVA-system, applies also to the p.u.-system. The calculation of short-circuit currents can be carried out using the %/MVA- or the p.u.-system as outlined in Section 2.7. The rated voltage of equipment U_r is chosen as reference voltage U_B for the calculation of the impedance of transformers, generators, etc. For system feeders and lines the rated voltage is not defined, therefore the nominal system voltage U_n is taken as reference voltage. The rated voltages of transformers in most cases are unequal to the nominal voltage of the power system, connected to the transformer. Figure 8.1 indicates a 110/10-kV-system as an example.

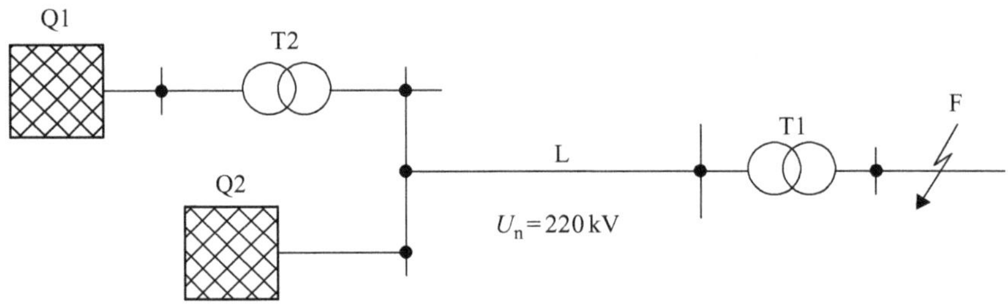

Figure 8.1 Equivalent circuit diagram of a power system with different voltage levels

Rated data of equipment are given below:

$U_{nQ1} = 110$ kV; $S''_{kQ1} = 1$ GVA; $U_{nQ2} = 220$ kV; $S''_{kQ2} = 3$ GVA
$S_{rT2} = 300$ MVA; $u_{krT2} = 15\%$; $t_{rT2} = 225$ kV/115 kV
$S_{rT1} = 250$ MVA; $u_{krT1} = 17\%$; $t_{rT1} = 220$ kV/115 kV
$R'_L = 0.03\ \Omega$/km; $X'_L = 0.12\ \Omega$/km; $l = 100$ km

The impedances of the equipment using the Ohm-system related to the short-circuit location (Column 2), the %/MVA-system (Column 3) and the p.u.-system (Column 4) are summarised in Table 8.1. The values in the %/MVA-system are converted to the Ohm-system by using the equations as per Table 2.3 (Column 5). The result (Column 6) indicates a difference of 6.3 per cent for the total short-circuit impedance.

The differences of the impedances result from the fact that the calculation in the Ohm-system takes account of the real transformation ratio of the transformers $t_r = U_{rTHV}/U_{rTMV}$, whereas the calculation in the %/MVA- and in the p.u.-system assumes a voltage of 100 per cent for all different voltage levels. The final conversion to the Ohm-system is done using the voltage at the short-circuit location U_{nF}, thus denying the differences between rated voltages of the transformers and nominal voltages of the connected power systems. These differences obviously will result in differences of the short-circuit currents as outlined in Table 8.1. A correction of the impedances is necessary in case the rated voltages of transformers and the nominal system voltages differ from each other.

Table 8.1 Impedances of equipment and short-circuit current as per Figure 8.1

1	2	3	4	5	6
Equipment	Impedance related to s.-c. location (Ω)	Impedance (%/MVA)	Impedance (p.u.)	Impedance (%/MVA) → (Ω) (p.u.) → (Ω)	Difference (%)
System Q1	13.929	0.11	0.11	13.31	4.7
Transformer T2	6.917	0.05	0.05	6.05	14.3
System Q2	4.849	0.037	0.037	4.48	8.2
Line L	3.279	0.0248	0.0248	3.0	9.3
Transformer T1	8.993	0.068	0.068	8.228	9.3
Total s.-c. impedance	16.206	0.126	0.126	15.246	6.3
Short-circuit current	4.31 kA	4.60 kA	10.12 p.u.	—	6.3

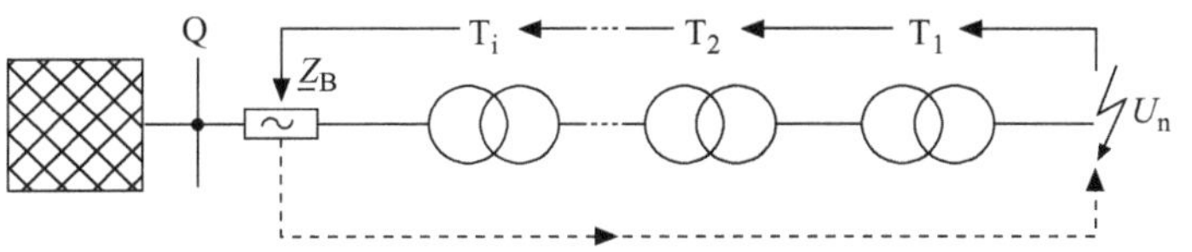

Figure 8.2 Equivalent circuit diagram for the calculation of impedance correction factor using %/MVA- or p.u.-system

The impedance correction factors (see Figure 8.2 for explanation) are obtained starting from the short-circuit location F indicated in Figure 8.1 by multiplying the ratios of the rated voltages of all transformers T_1–T_i between the short-circuit location F and the equipment under consideration B, then going back with the ratio of the rated voltage of the equipment U_{rB} and the nominal voltage U_n at the short-circuit location. For system feeders and lines the nominal system voltage at the very location U_{nB} has to be taken instead of the rated voltage, which is not defined for feeders and lines. The impedance correction factor K_B is calculated by

$$K_B = \left(\frac{U_{rT1E}}{U_{rT1A}} * \frac{U_{rT2E}}{U_{rT2A}} * \frac{U_{rT3E}}{U_{rT3A}} * \cdots \right)^2 * \left(\frac{U_{rB}}{U_{nF}} \right)^2 \tag{8.1}$$

The impedance correction factor using the %/MVA- or the p.u.-system must be applied for any equipment [33] except power station units for which special correction factors are valid.

Table 8.2 Impedances of equipment and short-circuit current using correction factor

1	2	3	4	5	6
Equipment	Impedance related to s.-c. location (Ω)	Impedance (%/MVA)	Correction factor K_B as per Equation (8.1)	Impedance (%/MVA) using K_B	Impedance (p.u.) using K_B
System Q1	13.929	0.11	1.046	0.115	0.115
Transformer T2	6.917	0.05	1.093	0.057	0.057
System Q2	4.849	0.037	1.143	0.04	0.04
Line L	3.279	0.0248	1.093	0.027	0.027
Transformer T1	8.993	0.068	1.093	0.0743	0.0743
Total s.-c. impedance	16.206	0.126	—	0.134	0.134
Short-circuit current	4.31 kA	—	—	4.31 A	9.48 p.u.

Applying the impedance correction factors as per Equation (8.1) the impedances calculated with the %/MVA- and the p.u.-system are identical to those obtained by using the Ohm-system as outlined in Table 8.2.

8.3 Impedance correction factors

Within this book the deduction of the impedance correction factor K_G for synchronous machines (generators) is given. The factor is valid for generators connected directly without unit transformers to the power system which is normally the fact in medium-voltage and low-voltage systems [37]. Assuming an overexcited turbine generator as per Figure 8.3 with voltage control at the terminal connection to $U_G = (1 \pm p_G) * U_{rG}$, the control range normally is set to $(1 + p_G) = 1.05$. Prior to fault the generator generates the apparent power $\underline{S}_G = P_G + jQ_G$ to be fed into the system.

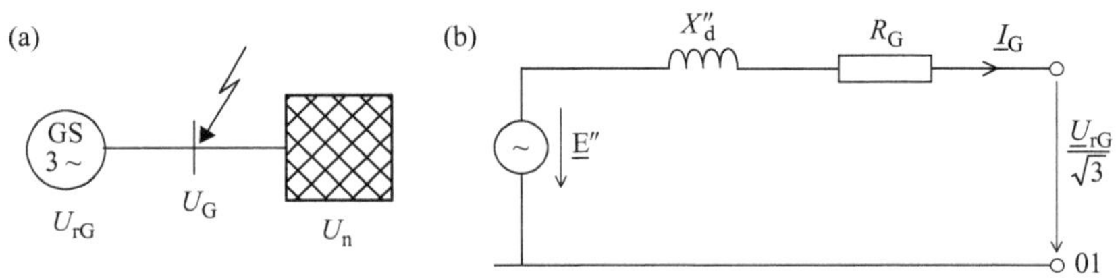

Figure 8.3 Generator directly connected to the power system. (a) Equivalent system diagram and (b) equivalent circuit diagram in the positive-sequence component

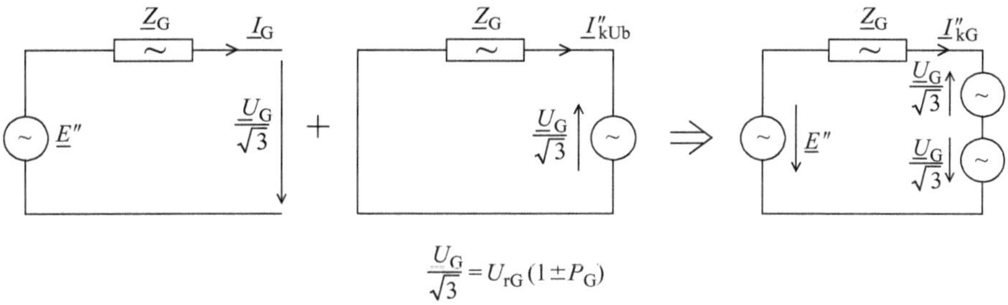

Figure 8.4 Determination of the short-circuit current by superposition

In the case of a short-circuit as indicated F in Figure 8.3(a) the short-circuit current can be calculated by superposition of the generator current $\underline{I}_G$ prior to fault and the short-circuit current $\underline{I}''_{kUb}$ based on the voltage $U_b = U_G = U_{rG}/\sqrt{3}$ prior to fault as outlined in Figure 8.4.

The generator current $\underline{I}_G$ prior to fault is given by

$$\underline{I}_G = \frac{\underline{E}'' - (\underline{U}_{rG}/\sqrt{3})}{\underline{Z}_G} \tag{8.2}$$

and the fault current $\underline{I}''_{kUb}$ can be calculated by

$$\underline{I}''_{kUb} = \frac{\underline{U}_{rG}/\sqrt{3}}{\underline{Z}_G} \tag{8.3}$$

The short-circuit current of the generator $\underline{I}''_{kG}$ is obtained by superposition of the two currents

$$\underline{I}''_{kG} = \underline{I}''_{kUb} + \underline{I}_G = \frac{\underline{U}_{rG}/\sqrt{3}}{\underline{Z}_G} + \frac{\underline{E}'' - (\underline{U}_{rG}/\sqrt{3})}{\underline{Z}_G} \tag{8.4}$$

where $\underline{E}''$ is the subtransient voltage of the generator, $\underline{U}_{rG}$ is the rated generator voltage and $\underline{Z}_G$ is the generator impedance.

If the method of the equivalent voltage source with the voltage $\underline{E}''$ is used the short-circuit current $\underline{I}''_{kG}$ is calculated by

$$\underline{I}''_{kG} = \frac{\underline{E}''}{\underline{Z}_G} \tag{8.5a}$$

which is different from the calculation as per Equation (8.4). If the voltage at the short-circuit location $\underline{E}'' = c * U_n/\sqrt{3}$ with the voltage factor c as per Table 4.1 and an impedance correction factor $\underline{K}_G$ are introduced, the short-circuit current $\underline{I}''_{kG}$ is found to be

$$\underline{I}''_{kG} = \frac{c * U_n}{\sqrt{3} * \underline{Z}_G * \underline{K}_G} = \frac{c * U_n}{\sqrt{3} * (R_G + jX''_d) * \underline{K}_G} \tag{8.5b}$$

where $\underline{K}_G$ is the impedance correction factor (to be determined), U_n is the nominal system voltage, $\underline{Z}_G$ is the generator impedance, R_G is the stator resistance of the generator and X''_d is the subtransient reactance of the generator.

Equations (8.5a) and (8.5b) are set to be equal. The unknown internal subtransient voltage of the generator $\underline{E}''$ can be found in accordance with Figure 8.3(b).

$$\underline{E}'' = \frac{\underline{U}_{rG}}{\sqrt{3}} + I_G * (\cos\varphi_G - j\sin\varphi_G) * (R_G + jX''_d) \tag{8.6}$$

where $\underline{U}_{rG}$ is the rated voltage of the generator and φ_G is the phase angle of the generator current (power factor: $\cos\varphi$) and other quantities are explained above.

The impedance correction factor $\underline{K}_G$ is then calculated by

$$\underline{K}_G = \frac{U_n}{U_{rG}} * c * \left[1 + \frac{\sqrt{3} * I_G}{U_{rG}} * (R_G * \cos\varphi_G + X''_d * \sin\varphi_G) + j\frac{\sqrt{3} * I_G}{U_{rG}} * (X''_d * \cos\varphi_G - R_G * \sin\varphi_G)\right]^{-1} \tag{8.7}$$

The resistance R_G normally can be neglected against the subtransient reactance X''_d of the generator; the correction factor then results in

$$K_G \approx \frac{U_n}{U_{rG}} * \frac{c}{1 + (I_G/I_{rG}) * x''_d * \sin\varphi_G} \tag{8.8}$$

whereas the subtransient reactance is introduced as a p.u.-value $x''_d = (X''_d/U^2_{rG}) * S_{rG}$. The correction factor is maximal when the maximal voltage factor c_{max} and rated operating conditions with $I_G = I_{rG}$ and $\varphi_G = \varphi_{rG}$ are applied. The impedance correction factor K_G is given by

$$K_G \approx \frac{U_n}{U_{rG}} * \frac{c_{max}}{1 + x''_d * \sin\varphi_G} \tag{8.9}$$

For other equipment such as power station units and transformers impedance correction factors can be deducted in a similar way [35] as explained for the correction factor of the generator. Details can be found in IEC 60909-1.

8.4 Factor κ for peak short-circuit current

The peak short-circuit current is the maximal instant value of the short-circuit current which occurs normally within the first few milliseconds after the occurrence of the short-circuit. The system configuration with a single-fed three-phase short-circuit is dealt with in Figure 8.5.

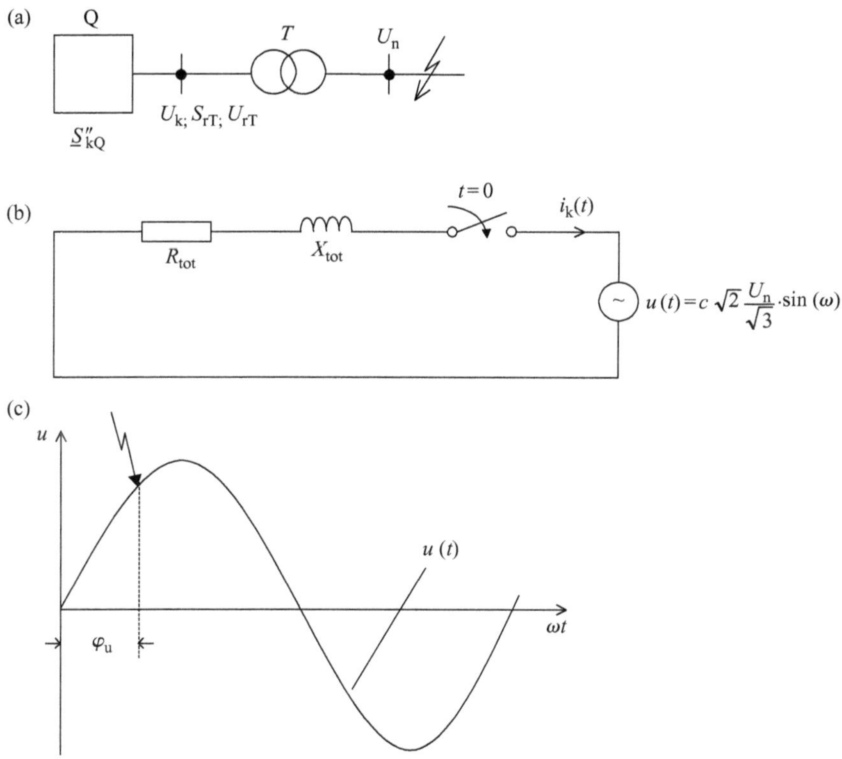

Figure 8.5 *Equivalent circuit diagram of a power system with three-phase short-circuit. (a) Circuit diagram, (b) simplified diagram of a single-fed three-phase short-circuit and (c) time course of voltage with voltage angle* φ_U

The time course of the short-circuit current $i_k(t)$ is calculated from the differential equation

$$L * \frac{di_k(t)}{dt} + R * i_k(t) = \sqrt{2} * \frac{c * U_n}{\sqrt{3}} * \sin(\omega t + \varphi_U) \tag{8.10}$$

The solution of the differential equation is given by

$$i_k(t) = \sqrt{2} * \frac{c * U_n}{\sqrt{3}} * \frac{1}{\sqrt{R^2 + X^2}} * (\sin(\omega t + \varphi_U - \gamma) - e^{-t/T} * \sin(\varphi_U - \gamma)) \tag{8.11}$$

where U_n is the nominal system voltage, φ_U is the angle of voltage related to zero crossing as per Figure 8.5, c is the voltage factor according to Table 4.1, T is the time constant: $T = L/R$, X is the reactance of the short-circuit impedance: $X = \omega L$, R is the resistance of the short-circuit impedance, ω is the angular velocity and γ the angle of the short-circuit impedance: $\gamma = \arctan(X/R)$.

The initial short-circuit current I''_{k} is equal to the first part (periodic term) of Equation (8.11), the second term is the aperiodic and decaying d.c.-component of the current. If the time course of the short-circuit current as per Equation (8.11) is related to the peak value of the initial short-circuit current

$$I''_{\mathrm{k}} = \sqrt{2} * \frac{c * U_{\mathrm{n}}}{\sqrt{3}} * \frac{1}{\sqrt{R^2 + X^2}} \tag{8.12}$$

the peak factor κ is obtained

$$\kappa(t) = \sin(\omega t + \varphi_{\mathrm{U}} - \gamma) - \mathrm{e}^{-t/T} * \sin(\varphi_{\mathrm{U}} - \gamma) \tag{8.13a}$$

The maximum of the peak factor κ determines the maximum of the short-circuit current (peak short-circuit current i_{p}) to be calculated by partial differentiation of Equation (8.13a) with respect to φ_{U} and t. The maximum of the peak factor always occurs for short-circuits at $\varphi_{\mathrm{U}} = 0$ and $t \leq 10$ ms (50 Hz), respectively $t \leq 8.33$ ms (60 Hz), whatever the ratio R/X might be

$$\kappa(t) = \sin(\omega t - \gamma) + \mathrm{e}^{-(R/X)*\omega t} * \sin\gamma \tag{8.13b}$$

A sufficient approximation of the peak factor κ is given by

$$\kappa = 1.02 + 0.98 * \mathrm{e}^{-3(R/X)} \tag{8.14}$$

Special attention for the calculation of peak short-circuit current must be given in the case of short-circuits in meshed systems or in systems having parallel lines with R/X-ratios different from each other [34]. A detailed analysis of these conditions is given in IEC 60909-1 and is mentioned in Chapter 4. The peak factor κ according to Equation (8.14) is outlined in Figure 4.7.

8.5 Factor μ for symmetrical short-circuit breaking current

The short-circuit current in the case of a near-to-generator short-circuit decays significantly during the first periods after initiation of the short-circuit due to the change of the rotor flux in the generator. This behaviour cannot be calculated exactly as eddy currents in the forged rotor of turbine generators, non-linearities of the iron and saturation effects especially in the stator tooth are difficult to be calculated. Furthermore, the decay of the short-circuit current and by this the breaking current depend on different generator and system parameters such as time constants of the generator itself, location of short-circuit in the system, operational condition prior to the fault, operation of excitation and voltage control device, tap-changer position of transformers, etc. which cause unpredictable deviations of the calculated results from those obtained from measurements. Detailed calculations with digital programmes are therefore only applicable in special cases if high safety requirements are to be met.

The time decay of the a.c. part of the short-circuit current is calculated by

$$I_{\mathrm{kac}}(t) = (I''_{\mathrm{k}} - I'_{\mathrm{k}}) * \mathrm{e}^{-t/T''_{\mathrm{N}}} + (I'_{\mathrm{k}} - I_{\mathrm{k}}) * \mathrm{e}^{-t/T'_{\mathrm{N}}} + I_{\mathrm{k}} \tag{8.15}$$

Equation (8.15) is composed of the initial short-circuit current

$$I''_{\mathrm{k}} = \frac{E''}{X''_{\mathrm{d}} + X_{\mathrm{N}}} \tag{8.16a}$$

the transient short-circuit current

$$I'_{\mathrm{k}} = \frac{E'}{X'_{\mathrm{d}} + X_{\mathrm{N}}} \tag{8.16b}$$

and the steady-state short-circuit current

$$I_{\mathrm{k}} = \frac{E}{X_{\mathrm{d}} + X_{\mathrm{N}}} \tag{8.16c}$$

where E; E'; E'' are the steady-state, transient and subtransient voltages, X_{d}; X'_{d}; X''_{d} is the steady-state, transient and subtransient reactance of the generator and X_{N} is the reactance between the generator and the short-circuit location, e.g., including the reactance of the unit transformer.

The individual components of the short-circuit currents as per Equations (8.15) and (8.16) are declining with different time constants, i.e., the subtransient time constant T''_{N} which can be set approximately equal to the subtransient time constant T''_{dN} in the direct axis. Typical values of the time constants are $T''_{\mathrm{N}} \approx$ 0.03–0.04 s and $T'_{\mathrm{N}} \approx$ 1.0–1.5 s and are calculated by

$$T''_{\mathrm{N}} = T'' * \frac{X'_{\mathrm{d}}}{X''_{\mathrm{d}}} * \frac{X''_{\mathrm{d}} + X_{\mathrm{N}}}{X'_{\mathrm{d}} + X_{\mathrm{N}}} \tag{8.17a}$$

$$T'_{\mathrm{N}} = T' * \frac{X_{\mathrm{d}}}{X'_{\mathrm{d}}} * \frac{X'_{\mathrm{d}} + X_{\mathrm{N}}}{X'_{\mathrm{d}} + X_{\mathrm{N}}} \tag{8.17b}$$

The units T'' and T' are the subtransient and transient time constants of the generator. Regarding the time course of the decaying a.c. part of the current as per Equation (8.15), the approximation $\mathrm{e}^{-t/T'_{\mathrm{N}}} \approx 1$ can be assumed if the time range (minimal time delay of circuit-breakers) $t = t_{\mathrm{min}} =$ 0.02–0.25 s is considered:

$$I_{\mathrm{kac}}(t) \approx I''_{\mathrm{k}} * \left[\left(1 - \frac{X''_{\mathrm{d}} + X_{\mathrm{N}}}{X'_{\mathrm{d}} + X_{\mathrm{N}}}\right) * \mathrm{e}^{-t_{\mathrm{min}}/T''_{\mathrm{N}}} + \frac{X''_{\mathrm{d}} + X_{\mathrm{N}}}{X'_{\mathrm{d}} + X_{\mathrm{N}}}\right] = I''_{\mathrm{k}} * \mu \tag{8.18}$$

with t_{min} being the minimal time for breaking the short-circuit current. The second part of Equation (8.18) is taken as the factor μ for the calculation of the breaking current.

As the exponent $t_{\min}/T''_{\mathrm{N}}$ may be greater or smaller than 1, the infinite progression for $\mathrm{e}^{-t/T''_{\mathrm{N}}}$ is applied. For no-load conditions, where $E'' \approx E' \approx E \approx U_{\mathrm{rG}}/\sqrt{3}$, the factor μ is obtained by

$$\mu \approx 1 - \frac{X'_{\mathrm{d}} - X''_{\mathrm{d}}}{T''_0 * (U_{\mathrm{rG}}/\sqrt{3})} * I''_{\mathrm{k}} * t_{\min} + \frac{X'_{\mathrm{d}} - X''_{\mathrm{d}}}{X'_{\mathrm{d}} + X_{\mathrm{N}}} * \left(\frac{t_{\min}}{T''_{\mathrm{N}}} + \mathrm{e}^{-t_{\min}/T''_{\mathrm{N}}} - 1\right) \tag{8.19}$$

with the no-load subtransient time constant

$$T''_0 = T'' * \frac{X'_{\mathrm{d}}}{X''_{\mathrm{d}}} \tag{8.17c}$$

When if $t_{\min} \ll T''_{\mathrm{N}}$ the last part of Equation (8.19) can be neglected. The factor μ can be presented depending on the minimal time delay of the circuit-breaker $t_{\min}$ and the ratio $I''_{\mathrm{k}}/I_{\mathrm{rG}}$ as outlined in Figure 4.10.

8.6 Factor λ for steady-state short-circuit current

The steady-state short-circuit currents of generators are determined by the method of excitation, the maximal possible excitation voltage, the type of voltage control and strongly by the saturation effects. As salient-pole and turbine generators differ significantly with respect to their reactances and are mostly equipped with different types of excitation, the steady-state short-circuit currents of both generators will differ even if all other conditions are equal. The calculation is carried out with the factor λ based on the rated current I_{rG} of the generator which is determined separately for minimal and maximal current.

$$I_{\mathrm{k\,max}} = \lambda_{\max} * I_{\mathrm{rG}} \tag{8.20a}$$

$$I_{\mathrm{k\,min}} = \lambda_{\min} * I_{\mathrm{rG}} \tag{8.20b}$$

The factor λ is found from the characteristic curve method as per Figure 8.6 defining Potier's reactance X_{P}.

$$I_{\mathrm{k}} = \frac{E_0(I_{\mathrm{f}})}{X_{\mathrm{P}} + X_{\mathrm{N}}} \tag{8.21}$$

where X_{P} is the Potier's reactance as per Figure 8.6, E_0 is the no-load voltage, I_{f} is the excitation current and X_{N} is the reactance between the generator and the short-circuit location, e.g., including the reactance of the unit transformer.

The value of Potier's reactance is between the transient reactance X'_{d} (pole saturation only) and the stator leakage reactance X_σ (teeth saturation only).

The method to determine Potier's reactance requires detailed knowledge of the saturation within the machine and is not practicable for the determination of the factor λ.

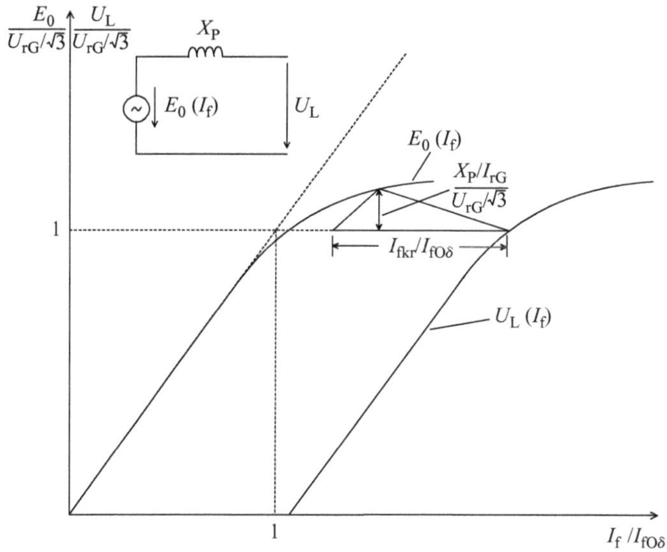

Figure 8.6 Characteristic saturation curve method for determination of Potier's reactance

IEC 60909-1 recommends a simplified method. Potier's reactance and the source voltage E_0 which is a function of the field current I_f are reduced due to the saturation as can be seen from Figure 8.6. Both effects compensate each other to a certain extent and are ignored therefore. The current I_k is calculated by

$$I_k = \frac{u_{f\,max} * E_r}{X_{dsat} + X_N} \tag{8.22}$$

where $u_{f\,max}$ is the highest possible excitation voltage (p.u.-value), E_r is the internal steady-state voltage of the generator at rated operating conditions, X_{dsat} is the saturated value of the synchronous reactance (equal to the reciprocal of the short-circuit ratio) and X_N is the reactances between the generator and the short-circuit location, e.g., including the reactance of the unit transformer.

Furthermore, the subtransient internal voltage is given as

$$E'' = I''_G * (X''_d + X_N) \tag{8.23}$$

By this the factor λ is determined to be

$$\lambda = \frac{I_k}{I_{rG}} = \frac{u_{f\,max} * E_r}{(X_{dsat} - X''_d) * I_{rG} + E'' * (I_{rG}/I''_{kG})} \tag{8.24}$$

Table 8.3 Assumed values of $u_{f\,max}$ for the calculation of the factor λ

	$u_{f\,max} = U_{f\,max}/U_{fr}$	
Type of synchronous machine	Series 1	Series 2
Salient-pole generator	1.3	1.6
Turbine generator	1.6	2.0

The voltages E'' and E_r can be determined if $R_G \ll X''_d$ by

$$E'' \approx \frac{U_{rG}}{\sqrt{3}} * (1 + x''_d * \sin\varphi_{rG}) \tag{8.25}$$

$$E_r \approx \frac{U_r}{\sqrt{3}} * \sqrt{1 + x^2_{dsat} + 2 * x_{dsat} * \sin\varphi_{rG}} \tag{8.26}$$

where $x''_d = X''_d * ((\sqrt{3} * I_{rG})/U_{rG})$ and $x_{dsat} = X_{dsat} * ((\sqrt{3} * I_{rG})/U_{rG})$.

The rated impedance is $Z_{rG} = U_{rG}/(\sqrt{3} * I_{rG})$. The values for λ_{max} and λ_{min} as per Figures 4.11 and 4.12 are calculated by Equations (8.23)–(8.25), $\cos\varphi_{rG} = 0.85$ and $X'_d = 0.2 * Z_{rG}$. The highest possible excitation voltage (p.u.-value) $u_{f\,max}$ is assumed for the calculation in accordance with the values as per Table 8.3.

8.7 Factor *q* for short-circuit breaking current of asynchronous motors

Asynchronous motors are contributing to the short-circuit current as outlined in Chapter 4. As the short-circuit current of asynchronous motors decays faster as compared with the short-circuit current of synchronous machines, the short-circuit breaking current is based on the initial short-circuit current of the motor I''_{kM} using the factor μ which is identical to the factor for the calculation of breaking current of synchronous generators and an additional factor q

$$I_{bM} = q * \mu * I''_{kM} \tag{8.27}$$

The factor q as per Figure 4.13 is derived from the results of transient calculations and measurements using 28 motors with different rating $P_{rM} = 11$–160 kW in the low-voltage range and up to $P_{rM} = 160$ kW–10 MW in the medium-voltage range. A detailed list of the rated data of the asynchronous motors is included in Table 2 of IEC 60909-1:1991 (similar to those given in Tables 13.6

and 13.7). The minimum time delay of the circuit-breakers was assumed in four steps $t_{min} = 0.02–0.05–0.1–\geq 0.25$ s. The results are outlined in Figure 8.7.

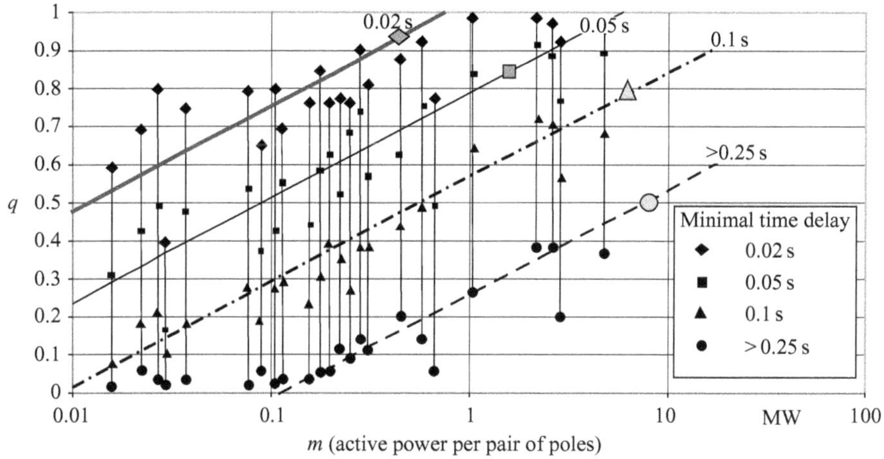

Figure 8.7 *Calculated and measured values of factor q for the calculation of short-circuit breaking current of asynchronous motors; values of q as per Figure 4.13 (According to Figure 20 of IEC 60909-1:1991.)*

As can be seen from Figure 8.7 the values of the factor q (approximation) as per Figure 4.13 are mean values of the calculated and measured ones with the 50 per cent-frequency deviation between the exact values and the approximated values in the range of $\Delta q < 5\%$.

Chapter 9

Calculation of short-circuit currents in d.c. auxiliary installations

9.1 General

The calculation of short-circuit currents in d.c. auxiliary installations, e.g., in power plants and substations is dealt with in IEC 61660-1. Contrary to the approach for the calculation of short-circuit currents in a.c. three-phase systems, the determination of the exact time course of the short-circuit current is needed besides the calculation of defined parameters [42]. The equipment as below contribute to short-circuit currents in d.c. installations:

- Smoothing capacitors
- Stationary batteries (normally of the lead-acid type)
- Rectifiers (IEC 61660-1 deals only with rectifiers in three-phase a.c. bridge connection for 50 Hz, parameters for 60 Hz are actually under consideration)
- d.c. motors with independent excitation.

The branch short-circuit currents from the equipment mentioned above are characterised by different time course, depending on the ohmic, inductive and capacitive parts, the d.c. voltage source and other parameters. Figure 9.1 presents the equivalent circuit diagrams of the equipment and the typical time course of the short-circuit current.

d.c. installations in auxiliary supply systems include several pieces of equipment; the total short-circuit current at the short-circuit location is the superposition of the individual branch short-circuit currents from the different equipment. In principle the short-circuit current can be defined by a time function $i_1(t)$, describing the time span t_p from short-circuit initiation till the maximal short-circuit current i_p (peak short-circuit current) and a time function $i_2(t)$, describing the decaying time course to the

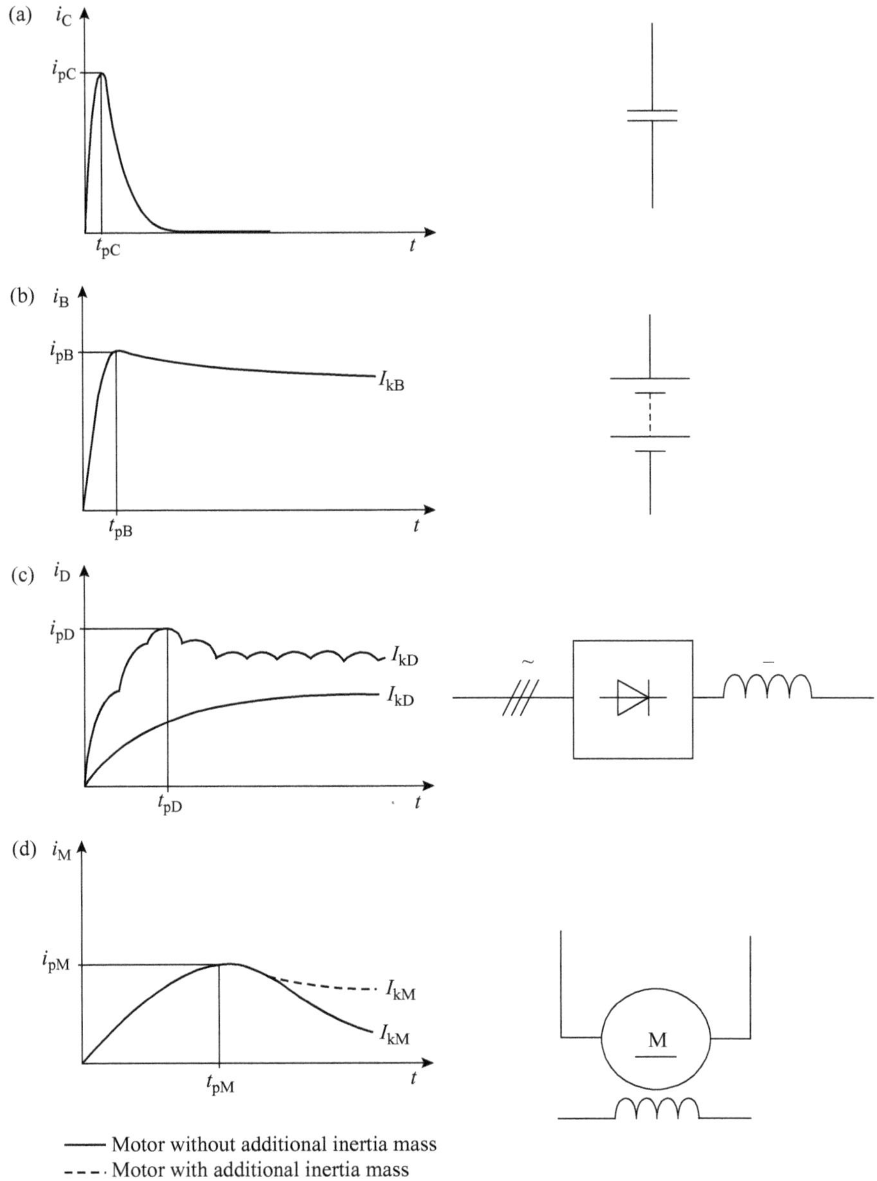

Figure 9.1 Equivalent circuit diagrams of equipment in d.c. auxiliary installations; typical time course of short-circuit current (according to Figure 1 of DIN EN 61660-1 (VDE 0102 Teil 10)). (a) Capacitor, (b) battery, (c) rectifier in three-phase a.c. bridge connection and (d) d.c. motor with independent excitation

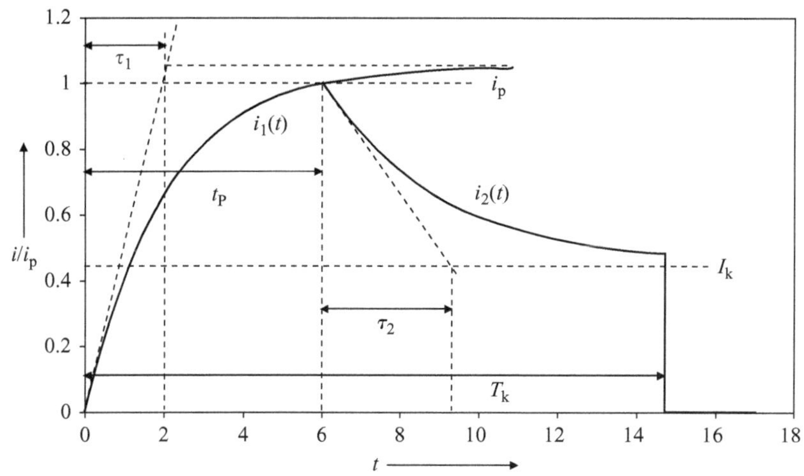

Figure 9.2 Standard approximation function of the short-circuit current (according to Figure 2 of IEC 61660-1:1997)

quasi steady-state short-circuit current I_k as outlined in Figure 9.2. Time functions can be calculated according to [50] by

$$i_1(t) = i_p * \frac{1 - e^{-t/\tau_1}}{1 - e^{-t_p/\tau_1}} \quad \text{for } 0 \leq t \leq t_p \tag{9.1a}$$

$$i_2(t) = i_p \left(\left(1 - \frac{I_k}{i_p} \right) * e^{-(t-t_p)/\tau_2} + \frac{I_k}{i_p} \right) \quad \text{for } t_p \leq t \leq T_k \tag{9.1b}$$

If no distinct maximum of the short-circuit current is present, the peak short-circuit current i_p and the quasi steady-state short-circuit current I_k are equal. This time course as well is described by the standard approximation function according to Figure 9.2.

The approach to calculate the parameter i_p and I_k, the time constants τ_1 and τ_2 and the time to peak t_p is explained below:

- The parameter of the short-circuit current will be calculated for each branch separately, i.e., for each individual equipment contributing to the short-circuit current.
- In case of several sources, the short-circuit parameters will be calculated by superposition of the branch short-circuit currents by:
  - Calculation of the branch short-circuit currents taking account of the common branch, i.e., that branch in the installation carrying branch short-circuit currents from different sources.
  - Correction of the branch short-circuit currents by a factor σ, which depends on the different resistances.

- Calculation of the time course of the branch short-circuit currents with the corrected impedances.
- Superposition of the calculated time functions of the branch short-circuit currents to determine the total short-circuit current.
- The thermal and electromagnetic effects of short-circuit currents are calculated using the standard approximation function.

The calculation of impedances of the sources in d.c. auxiliary installations feeding the short-circuit current was dealt with in Section 3.3. The calculation of the short-circuit parameters is carried out as mentioned above and explained below. For the calculation of maximal short-circuit currents in d.c. auxiliary installations the items as below had to be taken into account:

- Short-circuit impedance of system shall be minimal (Z_{Qmin}), so that the contribution to the short-circuit current is maximal.
- Resistance of lines is to be calculated for a temperature of 20°C.
- System topology leading to the maximal short-circuit currents shall be taken into account.
- Joint resistance of busbars shall be neglected.
- Control circuits to limit the contribution of rectifiers are disconnected, i.e., limitation is not active.
- Batteries are fully charged.
- Systems coupled by diodes are regarded as directly connected.

For the calculation of minimal short-circuit currents in d.c. auxiliary installations the items as stated below shall be considered:

- Short-circuit impedance of system shall be maximal (Z_{Qmax}), so that the contribution to the short-circuit current is minimal.
- System topology leading to the minimal short-circuit currents shall be taken into account.
- Resistance of lines shall be calculated for maximal permissible operating temperature.
- Joint resistance of busbar has to be taken into account.
- Contribution of rectifiers to the short-circuit current is limited to the rated value of the current limiter.
- Voltage of batteries shall be defined equal to the minimal discharge voltage as per manufacturer's information.
- Systems coupled by diodes are regarded as disconnected.

The current limiting characteristic of fuses and switchgear in d.c. auxiliary installations has to be taken into account for the calculation of both maximal and minimal short-circuit current.

9.2 Short-circuit currents from capacitors

The quasi steady-state short-circuit current of a capacitor is zero

$$I_{kC} = 0 \tag{9.2}$$

The peak short-circuit current i_{pC} is calculated using

$$i_{pC} = \kappa_C * \frac{E_C}{R_{CBr}} \tag{9.3}$$

where E_C is the capacitor voltage at the instant of short-circuit and R_{CBr} is the resistance of capacitor including connection and common branch (see Section 3.3.1 and Table 3.13).

The factor κ_C depends on the eigen-frequency ω_0

$$\omega_0 = \frac{1}{\sqrt{L_{CBr} * C}} \tag{9.4a}$$

and on the decay coefficient δ

$$\frac{1}{\delta} = \frac{2 * L_{CBr}}{R_{CBr}} \tag{9.4b}$$

where L_{CBr} is the inductance of the capacitor connection including common branch (see Table 3.13) and C is the capacitance.

The reasonable range of values for κ_C is outlined in Figure 9.3; equations for the calculation of κ_C are included in the Annex of IEC 61660-1:1997 and are not repeated here.

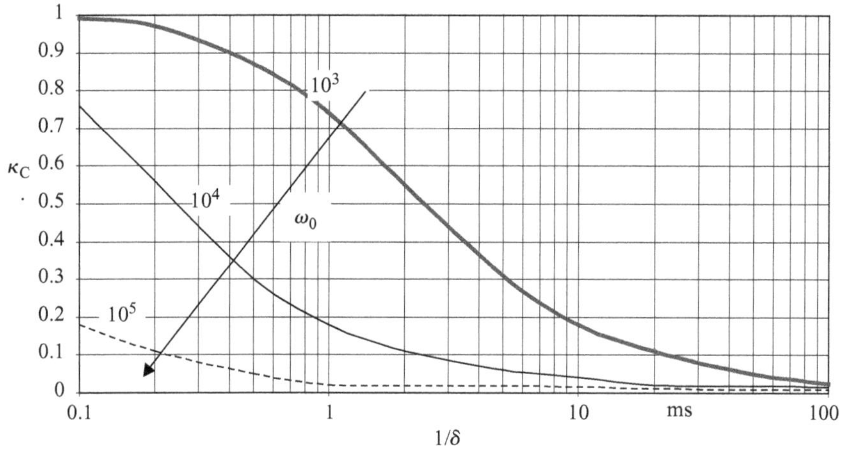

Figure 9.3 Factor κ_C for the calculation of peak short-circuit current of capacitors (according to Figure 12 of IEC 61660-1:1997)

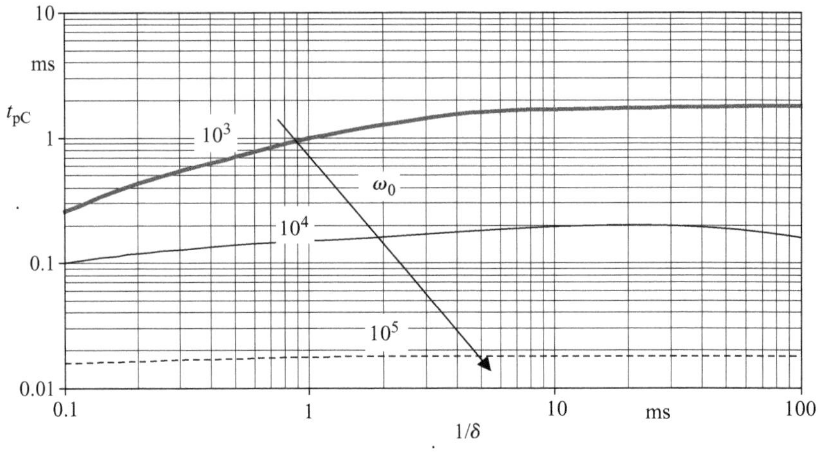

Figure 9.4 Time-to-peak t_{pC} for the calculation of short-circuit currents of capacitors (according to Figure 13 of IEC 61660-1:1997)

The time-to-peak t_{pC} also depends on the eigen-frequency ω_0 and the decay coefficient δ. The reasonable range of values for t_{pC} is outlined in Figure 9.4; equations for the calculation of t_{pC} are included in the Annex of IEC 61660-1:1997.

The rise-time constant τ_{1C}

$$\tau_{1C} = k_{1C} * t_{pC} \tag{9.5a}$$

and the decay-time constant τ_{2C}

$$\tau_{2C} = k_{2C} * R_{CBr} * C \tag{9.5b}$$

depend upon the factors k_{1C} and k_{2C}, i.e., upon eigen-frequency and decay component. Reasonable ranges of values are outlined in Figures 9.5 and 9.6. Quantities of Equations (9.5) are explained above.

Calculation equation for the factors k_{1C} and k_{2C} are not included in IEC 61660-1:1997.

9.3 Short-circuit currents from batteries

As rise-time constants of the short-circuit current of batteries are always below 100 ms, the quasi steady-state short-circuit current I_k is calculated for the time instant of 1 s after initiation of the short-circuit.

$$I_{kB} = \frac{0.95 * E_B}{R_{BBr} + 0.1 * R_B} \tag{9.6}$$

where E_B is the open-circuit voltage of the battery, R_{BBr} is the resistance of the battery including connection and common branch (see Section 3.3.1 and Table 3.14) and R_B is the resistance of the charged battery.

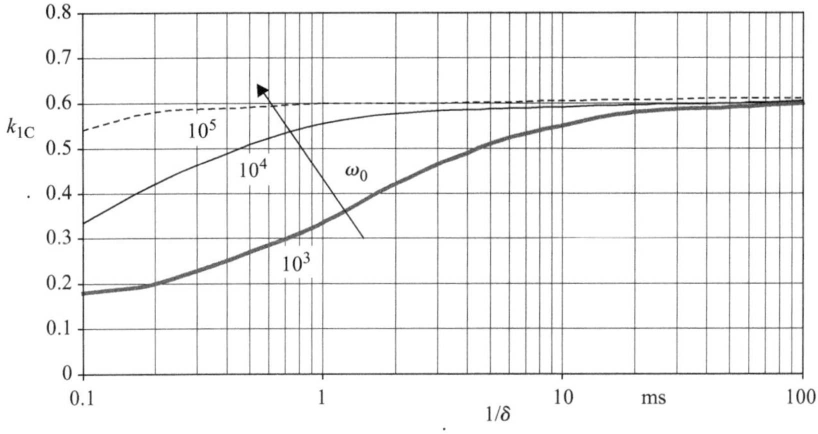

Figure 9.5 *Factor k_{1C} for the calculation of rise-time constant (according to Figure 14 of IEC 61660-1:1997)*

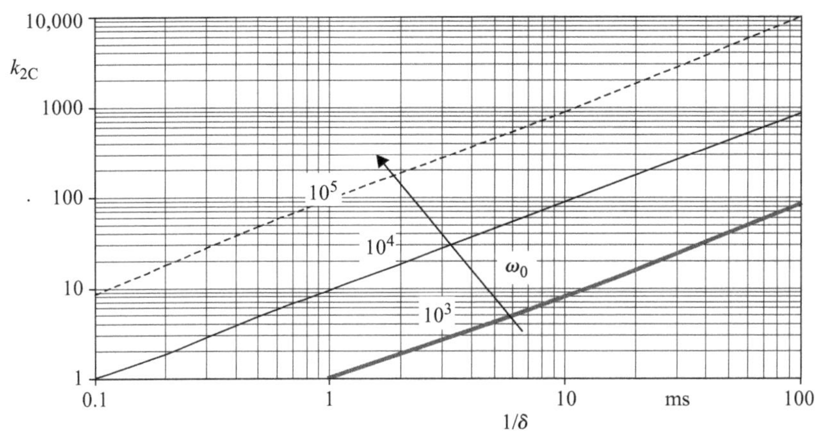

Figure 9.6 *Factor k_{2C} for the calculation of decay-time constant (according to Figure 15 of IEC 61660-1:1997)*

The peak short-circuit current i_{pB} is calculated using the battery voltage E_B by

$$i_{pB} = \frac{E_B}{R_{BBr}} \tag{9.7}$$

Reasonable ranges of the values for the rise-time constant τ_{1B} and the time-to-peak t_{pB}, both depending on the decay component δ

$$\frac{1}{\delta} = \frac{2}{R_{BBr}/L_{BBr} + 1/T_B} \tag{9.8}$$

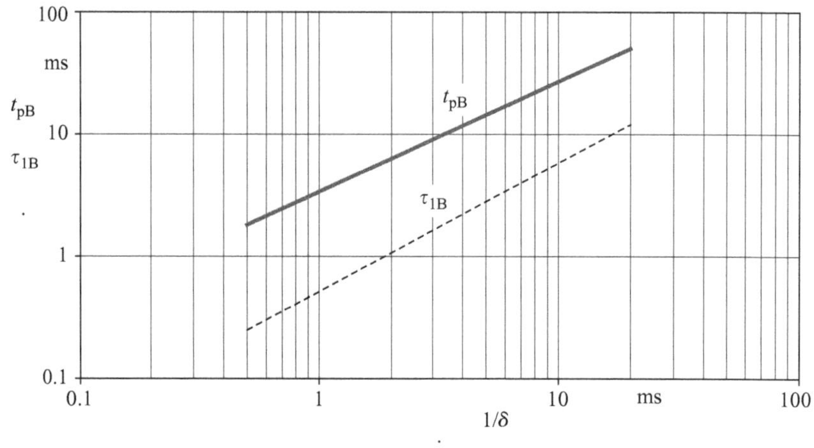

Figure 9.7 Rise-time constant τ_{1B} and time to peak t_{pB} of short-circuit currents of batteries (according to Figure 10 of IEC 61660-1:1997)

where L_{BBr} is the reactance of the battery including connection and common branch (see Table 3.14) and T_B is the time constant of the battery assumed to be $T_B = 30$ ms (outlined in Figure 9.7). The decay-time constant is set to $\tau_{2B} = 100$ ms.

More details of the calculation of short-circuit currents fed from batteries are included in [56].

9.4 Short-circuit currents from rectifiers

The quasi steady-state short-circuit current I_{kD} of a rectifier in three-phase a.c. bridge connection is

$$I_{kD} = \lambda_D * \frac{3 * \sqrt{2}}{\pi} * \frac{c * U_n}{\sqrt{3} * Z_N} * \frac{U_{rTLV}}{U_{rTHV}} \tag{9.9}$$

where U_n is the nominal system voltage on a.c. side of rectifier, Z_N is the network impedance a.c. side, U_{rTLV} is the rated voltage at low voltage side of transformer (a.c. side) and U_{rTHV} is the rated voltage at high voltage side of transformer (a.c. side).

The factor λ_D depends on the ratio R_N/X_N of the a.c. side of the rectifier as well as on the ratio of the resistances R_{DBr}/R_N (ratio of resistance d.c. side to resistance a.c. side). A reasonable range of values of the factor λ_D is outlined in Figure 9.8. Equations for the calculation of λ_D are included in IEC 61660-1:1997.

The peak short-circuit current is calculated using

$$i_{pD} = \kappa_D * I_{kD} \tag{9.10}$$

whereas the factor κ_D depends on the ratio of the inductivities L_{DBr}/L_N (ratio of inductance d.c. side to inductance a.c. side) and on the resistances and reactances

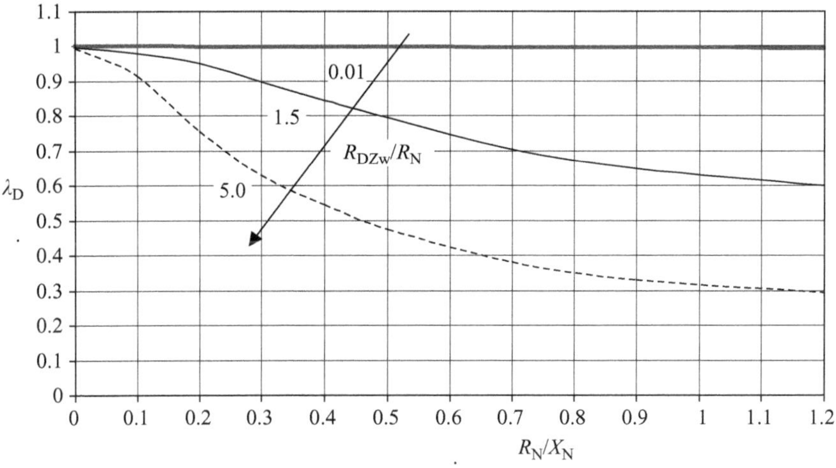

Figure 9.8 *Factor λ_D for the calculation of quasi steady-state short-circuit current of rectifiers (according to Figure 7 of IEC 61660-1:1997)*

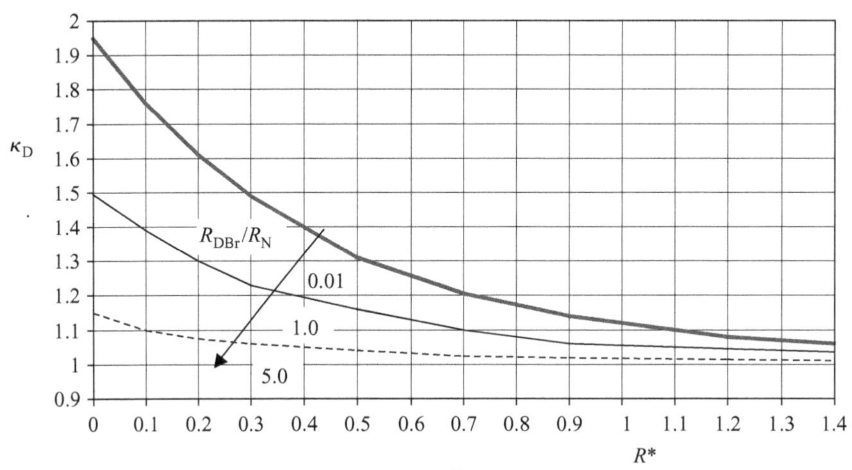

Figure 9.9 *Factor κ_D for the calculation of peak short-circuit currents of rectifiers. Factor: $R^* = (R_N/X_N)(1 + 2R_{DBr}/3R_N)$ (according to Figure 8 of IEC 61660-1:1997)*

R_N, R_{BBr} and X_N. The reasonable range of values of the factor κ_D is outlined in Figure 9.9. Equations for the calculation are included in IEC 61660-1:1997.

In the case of $\kappa_D < 1.0$ no distinct maximum of the short-circuit current is present, thus the peak short-circuit current can be neglected as the quasi steady-state short-circuit current will be the maximal value of the current. IEC 61660-1 determines for

$\kappa_D < 1.05$ the time to peak t_{pD} equal to the duration of the short-circuit T_k. For all other cases ($\kappa_D \geq 1.05$) the time to peak is calculated using

$$t_{pD} = (3 * \kappa_D + 6)\ \text{ms} \qquad \text{for } \frac{L_{DBr}}{L_N} \leq 1 \tag{9.11a}$$

$$t_{pD} = \left((3 * \kappa_D + 6) + 4\left(\frac{L_{DBr}}{L_N} - 1\right)\right)\ \text{ms} \quad \text{for } \frac{L_{DBr}}{L_N} > 1 \tag{9.11b}$$

The rise-time constant τ_{1D} for rectifiers fed from 50 Hz (60-Hz-values are actually under consideration) is

$$\tau_{1D} = \left(2 + (\kappa_D - 0.9) * \left(2.5 + 9 * \frac{L_{DBr}}{L_N}\right)\right)\ \text{ms} \quad \text{for } \kappa_D \geq 1.05 \tag{9.12a}$$

$$\tau_{1D} = \left(0.7 + \left(7 - \frac{R_N}{X_N} * \left(1 + \frac{2}{3} * \frac{L_{DBr}}{L_N}\right)\right)\right) * \left(0.1 + 0.2 * \frac{L_{DBr}}{L_N}\right)\ \text{ms} \quad \text{for } \kappa_D < 1.05 \tag{9.12b}$$

A suitable approximation, giving results on the safe side is

$$\tau_{1D} = \frac{t_{pD}}{3} \tag{9.12c}$$

The decay-time constant τ_{2D} for 50 Hz is calculated using

$$\tau_{2D} = \frac{2}{R_N/X_N * (0.6 + 0.9 * (R_{DBr}/R_N))}\ \text{ms} \tag{9.13}$$

Quantities as per Equations (9.11)–(9.13) are: X_N is the inductance of system N (a.c. side) $X_N = \omega L_N$, R_N is the resistance of system N (a.c. side), R_{DBr} is the resistance of d.c. side including smoothing reactor, connection and common branch (see Section 3.3.1 and Table 3.15), L_{DBr} is the inductance of d.c. side including smoothing reactor (saturated value), connection and common branch (see Table 3.15) and κ_D the factor as per Figure 9.9.

More details of the calculation of short-circuit currents fed from rectifiers are included in [54].

9.5 Short-circuit currents from d.c. motors with independent excitation

Branch short-circuit currents from d.c. motors with independent excitation are only considered if the total sum of the rated currents of all d.c. motors is greater than 1 per cent of the branch short-circuit currents of one rectifier.

The parameters of the motor, i.e., rated voltage U_{rM}, rated current I_{rM}, saturated inductance of the field circuit at short-circuit L_F and the unsaturated inductance of the field circuit at no-load L_{OF}, determine the quasi steady-state short-circuit current I_{kM} which is calculated using

$$I_{kM} = \frac{L_F}{L_{OF}} * \left(\frac{U_{rM} - I_{rM} * R_M}{R_{MBr}}\right) \tag{9.14}$$

where L_F is the field inductance, L_{OF} is field inductance at no-load, U_{rM} is the rated voltage, I_{rM} is the rated current of the motor, R_M is the resistance of the motor and R_{MBr} the resistance of the motor including connection and common branch (see Section 3.3.1 and Table 3.16).

Equation (9.14) is valid when the speed of the motor remains constant during the duration of the short-circuit. When the speed of the motor decreases to zero ($n \to 0$) the quasi steady-state short-circuit current is $I_{kM} = 0$.

The mechanical time constant τ_{Mec}, the time constants of the field circuit τ_F and the armature circuit τ_M determine the peak short-circuit current i_{pM}, which is calculated using

$$i_{pM} = \kappa_M * \frac{U_{rM} - I_{rM} * R_M}{R_{MBr}} \tag{9.15}$$

The factor κ_M is equal to One in the case of nominal speed and in all cases where $\tau_{Mec} \geq 10^* t_F$, else ($\tau_{Mec} < 10 * \tau_F$). κ_M depends on the eigen-frequency ω_0,

$$\omega_0 = \sqrt{\frac{1}{\tau_{Mec} * \tau_M} * \left(1 - \frac{R_M * I_{rM}}{U_{rM}}\right)} \tag{9.16a}$$

where τ_{Mec} is the mechanical time constant (see Table 3.16) and τ_M the time constant of armature circuit up to short-circuit location (see Table 3.16); and on the decay coefficient δ.

$$\frac{1}{\delta} = 2 * \tau_M \tag{9.16b}$$

The reasonable range of the factor κ_M is outlined in Figure 9.10.

The time to peak t_{pM}, the rise-time and the decay-time constants τ_{1M} and τ_{2M} depend on the value of the mechanical time constant τ_{Mec}. IEC 61660-1:1997 defines four factors k_{1M}, k_{2M}, k_{3M} and k_{4M} which are outlined in Figures 9.12 to 9.15. Calculation equations are not included in IEC 61000-1:1997.

When the motor speed remains constant or in all cases when $\tau_{Mec} \geq 10 * \tau_F$ the time to peak t_{pM} is calculated using

$$t_{pM} = k_{1M} * \tau_M \tag{9.17}$$

the rise-time constant τ_{1M}

$$\tau_{1M} = k_{2M} * \tau_M \tag{9.18}$$

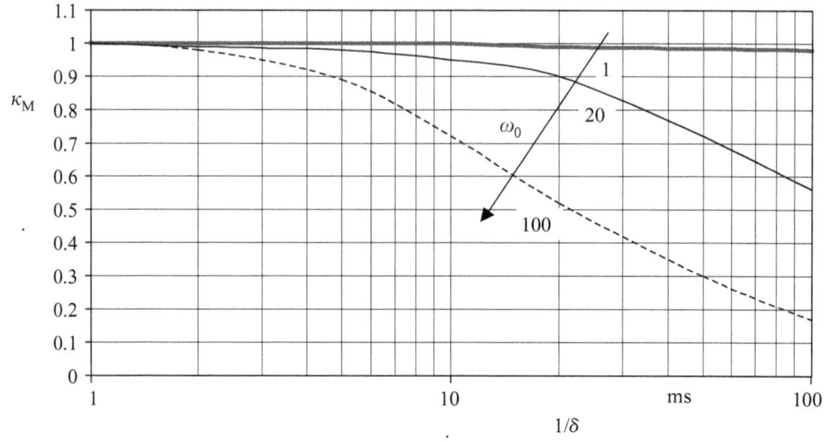

Figure 9.10 Factor κ_M for the calculation of peak short-circuit current of d.c. motors with independent excitation (according to Figure 17 of IEC 61660-1:1997)

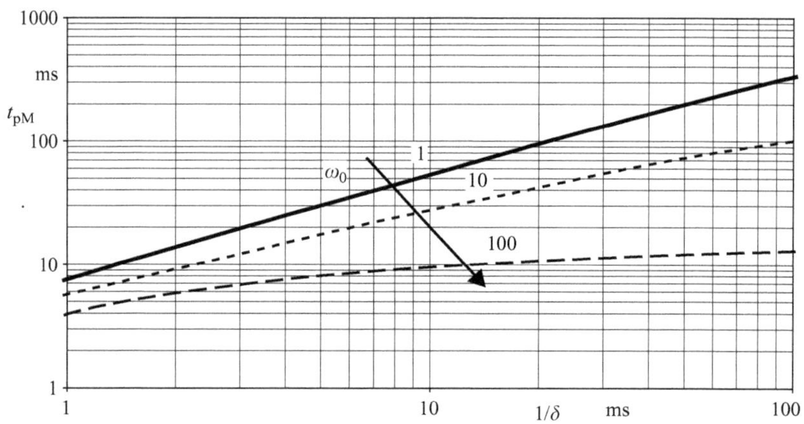

Figure 9.11 Time to peak of short-circuit currents for d.c. motors with independent excitation and $\tau_{Mec} < 10 * \tau_F$ (according to Figure 19 of IEC 61660-1:1997)

and the decay-time constant τ_{2M}

$$\tau_{2M} = \tau_F \qquad n = n_n = \text{const.} \tag{9.19a}$$

$$\tau_{2M} = k_{4M} * \frac{L_{OF}}{L_F} * \tau_{Mec} \quad n \to 0 \tag{9.19b}$$

For all other cases, i.e., $\tau_{Mec} < 10 * \tau_F$ the reasonable range of the time to peak t_{pM} is outlined in Figure 9.11.

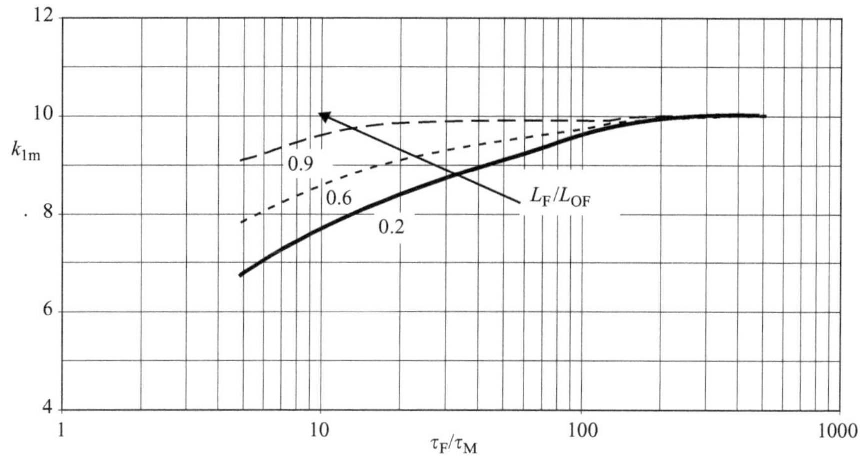

Figure 9.12 *Factor* k_{1M} *in the case of d.c. motors with independent excitation and* $\tau_{Mec} \geq 10 * \tau_F$ *(according to Figure 18 of IEC 61660-1:1997)*

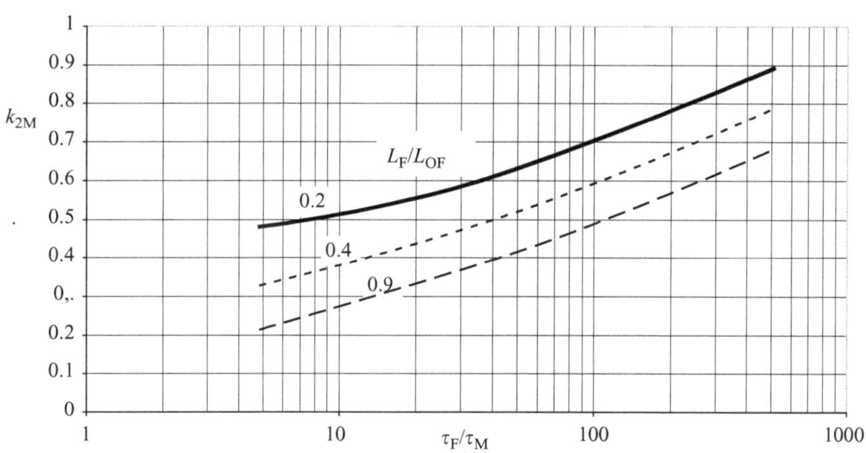

Figure 9.13 *Factor* k_{2M} *in the case of d.c. motors with independent excitation and* $\tau_{Mec} < 10 * \tau_F$ *(according to Figure 18 of IEC 61660-1:1997)*

The rise-time constant and the decay-time constant τ_{1M} and τ_{2M} are calculated using

$$\tau_{1M} = k_{3M} * \tau_M \tag{9.20a}$$

$$\tau_{2M} = k_{4M} * \tau_{Mec} \tag{9.20b}$$

The factors k_{1M}, k_{2M}, k_{3M} and k_{4M} are outlined in Figures 9.12 to 9.15.

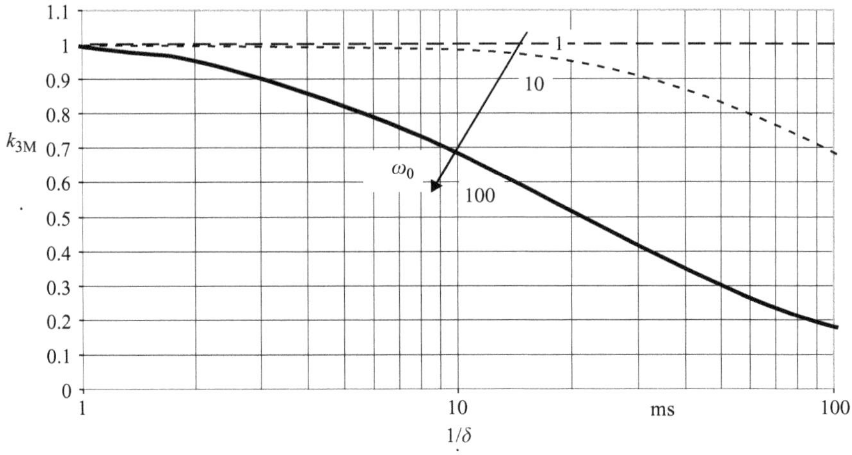

Figure 9.14 Factor k_{3M} in the case of d.c. motors with independent excitation and $\tau_{Mec} < 10 * \tau_F$ (according to Figure 20 of IEC 61660-1:1997)

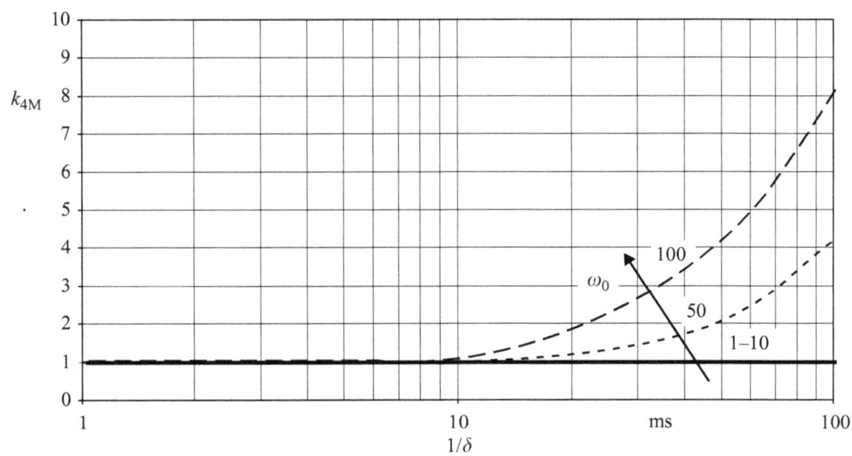

Figure 9.15 Factor k_{4M} in the case of d.c. motors with independent excitation and $\tau_{Mec} < 10 * \tau_F$ (according to Figure 21 of IEC 61660-1:1997)

More details of the calculation of short-circuit currents fed from motors with independent excitation are included in [52].

9.6 Total short-circuit current

The calculation of the total short-circuit current is carried out taking into account the common branches of the d.c. installation. A common branch is a branch, e.g., conductor circuit, which carries branch short-circuit currents from different sources.

When no common branch exists in the d.c. installation, the total short-circuit current is calculated by adding the branch (partial) short-circuit currents. Otherwise, the partial short-circuit currents of the different sources (Index j) are to be corrected (Index kor) with factors σ_j, which reflect the different resistances of the sources and the common branches R_Y. The peak short-circuit current and quasi steady-state short-circuit current are calculated by

$$i_{\mathrm{pkor}j} = \sigma_\mathrm{j} * i_\mathrm{pj} \tag{9.21a}$$

$$I_{\mathrm{kkor}j} = \sigma_\mathrm{j} * I_\mathrm{kj} \tag{9.21b}$$

with the correction factor σ_j for each source j

$$\sigma_\mathrm{j} = \frac{R_{\mathrm{res}j} * (R_\mathrm{ij} + R_\mathrm{Y})}{R_{\mathrm{res}j} * R_\mathrm{ij} + R_\mathrm{ij} * R_\mathrm{Y} + R_{\mathrm{res}j} * R_\mathrm{Y}} \tag{9.22}$$

The resistance up to the common branch of a source is named R_ij and the equivalent resistance of the other sources up to the common branch which contributes to the short-circuit current is named $R_{\mathrm{res}j}$. Resistances of capacitors are neglected and the resistance of motors shall only be taken into account if the motor contributes to the quasi steady-state short-circuit current.

The calculation equations of the resistances R_ij and equivalent resistances $R_{\mathrm{res}j}$ are outlined in Table 9.1. It is assumed, that all four sources are contributing to the short-circuit current through one common branch, as outlined in Figure 9.16.

Table 9.1 Resistances R_ij and equivalent resistances $R_{\mathrm{res}j}$ for the calculation of correction factors; U: Voltage at short-circuit location prior to the short-circuit [42,50,57]

Source j	Resistance R_ij	Equivalent resistance $R_{\mathrm{res}j}$
Capacitor	$R_\mathrm{iC} = R_\mathrm{C} + R_\mathrm{CL}$	$R_\mathrm{resC} = \dfrac{1}{1/R_\mathrm{iD} + 1/R_\mathrm{iB} + 1/R_\mathrm{iM}}$
Battery	$R_\mathrm{iB} = R_\mathrm{B} + R_\mathrm{BL}$	$R_\mathrm{resB} = \dfrac{1}{1/R_\mathrm{iD} + 1/R_\mathrm{iM}}$
Rectifier	$R_\mathrm{iD} = \dfrac{U}{I_\mathrm{kD}} - R_\mathrm{Y}$	$R_\mathrm{resD} = \dfrac{1}{1/R_\mathrm{iB} + 1/R_\mathrm{iM}}$
d.c. motor with independent excitation	$R_\mathrm{iM} = R_\mathrm{M} + R_\mathrm{ML}$	$R_\mathrm{resM} = \dfrac{1}{1/R_\mathrm{iD} + 1/R_\mathrm{iB}}$

Remarks: Index L – Connection of equipment; index Y – Common branch (see Section 3.3.1); and indices C; B; D; M – Capacitor; battery; rectifier; motor

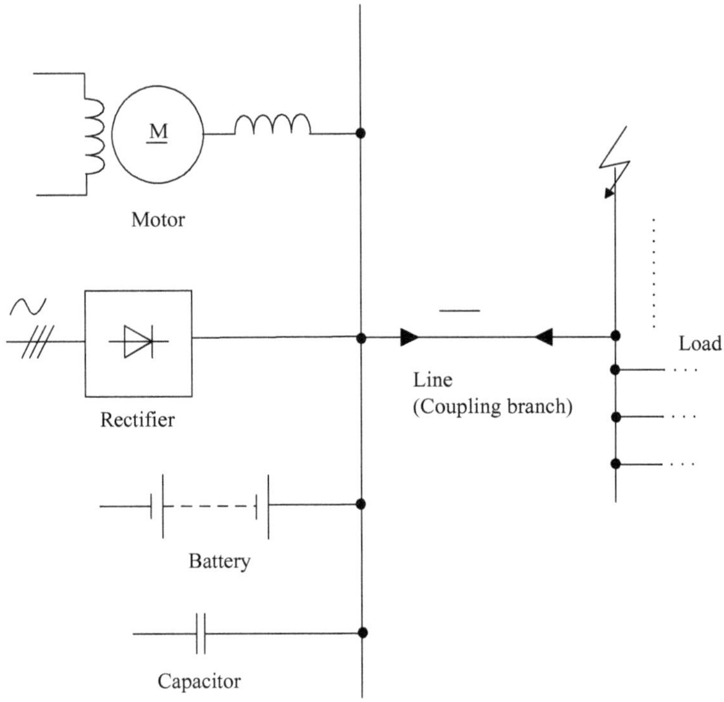

Figure 9.16 Equivalent circuit diagram of a d.c. auxiliary installation

Rise-time and decay-time constants τ_{1M} and τ_{2M} of the partial short-circuit currents are not corrected.

The total short-circuit current is calculated by superposition of the corrected partial short-circuit currents of the different sources.

$$i_1(t) = \sum_{j=1}^{m} i_{\text{pkor}\,j} * \frac{1 - e^{-t/\tau_{1j}}}{1 - e^{-t_{\text{p}j}/\tau_{1j}}} \qquad \text{for } 0 \le t \le t_{\text{p}j} \tag{9.23a}$$

$$i_2(t) = \sum_{j=1}^{m} i_{\text{pkor}\,j} \left(\left(1 - \frac{I_{\text{kkor}\,j}}{i_{\text{pkor}\,j}} \right) * e^{-(t-t_{\text{p}j})/\tau_{2j}} + \frac{I_{\text{kkor}\,j}}{i_{\text{pkor}\,j}} \right) \qquad \text{for } t_{\text{p}j} \le t \le T_{\text{k}} \tag{9.23b}$$

The calculation of the thermal and electromagnetic (mechanical) effects of short-circuit currents as per IEC 61660-2 is based on the standard approximation function. Typical time-curves of total short-circuit currents are outlined in Figure 9.17.

The peak short-circuit current i_{p}, the quasi steady-state short-circuit current I_{k} and the decay-time constant τ_2 are determined graphically from the time curve of the total short-circuit current. The rise-time constant is calculated in accordance with

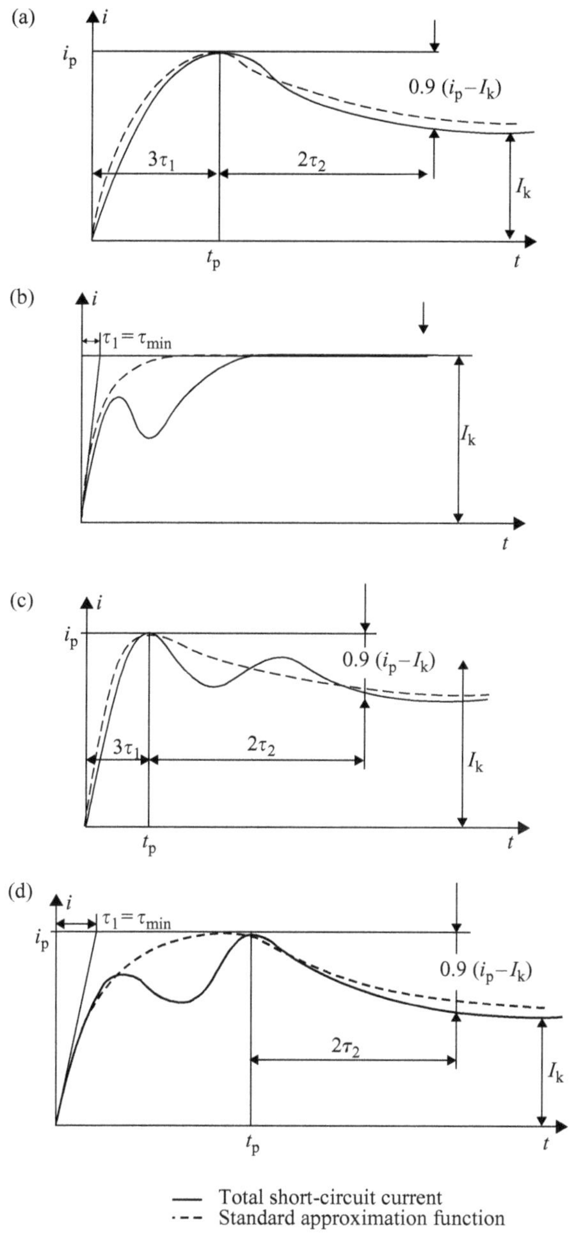

Figure 9.17 Typical time curves of total short-circuit current in d.c. auxiliary installations, e.g., (a) with dominating part of motors, (b) with dominating part of rectifiers, (c) with dominating part of batteries and (d) in the case of low rectifier load (according to Figure 22 of DIN EN 61660-1 (VDE 0102 Teil 10))

Figure 9.17(a) and (c), respectively by

$$\tau_1 = \frac{t_p}{3} \tag{9.24a}$$

and for time curves as per Figure 9.17(b) or (d)

$$\tau_1 = \mathrm{MIN}\{\tau_{1C}; \tau_{1B}; \tau_{1D}; \tau_{1M}\} \tag{9.24b}$$

The rise-time constant τ_{1C} of capacitors is neglected in the case where

$$i_{pC} \leq 0.5 * i_p \tag{9.25}$$

The decay-time constant τ_2 is equal to 50 per cent of that time span in which the short-circuit current $i_2(t)$ is reduced to $0.9 * (i_p–I_k)$, i.e., the short-circuit current has the value of $(I_k + 0.1 * i_p)$. Reference is made to Figure 9.17.

$$i_2(t_p + 2\tau_2) = 0.9 * (i_p - I_k) \tag{9.26a}$$

and

$$i_2(t_p + 2\tau_2) = I_k + 0.1 * i_p \tag{9.26b}$$

9.7 Example

The short-circuit currents at location F at the main busbar of the auxiliary supply installation of a power station are to be calculated. The 220-V-installation as outlined in Figure 9.18 include a battery of 1100 Ah, feeding from an a.c. LV-system through rectifier with smoothing capacitor and a d.c. motor. The data and parameter of the equipment are given below:

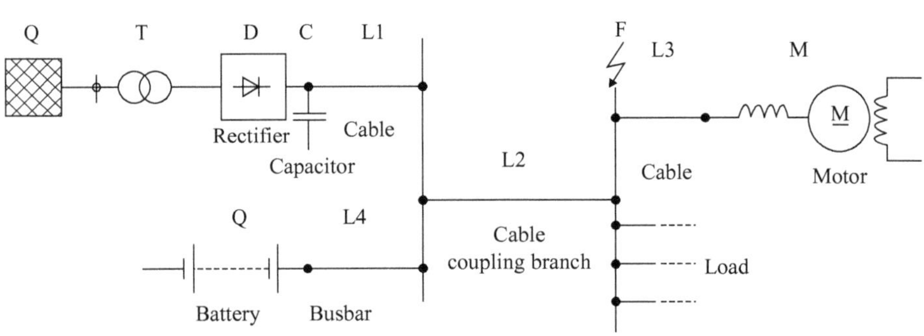

Figure 9.18 Equivalent circuit diagram of the d.c. auxiliary installation (220 V), e.g., of a power station

Equipment	Quantity	Parameter	Remarks
Battery B	Nominal voltage	$U_{nB} = 225$ V	Table 3.14
	Capacity for 10 h-discharge	$K_{10} = 1100$ Ah	
	Resistance per cell	$R_{BZ} = 0.12$ mΩ	
	Inductance per cell	$L_{BZ} = 0.2$ μH	
	Number of cells	$n = 109$	
Power system	Nominal voltage	$U_{nQ} = 660$ V	Table 3.1
	Initial short-circuit power	$S''_{kQ} = 33.3$ MVA	
	Resistance	$R_Q/X_Q = 0.28$	
AC/DC-transformer	Rated voltages	$U_{rTHV}/U_{rTLV} =$ 660 V/240 V	Table 3.2
	Rated power	$S_{rT} = 380$ kVA	
	Impedance voltage	$u_{kT} = 3.2\%$	
	Losses; resistance voltage	$P_{Cu} = 4$ kW; $u_{RT} = 0.5\%$	
Rectifier D	Rated voltage	$U_{rD} = 220$ V	Table 3.15
	Rated current	$I_{rD} = 1000$ A	
Smoothing reactor	Resistance	$R_{DBr} = 0.87$ mΩ	Table 3.15
	Reactance	$X_{DBr} = 29$ μH	
Smoothing capacitor C	a.c. capacitance	$C_{ac} = 70$ mF	Table 3.13
	Resistance	$R_C = 8$ mΩ	
d.c. motor M	Rated voltage	$U_{rM} = 220$ V	Table 3.16
	Rated power	$P_{rM} = 95$ kW	
	Rated current	$I_{rM} = 432$ A	
	No-load speed	$n_0 = 25\ s^{-1}$	
	Inductance of armature circuit	$L_M = 0.4$ mH	
	Resistance of armature	$R_M = 41.9$ mΩ	
	Inductance of field circuit	$L_F = 9.9$ H	
	Resistance of field circuit	$R_F = 9.9$ Ω	
	Moment of inertia	$J = 6.5\ kg\,m^2$	
	Motor speed $T_k > 1$ s	$n \rightarrow 0$	

Equipment	Quantity	Parameter	Remarks
Cables L1; L2; L3	Cross section	$q_n = 300\ \text{mm}^2$	Table 3.12
	Number of cables	3 in parallel (triangle)	
	Specific resistance	$\rho = 0.0173\ \Omega\text{mm}^2/\text{m}$	
	Radius of conductor	$r = 10.5\ \text{mm}$	
	Length of cables	$L1 = 2\ \text{m}$; $L2 = 6\ \text{m}$; $L3 = 20\ \text{m}$	
Conductor bars L4	Cross section	$q_n = 400\ \text{mm}^2$	Table 3.12
	Number of bars	2 in parallel	
	Width and height	$d \times b = 10\ \text{mm} \times 40\ \text{mm}$	
	Specific resistance	$\rho = 0.0173\ \Omega\text{mm}^2/\text{m}$	
	Distance between sub-conductors	$a_S = 10\ \text{mm}$	
	Distance of bars	$a_m = 75\ \text{mm}$	
	Length of bar	$L4 = 14\ \text{m}$	

The calculation of the short-circuit current is as follows:

1 Calculation of the impedances of cables and busbar conductors (Section 9.7.1).
2 Calculation of the short-circuit currents of the individual equipment (Section 9.7.2).
3 Calculation of the correction factors (Section 9.7.3).
4 Calculation of partial short-circuit currents (Section 9.7.4).
5 Calculation of total short-circuit current (Section 9.7.5).

9.7.1 *Calculation of the impedances of cables and busbar conductors*

Three cables are laid in triangle arrangement. Specific resistance and inductance are calculated according to Table 3.12.

$$R' = 2\frac{\rho}{3q_n} = 2\frac{0.0173\ \Omega\text{mm}^2/\text{m}}{3 * 300\ \text{mm}^2} = 0.0384\ \text{m}\Omega/\text{m}$$

$$L' = \frac{\mu_0}{\pi} * \left(\ln\frac{a}{r_B} + \frac{1}{4n}\right)$$

$$\text{with } r_B = \sqrt[n]{n * r_T * r^{n-1}} = \sqrt[3]{3 * 17.3\ \text{mm} * (10.5\ \text{mm})^2} = 17.9\ \text{mm}$$

$$L' = \frac{4\pi * 10^{-7}\ \text{H/m}}{\pi}\left(\ln\frac{75\ \text{mm}}{17.9\ \text{mm}} + \frac{1}{4 * 3}\right) = 0.607\ \mu\text{H/m}$$

Two rectangular busbars are laid in parallel. Specific resistance and inductance are calculated according to Table 3.12.

$$R' = 2\frac{\rho}{2q_{\mathrm{n}}} = 2 * \frac{0.0173\ \Omega\mathrm{mm}^2/\mathrm{m}}{2 * 400\ \mathrm{mm}^2} = 0.0433\ \mathrm{m}\Omega/\mathrm{m}$$

$$L' = \frac{\mu_0}{\pi} * \left(\ln \frac{a}{0.223 * (2d + d_S + b)}\right)$$

$$= \frac{4\pi * 10^{-7}\ \mathrm{H/m}}{\pi} * \left(\ln \frac{75\ \mathrm{mm}}{0.223 * (30 + 40)\ \mathrm{mm}}\right) = 0.628\ \mu\mathrm{H/m}$$

The resistances and inductances of the individual connections by cables and bars as per Figure 9.18 are summarised below:

Connection	L1	L2	L3	L4
Length(m)	2	6	20	14
Resistance R(mΩ)	0.0768	0.2304	0.768	0.606
Inductance L(μH)	1.212	3.636	12.12	8.792

9.7.2 Calculation of the short-circuit currents of the individual equipment

9.7.2.1 Short-circuit current from capacitor

Total impedance of the capacitor according to Table 3.13 with common branch (cable L2) and capacitor connection (cable L1)

$$R_{\mathrm{CBr}} = R_{\mathrm{C}} + R_{\mathrm{L1}} + R_{\mathrm{L2}} = (8 + 0.0708 + 0.2304)\ \mathrm{m}\Omega = 8.301\ \mathrm{m}\Omega$$

$$L_{\mathrm{CBr}} = L_{\mathrm{L1}} + L_{\mathrm{L2}} = (1.212 + 3.636)\ \mu\mathrm{H} = 4.848\ \mu\mathrm{H}$$

Peak short-circuit current as per Equation (9.3)

$$i_{\mathrm{pC}} = \kappa_{\mathrm{C}} * \frac{E_{\mathrm{C}}}{R_{\mathrm{CBr}}}$$

Factor κ_{C} depends on the eigen-frequency ω_0

$$\omega_0 = \frac{1}{\sqrt{L_{\mathrm{CBr}} * C}} = \frac{1}{\sqrt{4.848\ \mu\mathrm{H} * 70\ \mathrm{mF}}} = 1.717 * 10^3/\mathrm{s}$$

and on the decay coefficient δ

$$\frac{1}{\delta} = \frac{2 * L_{\mathrm{CBr}}}{R_{\mathrm{CBr}}} = \frac{2 * 4.848\ \mu\mathrm{H}}{8.301\ \mathrm{m}\Omega} = 1.168\ \mathrm{ms}$$

$$\rightarrow \kappa_{\mathrm{C}} = 0.65 \text{ as per Figure 9.3}$$

$$i_{\mathrm{pC}} = \kappa_{\mathrm{C}} * \frac{E_{\mathrm{C}}}{R_{\mathrm{CBr}}} = 0.65 * \frac{225\ \mathrm{V}}{8.301\ \mathrm{m}\Omega} = 17.62\ \mathrm{kA}$$

Time-to-peak $t_{\mathrm{pC}} = 1.1$ ms as per Figure 9.4.

Rise-time constant and decay-time constant as per Equation (9.5) are

$$\tau_{1C} = k_{1C} * t_{pC}$$
$$\tau_{2C} = k_{2C} * R_{CBr} * C$$

With factors k_{1C} and k_{2C} depending on eigen-frequency and decay component $k_{1C} \approx 0.38$ as per Figure 9.5 and $k_{2C} \approx 1.4$ as per Figure 9.6

$$\tau_{1C} = 0.38 * 11\ \text{ms} = 0.42\ \text{ms}$$
$$\tau_{2C} = 1.4 * 8.301\ \text{m}\Omega * 70\ \text{mF} = 0.813\ \text{ms}$$

9.7.2.2 Short-circuit current from battery

Total impedance of the battery as per Table 3.14 with common branch (cable L2) and battery connection (cable L4). The battery consists of 109 cells.

$$\begin{aligned} R_{BBr} &= 0.9 * R_B + R_{L4} + R_{L2} \\ &= (0.9 * 109 * 0.12 + 0.606 + 0.2304)\ \text{m}\Omega = 12.608\ \text{m}\Omega \\ L_{BBr} &= L_B + L_{L4} + L_{L2} = (109 * 0.2 + 8.792 + 3.636)\ \mu\text{H} = 34.228\ \mu\text{H} \end{aligned}$$

Quasi steady-state short-circuit current as per Equation (9.6)

$$I_{kB} = \frac{0.95 * E_B}{R_{BBr} + 0.1 * R_B} = \frac{0.95 * 1.05 * 225\ \text{V}}{(12.608 + 0.1 * 109 * 0.12)\ \text{m}\Omega} = 16.13\ \text{kA}$$

Peak short-circuit current according to Equation (9.7) is

$$i_{pB} = \frac{E_B}{R_{BBr}} = \frac{1.05 * 225\ \text{V}}{12.608\ \text{m}\Omega} = 18.74\ \text{kA}$$

The decay component δ is

$$\frac{1}{\delta} = \frac{2}{R_{BBr}/L_{BBr} + 1/T_B} = \frac{2}{12.608\ \text{m}\Omega/34.228\ \mu\text{H} + 1/30\ \text{ms}} = 4.98\ \text{ms}$$

Time-to-peak and rise-time constant as per Figure 9.7 are $t_{pB} = 13.7$ ms and $\tau_{1B} = 2.6$ ms. The decay-time constant is $\tau_{2B} = 100$ ms per definition.

9.7.2.3 Short-circuit current from rectifier

Impedance of system feeder Q related to 220 V as per Table 3.1

$$Z_Q = \frac{c * U_{nQ}^2}{S''_{kQ}} = \frac{1.0 * (220\ \text{V})^2}{33.3\ \text{MVA}} = 1.73\ \text{m}\Omega$$

Resistance and reactance due to the ratio of R_Q/X_Q

$$X_Q = 0.963 * Z_Q = 1.666\ \text{m}\Omega$$

$$R_Q = \sqrt{Z_Q^2 - X_Q^2} = 0.466\ \text{m}\Omega$$

Impedance of transformer according to Table 3.2

$$Z_T = \frac{u_{kT}}{S_{rT}} * \frac{U_{rTLV}^2}{100\%} = \frac{3.2\%}{380\ \text{kVA}} * \frac{(240\ \text{V})^2}{100\%} = 4.85\ \text{m}\Omega$$

$$R_T = \frac{u_{RT}}{S_{rT}} * \frac{U_{rTLV}^2}{100\%} = \frac{1.05\%}{380\ \text{kVA}} * \frac{(240\ \text{V})^2}{100\%} = 1.592\ \text{m}\Omega$$

$$X_T = \sqrt{Z_T^2 - R_T^2} = 4.581\ \text{m}\Omega$$

Total impedance a.c. side according to Table 3.15

$$R_N = R_Q + R_T = (0.466 + 1.592)\ \text{m}\Omega = 2.058\ \text{m}\Omega$$

$$X_N = X_Q + X_T = (1.666 + 4.581)\ \text{m}\Omega = 6.247\ \text{m}\Omega$$

and inductance

$$L_N = X_N/\omega = (6.247/100 * \pi)\ \mu\text{H} = 4.581\ \mu\text{H}$$

Total impedance d.c. side according to Table 3.15 with common branch (cable L2)

$$R_{DBr} = R_S + R_{L1} + R_{L2} = (0.87 + 0.0768 + 0.2304)\ \text{m}\Omega = 1.177\ \text{m}\Omega$$

$$L_{DBr} = L_S + L_{L1} + L_{L2} = (29 + 1.212 + 3.636)\ \mu\text{H} = 33.85\ \mu\text{H}$$

Quasi steady-state short-circuit current of the rectifier as per Equation (9.9)

$$I_{kD} = \lambda_D * \frac{3 * \sqrt{2}}{\pi} * \frac{c * U_n}{\sqrt{3} * Z_N} * \frac{U_{rTLV}}{U_{rTHV}}$$

Factor λ_D depending on R_N/X_N and R_{DBr}/R_N

$$R_N/X_N = 0.33 \text{ and } R_{DBr}/R_N = 0.572 \rightarrow \lambda_D = 0.92 \text{ as per Figure 9.8}$$

$$I_{kD} = 0.92 * \frac{3 * \sqrt{2}}{\pi} * \frac{1.0 * 240\ \text{V}}{\sqrt{3} * 6.58\ \text{m}\Omega} * \frac{240\ \text{V}}{660\ \text{V}} = 26.44\ \text{kA}$$

Peak short-circuit current as per Equation (9.10)

$$i_{pD} = \kappa_D * I_{kD}$$

Factor κ_D depends on L_{DBr}/L_N and R_N, R_{BBr} and X_N

$$L_{DBr}/L_N = 1.702 \quad \text{and} \quad \frac{R_N}{X_N}\left(1 + \frac{2}{3} * \frac{R_{DBr}}{R_N}\right) = 0.455 \rightarrow \kappa_D = 1.12$$

as per Figure 9.9

$$i_{pD} = 1.12 * 26.44\ \text{kA} = 29.61\ \text{kA}$$

Time to peak according to Equation (9.11b) as $\kappa_D = 1.12 \geq 1.05$ and $L_{DBr}/L_N = 1.702 > 1$

$$t_{pD} = \left((3 * \kappa_D + 6) + 4\left(\frac{L_{DBr}}{L_N} - 1\right)\right)$$

$$= \left((3 * 1.12 + 6) + 4\left(\frac{33.85}{19.89} - 1\right)\right)\ \text{ms} = 12.17\ \text{ms}$$

Rise-time constant τ_{1D} as per Equation (9.12a)

$$\tau_{1D} = \left(2 + (\kappa_D - 0.9) * \left(2.5 + 9 * \frac{L_{DBr}}{L_N}\right)\right)$$

$$= \left(2 + (1.12 - 0.9) * \left(2.5 + 9 * \frac{33.85}{19.89}\right)\right)\ \text{ms} = 5.92\ \text{ms}$$

Decay-time constant τ_{2D} as per Equation (9.13)

$$\tau_{2D} = \frac{2}{R_N/X_N * (0.6 + 0.9 * R_{DBr}/R_N)}$$

$$= \frac{2}{2.058/6.247 * (0.6 + 0.9 * 1.177/2.058)}\ \text{ms} = 5.44\ \text{ms}$$

9.7.2.4 Short-circuit currents from d.c. motor

Total impedance of the motor as per Table 3.16 with connection cable L3, common branch is neglected in this case

$$R_{MBr} = R_M + R_{L3} = (41.9 + 0.768)\ \text{m}\Omega$$

$$L_{MBr} = L_M + L_{L3} = (400 + 12.12)\ \mu\text{H} = 412.12\ \mu\text{H}$$

Quasi steady-state short-circuit current ($n \to 0$)

$$I_{kM} = 0$$

Peak short-circuit current according to Equation (9.15)

$$i_{pM} = \kappa_M * \frac{U_{rM} - I_{rM} * R_M}{R_{MBr}}$$

With factor κ_M depending on the eigen-frequency ω_0 and on the decay coefficient δ as per Equation (9.16)

$$\omega_0 = \sqrt{\frac{1}{\tau_{Mec} * \tau_M} * \left(1 - \frac{R_M * I_{rM}}{U_{rM}}\right)}$$

$$\frac{1}{\delta} = 2 * \tau_M$$

Mechanical time constant as per Table 3.16

$$\tau_{\text{Mec}} = \frac{2\pi * n_0 J * R_{\text{MZw}} * I_{\text{rM}}}{M_{\text{r}} * U_{\text{rM}}} = \frac{2\pi * n_0 * J * R_{\text{MZw}} * I_{\text{rM}}}{(P_{\text{rM}}/2\pi n_0) * U_{\text{rM}}}$$

$$\tau_{\text{Mec}} = \frac{2\pi * 25/\text{s} * 6.5\ \text{kgm}^2 * 42.67\ \text{m}\Omega * 432\ \text{A}}{(100\ \text{kW}/2\pi * 25/\text{s}) * 220\ \text{V}} = 134.4\ \text{ms}$$

Time constant of armature circuit as per Table 3.16

$$\tau_{\text{M}} = \frac{L_{\text{MBr}}}{R_{\text{MBr}}} = \frac{412.12\ \mu\text{H}}{42.67\ \text{m}\Omega} = 9.66\ \text{ms}$$

Time constant of the field circuit

$$\tau_{\text{F}} = \frac{L_{\text{F}}}{R_{\text{F}}} = \frac{9.9\ \text{H}}{9.9\ \Omega} = 1\ \text{s}$$

Giving the eigen-frequency

$$\omega_0 = \sqrt{\frac{1}{134.4\ \text{ms} * 9.66\ \text{ms}} * \left(1 - \frac{41.9\ \text{m}\Omega * 432\ \text{A}}{220\ \text{V}}\right)} = 37.61/\text{s}$$

and the decay coefficient $1/\delta = 2 * 9.66\ \text{ms} = 19.32\ \text{ms}$.
As $\tau_{\text{Mec}} = 134.4\ \text{ms} < 10^* t_{\text{F}} = 100\ \text{s} \rightarrow \kappa_{\text{M}} = 0.81$ as per Figure 9.10

$$i_{\text{pM}} = 0.81 * \frac{220\ \text{V} - 432\ \text{A} * 41.9\ \text{m}\Omega}{42.67\ \text{m}\Omega} = 3.83\ \text{kA}$$

Time to peak as $\tau_{\text{Mec}} < 10 * \tau_{\text{F}}$

$$t_{\text{pM}} = 28\ \text{ms} \quad \text{according to Figure 9.11}$$

Rise-time and decay-time as per Equation (9.20) constants depend on the mechanical time constant as per Equations (9.18) to (9.19)

$$\tau_{1\text{M}} = k_{3\text{M}} * \tau_{\text{M}}$$

$$\tau_{2\text{M}} = k_{4\text{M}} * \tau_{\text{Mec}}$$

with factors $k_{3\text{M}}$ and $k_{4\text{M}}$

$$k_{3\text{M}} = 0.78 \text{ according to Figure 9.14}$$

$$k_{4\text{M}} = 1.05 \text{ according to Figure 9.15}$$

$$\tau_{1\text{M}} = 0.78 * 9.66\ \text{ms} = 7.53\ \text{ms}$$

$$\tau_{2\text{M}} = 1.05 * 134.4\ \text{ms} = 141.1\ \text{ms}$$

9.7.2.5 Summary of results

The result of the short-circuit currents of the individual equipment are summarised below

Symbol	Equipment	i_p (kA)	I_k (kA)	t_p (ms)	τ_1 (ms)	τ_2 (ms)
C	Capacitor	17.62	0.0	1.1	0.42	0.98
B	Battery	18.74	16.13	13.7	2.6	100
D	Rectifier	29.61	26.44	12.17	5.92	5.44
M	Motor	3.83	0.0	28	7.53	141.1

9.7.3 *Calculation of the correction factors and corrected parameters*

Correction factors σ as per Table 9.1 are only to be calculated for rectifier, capacitor and battery, as the motor feeds the short-circuit directly.

Source j	Resistance R_{ij}	Equivalent resistance $R_{res\,j}$
Capacitor	$R_{iC} = R_C + R_{CL}$	$R_{resC} = \dfrac{1}{(1/R_{iD}) + (1/R_{iB}) + (1/R_{iM})}$
	$R_{iC} = R_C$	$R_{resC} = \dfrac{1}{(1/8.29) + (1/13.686)}\ \text{m}\Omega$
	$R_{iC} = 8\ \text{m}\Omega$	$R_{resC} = 5.16\ \text{m}\Omega$
Battery	$R_{iB} = R_B + R_{L4}$	$R_{resB} = \dfrac{1}{(1/R_{iD}) + (1/R_{iM})}$
	$R_{iB} = (109 * 0.12 + 0.606)\ \text{m}\Omega$	$R_{resB} = R_{iD}$
	$R_{iB} = 13.686\ \text{m}\Omega$	$R_{resB} = 8.29\ \text{m}\Omega$
Rectifier	$R_{iD} = \dfrac{U}{I_{kD}} - R_{L2}$	$R_{resD} = \dfrac{1}{(1/R_{iB)} + (1/R_{iM})}$
	$R_{iD} = \dfrac{225\ \text{V}}{26.44\ \text{kA}} - 0.2304\ \text{m}\Omega$	$R_{resD} = R_{iB}$
	$R_{iD} = 8.29\ \text{m}\Omega$	$R_{resD} = 13.686\ \text{m}\Omega$
Motor	$R_{iM} = R_M + R_{L3}$	As the motor is feeding
	$R_{iM} = (41.9 + 0.768)\ \text{m}\Omega$	the short-circuit directly, R_{iM} is
	$R_{iM} = 42.668\ \text{m}\Omega$	neglected

Calculation of correction factors σ as per Equation (9.22)

- *Capacitor*

$$\sigma_C = \frac{R_{resC} * (R_{iC} + R_{L2})}{R_{resC} * R_{iC} + R_{iC} * R_{L2} + R_{resC} * R_{L2}}$$

$$\sigma_C = \frac{5.16 * (8 + 0.2304)}{5.16 * 8 + 8 * 0.2304 + 5.16 * 0.2304} = 0.958$$

- *Battery*

$$\sigma_B = \frac{R_{resB} * (R_{iB} + R_{L2})}{R_{resC} * R_{iB} + R_{iB} * R_{L2} + R_{resB} * R_{L2}}$$

$$\sigma_B = \frac{8.29 * (13.686 + 0.2304)}{8.29 * 13.686 + 13.686 * 0.2304 + 8.29 * 0.2304} = 0.973$$

- *Rectifier*

$$\sigma_D = \frac{R_{resD} * (R_{iD} + R_{L2})}{R_{resD} * R_{iD} + R_{iD} * R_{L2} + R_{resD} * R_{L2}}$$

$$\sigma_D = \frac{13.686 * (8.29 + 0.2304)}{13.686 * 8.29 + 8.29 * 0.2304 + 13.686 * 0.2304} = 0.984$$

- *Motor*
  Correction factor of the motor is set to 1, as the motor is feeding the short-circuit directly.
- *Correction of parameters*
  Correction of peak short-circuit current and quasi steady-state short-circuit current is carried out for the individual results based on Equation (9.21)

$$i_{pkor\,j} = \sigma_j * i_{pj}$$

$$I_{kkor\,j} = \sigma_j * I_{kj}$$

Symbol	Equipment	σ	i_p (kA)	I_k (kA)	i_{pkor} (kA)	I_{kkor} (kA)	t_p (ms)	τ_1 (ms)	τ_2 (ms)
C	Capacitor	0.958	17.62	0.0	16.88	0.0	1.1	0.42	0.98
B	Battery	0.973	18.74	16.13	18.23	15.69	13.7	2.6	100
D	Rectifier	0.984	29.61	26.44	29.14	26.02	12.17	5.92	5.44
M	Motor	1	3.83	0.0	3.83	0.0	28	7.53	141.1

9.7.4 Calculation of partial short-circuit currents

The partial short-circuit currents are calculated based on Equation (9.1)

$$i_1(t) = i_p * \frac{1 - e^{-t/\tau_1}}{1 - e^{-t_p/\tau_1}} \qquad \text{for } 0 \le t \le t_p$$

$$i_2(t) = i_p \left(\left(1 - \frac{I_k}{i_p} \right) * e^{-(t-t_p)/\tau_2} + \frac{I_k}{i_p} \right) \qquad \text{for } t_p \le t \le T_k$$

- *Capacitor*

$$i_{1C}(t) = 16.88\ \text{kA} * \frac{1 - e^{-t/0.42}}{1 - e^{-1.1/0.42}}$$

$$i_{1C}(t) = 18.21\ \text{kA} * (1 - e^{-t/0.42}) \qquad \text{for } 0 \le t \le 1.1\ \text{ms}$$

$$i_{2C}(t) = 16.88\ \text{kA} * (1 - e^{-(t-1.1)/0.98}) \qquad \text{for } 1.1\ \text{ms} \le t \le T_k$$

- *Battery*

$$i_{1B}(t) = 18.23\ \text{kA} * \frac{1 - e^{-t/2.6}}{1 - e^{-13.7/2.6}}$$

$$i_{1B}(t) = 18.32\ \text{kA} * (1 - e^{-t/2.6}) \qquad \text{for } 0 \le t \le 13.7\ \text{ms}$$

$$i_{2B}(t) = 16.88\ \text{kA} * \left(\left(1 - \frac{15.69}{18.23}\right) * e^{-(t-13.7)/100} + \frac{15.69}{18.23}\right)$$

$$i_{2B}(t) = 16.88\ \text{kA} * (0.139 * e^{-(t-13.7)/100} + 0.861)$$

$$\text{for } 13.7\ \text{ms} \le t \le T_k$$

- *Rectifier*

$$i_{1D}(t) = 29.14\ \text{kA} * \frac{1 - e^{-t/5.92}}{1 - e^{-12.17/5.92}}$$

$$i_{1D}(t) = 33.42\ \text{kA} * (1 - e^{-t/5.2}) \qquad \text{for } 0 \le t \le 12.17\ \text{ms}$$

$$i_{2D}(t) = 29.14\ \text{kA} * \left(\left(1 - \frac{26.02}{29.14}\right) * e^{-(t-12.17)/5.44} + \frac{26.02}{29.14}\right)$$

$$i_{2D}(t) = 29.14\ \text{kA} * (0.107 * e^{-(t-12.17)/5.44} + 0.893)$$

$$\text{for } 12.17\ \text{ms} \le t \le T_k$$

- *Motor*

$$i_{1M}(t) = 3.83\ \text{kA} * \frac{1 - e^{-t/7.53}}{1 - e^{-28/7.53}}$$

$$i_{1M}(t) = 3.83\ \text{kA} * (1 - e^{-t/7.53}) \qquad \text{for } 0 \le t \le 28\ \text{ms}$$

$$i_{2M}(t) = 3.83\ \text{kA} * e^{-(t-28)/141.1} \qquad \text{for } 28\ \text{ms} \le t \le T_k$$

The corrected partial short-circuit currents of the different equipment (sources) are outlined in Figure 9.19.

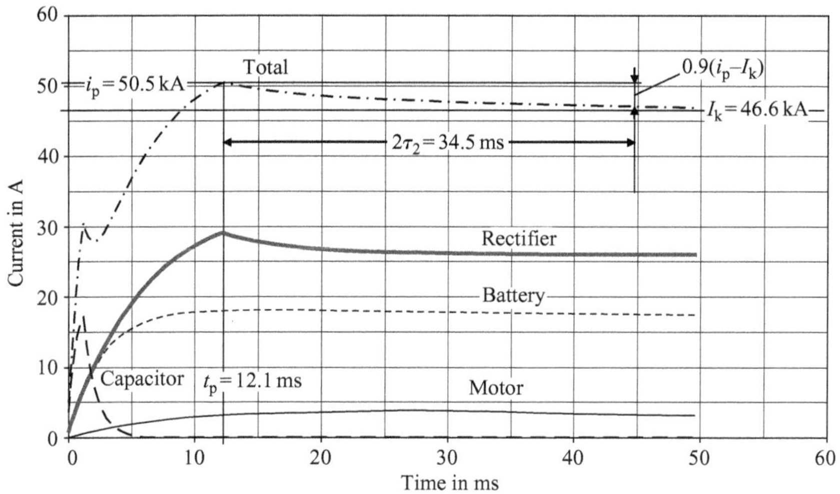

Figure 9.19 Partial short-circuit currents and total short-circuit current, d.c. auxiliary system as per Figure 9.18

9.7.5 Calculation of total short-circuit current

The total short-circuit current is calculated by superposition of the corrected partial short-circuit currents of the different sources as per Equation (9.23)

$$i_1(t) = \sum_{j=1}^{m} i_{\mathrm{pkor}\,j} * \frac{1 - \mathrm{e}^{-t/\tau_{1j}}}{1 - \mathrm{e}^{-t_{\mathrm{p}j}/\tau_{1j}}} \qquad \text{for } 0 \leq t \leq t_{\mathrm{p}j}$$

$$i_2(t) = \sum_{j=1}^{m} i_{\mathrm{pkor}\,j} \left(\left(1 - \frac{I_{\mathrm{kkor}\,j}}{i_{\mathrm{pkor}\,j}} \right) * \mathrm{e}^{-(t-t_{\mathrm{p}j})/\tau_{2j}} + \frac{I_{\mathrm{kkor}\,j}}{i_{\mathrm{pkor}\,j}} \right) \qquad \text{for } t_{\mathrm{p}j} \leq t \leq T_{\mathrm{K}}$$

The total short-circuit current obtained by superposition is outlined in Figure 9.19.

The peak short-circuit current

$$i_{\mathrm{p}} = 50.5 \text{ kA}$$

the quasi steady-state short-circuit current

$$I_{\mathrm{k}} = 46.6 \text{ kA}$$

the time-to-peak

$$t_{\mathrm{p}} = 12.1 \text{ ms}$$

and the decay-time constant

$$\tau_2 = 17.3 \text{ ms}$$

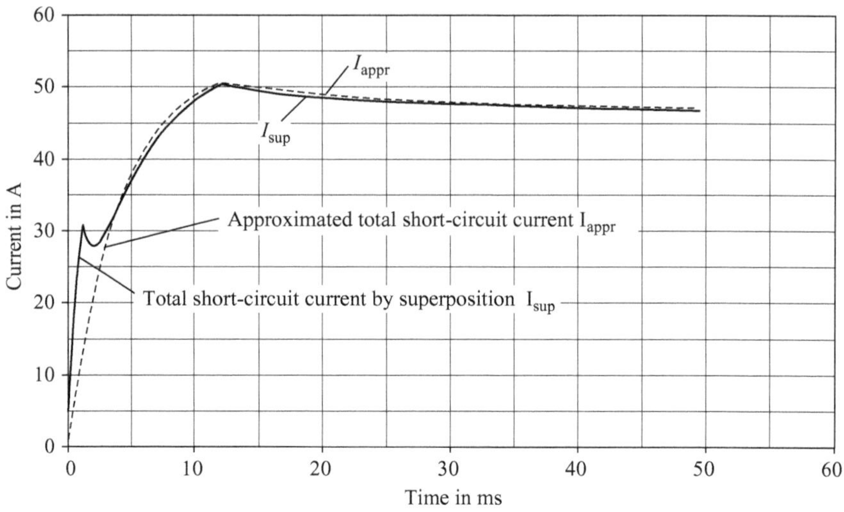

Figure 9.20 Total short-circuit current, obtained by superposition of the partial short-circuit currents and approximated short-circuit current, d.c. auxiliary system as per Figure 9.18

are obtained from Figure 9.19 as indicated. The rise-time constant can be estimated according to Equation (9.24) either by

$$\tau_1 = \frac{t_p}{3} = \frac{12.1 \text{ ms}}{3} = 4.03 \text{ ms} \quad \text{(time course as per Figure 9.17(a) and (c))}$$

or

$$\tau_1 = \text{MIN}\{\tau_{1C}; \tau_{1B}; \tau_{1D}; \tau_{1M}\} \quad \text{(time course as per Figure 9.17(b) and (d))}$$

$$\tau_1 = \text{MIN}\{5.92; 2.6; 0.42; 7.53\} \text{ ms} = 0.42 \text{ ms}$$

Comparison of the two different values revealed that the approximated time course of the total short-circuit current fits best to the superposition of the partial short-circuit currents, see Figure 9.20, if the rise-time constant is $\tau_1 = 4.03$ ms, which also is in line with the typical time course of the short-circuit current as per Figure 9.17(b) and (d).

Chapter 10
Effects of short-circuit currents

10.1 General

Calculation methods for the thermal and electromagnetic effects of short-circuit currents are dealt with within IEC 61660-2, which is applicable to short-circuit currents in d.c. auxiliary installations in power plants and substations and IEC 60865-1, related to three-phase a.c. systems.

10.2 a.c. systems

10.2.1 Thermal effects and thermal short-circuit strength

The thermal withstand capability (thermal short-circuit strength) of equipment is determined by the maximal permissible conductor temperature prior to the short-circuit, the duration of the short-circuit and the short-circuit current itself. The maximal permissible temperature of conductors under normal operating conditions and in the case of short-circuits, e.g., as per DIN VDE 0276, is summarised in Table 10.1. Figures are given for a short-circuit duration of $T_k = 5$ s. It is assumed that no heat transfer is taking place during the short-circuit duration (adiabatic heating). Skin- and proximity-effects are neglected, the specific caloric heat of the conductor and insulation is constant and the relation resistance-to-temperature is linear. Special consideration is to be taken for conductors in a.c. installations with cross-section above 600 mm^2, as the skin-effect has to be taken into account. Additional requirements according to IEC 60986:1989 and IEC 60949:1988 for cables and isolated conductors are to be met.

10.2.1.1 Conductors and equipment

The analysis is based on the calculation of the thermal equivalent short-time current I_{th}

$$I_{th} = \sqrt{\frac{Q}{R * T_k}} = \sqrt{\frac{\int_0^{T_k} i_k^2(t)\,dt}{R * T_k}} \tag{10.1}$$

Table 10.1 Maximal permissible conductor temperature and rated short-time current density; 1) – Normal operating condition; 2) – Short-circuit condition

Type of cable	Maximum permissible temperature (°C) 1)	2)	Temperature at beginning of short-circuit (°C) 90	80	70	65	60	50	40	30	20
			Rated short-time current density in A/mm^2; $t_{kr} = 1$ s								
Copper conductor											
Cables with soft-soldering	—	160	100	108	115	119	122	129	136	143	150
XPE-cables	90	250	143	149	154	157	159	165	170	176	181
PVC-cables (mm^2)											
≤300	70	160	—	—	115	119	122	129	136	143	150
>300	70	140	—	—	103	107	111	118	126	133	140
Mass-impregnated cables (kV)											
0.6/1	80	250	—	149	154	157	159	165	170	176	181
3.6/6	80	170	—	113	120	124	127	134	141	147	154
6/10	70	170	—	—	120	124	127	134	141	147	154
12/20	65	170	—	—	—	124	127	134	141	147	154
18/30	60	150	—	—	—	—	117	124	131	138	145
Radial-screen cable 12/20	65	170	—	—	—	124	127	134	141	147	154
Aluminium conductor											
XPE-cables	90	250	94	98	102	104	105	109	113	116	120
PVC-cables (mm^2)											
≤300	70	160	—	—	76	78	81	85	90	95	99
>300	70	140	—	—	68	71	73	78	83	88	93
Mass-impregnated cables (kV)											
0.6/1	80	250	—	98	102	104	105	109	113	116	120
3.6/6	80	170	—	75	80	82	84	89	93	97	102
6/10	70	170	—	—	80	82	84	89	93	97	102
12/20	65	170	—	—	—	82	84	89	93	97	102
18/30	60	150	—	—	—	—	77	82	87	91	96
Radial-screen cable 12/20	65	170	—	—	—	82	84	89	93	97	102

where Q is the thermal heat produced by the short-circuit current, R the resistance of the equipment, T_k the short-circuit duration and $i_k(t)$ the time course of short-circuit current which produces the same thermal heat Q within the conductor as the short-circuit current $i_k(t)$ during the short-circuit duration T_k. The thermal equivalent short-time current is calculated from the initial short-circuit current I_k'' by using

$$I_{th} = I_k'' * \sqrt{m+n} \tag{10.2}$$

The factors m and n represent the heat dissipation of the d.c. component and the a.c. component of the short-circuit current [38]. Suitable ranges of values for m and n are outlined in Figures 10.1 and 10.2. If an interrupting short-circuit is present or if multiple short-circuits (number n) occur with different duration T_{ki} and current I_{thi} the resulting thermal equivalent rated short-time current is

$$I_{th} = \sqrt{\frac{1}{T_k}\sum_{i=1}^{n} I_{thi}^2 * T_{ki}} \tag{10.3}$$

where

$$T_k = \sum_{i=1}^{n} T_{ki}$$

IEC 60909-0 includes calculation equations for the factors m and n.

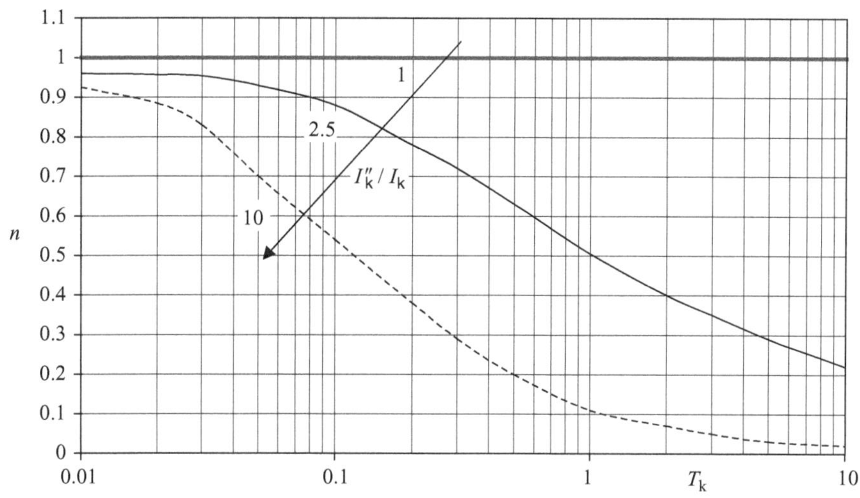

Figure 10.1 Factor n for the calculation of thermal short-time current (heat dissipation of a.c. component) (according to Figure 22 of IEC 60909-0:2001)

According to IEC 60865-1 separate considerations have to be taken concerning the thermal strength of equipment, i.e., transformers, transducers, etc., and conductors, i.e., busbars, cables, etc.

Equipment have a suitable thermal short-circuit strength if the rated short-time short-circuit current I_{thr} (as per manufacturer's data) for the short-circuit duration $T_k < T_{kr}$ (e.g., $T_{kr} = 1$ s) is above the thermal equivalent short-circuit current I_{th}.

$$I_{th} \le I_{thr} \quad \text{for } T_k \le T_{kr} \tag{10.4a}$$

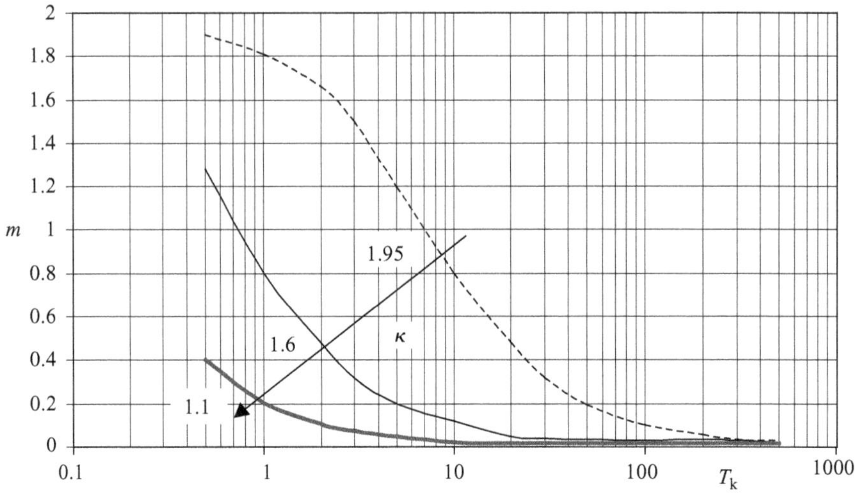

Figure 10.2 Factor m for the calculation of thermal short-time current (heat dissipation of d.c. component) (according to Figure 21 of IEC 60909-0:2001)

In case the short-circuit duration T_k is longer than the rated short-circuit duration T_{kr}, the thermal short-circuit strength is fulfilled if

$$I_{th} \leq I_{thr} * \sqrt{\frac{T_{kr}}{T_k}} \quad \text{for } T_k \geq T_{kr} \tag{10.4b}$$

The maximal permissible values for conductor material can be obtained from Figure 10.3.

In the case of bare conductors the thermal short-time current density J_{th} is calculated on the basis of the thermal equivalent short-time current and the conductor cross-section q_n.

$$J_{th} = \frac{I_{th}}{q_n} \tag{10.5}$$

where q_n is the nominal cross-section of conductor and I_{th} is the equivalent short-circuit current.

In the case of overhead line conductors of the Al/St-type, only the cross-section of the aluminium part is considered. Conductors have sufficient thermal short-circuit strength if

$$J_{th} \leq J_{thr} * \sqrt{\frac{T_{kr}}{T_k}} \tag{10.6}$$

Values for the rated short-time current density are included in Table 10.1 and Figure 10.3. In IEC 60865 equations for the calculation of rated short-time current density are included.

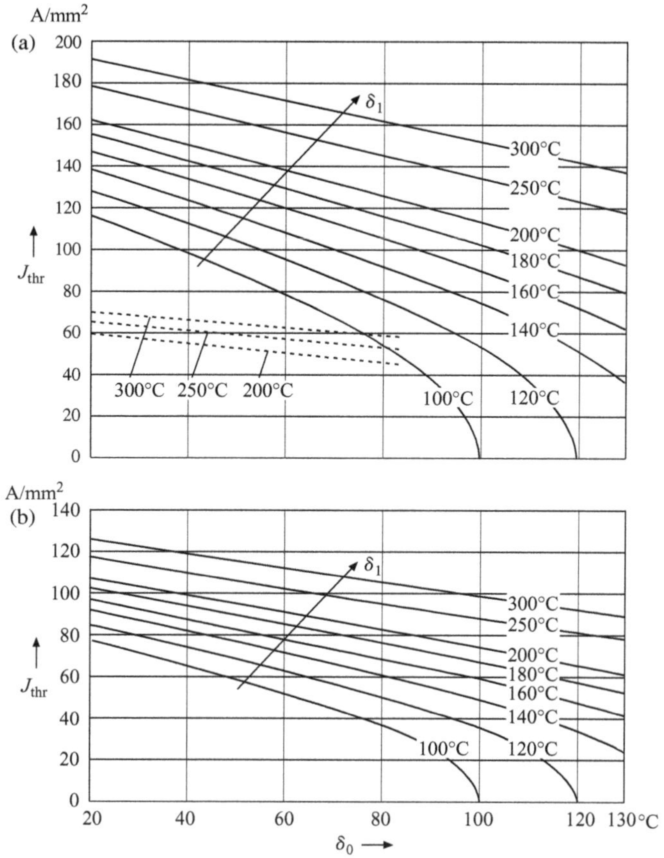

Figure 10.3 Rated short-time current density of conductors. δ_0 is the temperature at beginning of short-circuit and δ_1 is the temperature at end of short-circuit [1]. (a) ——: Copper; - - - -: unalloyed steel and steel cables and (b) Al, aluminium alloy, ACSR

Regarding non-insulated conductors, e.g., bare conductors and busbars, the thermal equivalent short-time current density is allowed to exceed the rated short-time current density in the case $T_k < T_{kr}$.

Manufacturer's cable lists usually include data on the maximal permissible thermal short-circuit currents I_{thz}. An example is outlined in Figure 10.4. The rated short-time current is given for a short-circuit duration of $T_k = 1$ s.

10.2.1.2 Cable screening, armouring and sheath

Sheaths, screening and armouring of cables carry parts of the short-circuit current in the case of asymmetrical short-circuits. Depending on the type of the short-circuit and the method of cable-laying, this current can be in the range of the short-circuit current itself, e.g., when the cables are laid in air or on wall-racks. In case cables are laid in earth the part of the short-circuit current through the sheaths, armouring and

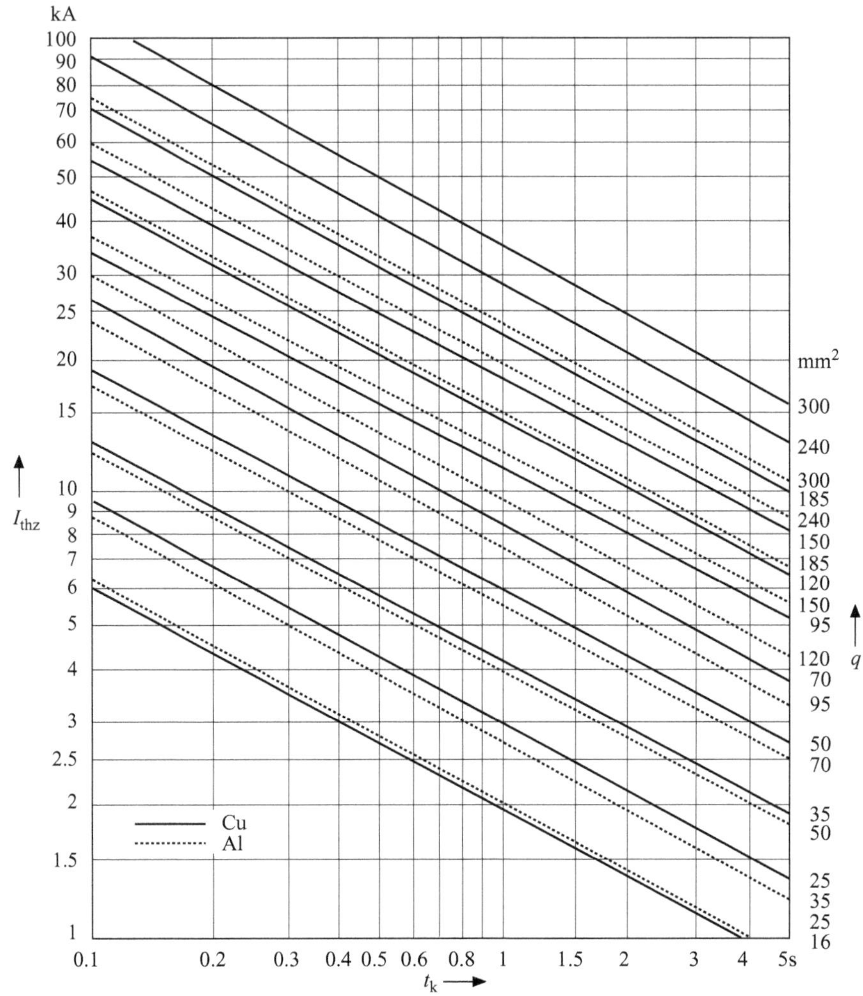

Figure 10.4 *Maximal permissible thermal short-circuit current for impregnated paper-insulated cables U_n up to 10 kV*

Source: KABELRHEYDT

screening is lower than the short-circuit current, as one part is flowing through earth as well. Due to the comparatively high specific resistance of lead (used for sheaths) and steel (used for armouring), the short-circuit current preferably flows through the screening made from copper or aluminium. Data of the different materials are outlined in Table 10.2. Due to different production processes and degree of purity of the material, data obtained from other tables can be slightly different.

The maximal permissible temperature δ_0 or screening and sheaths of cables are to be observed in the case of short-circuits. It is assumed for the analysis that the heat

Table 10.2 Data of materials for screening, armouring and sheaths of cables

Material	Specific caloric heat (J/K * mm^3)	Specific resistance (Ω * mm^2/m)	Temperature coefficient of resistance (K^{-1})
Copper	$3.48 * 10^{-3}$	$17.28 * 10^{-6}$	$3.8 * 10^{-3}$
Aluminium	$2.39 * 10^{-3}$	$28.6 * 10^{-6}$	$4.0 * 10^{-3}$
Lead	$1.45 * 10^{-3}$	$214 * 10^{-6}$	$4.35 * 10^{-3}$
Steel	$3.56 * 10^{-3}$	$143 * 10^{-6}$	$4.95 * 10^{-3}$

production within the cable during the short-circuit duration is an adiabatic process. The heat is dissipated to the surroundings only after the short-circuit is switched-off. The maximal permissible short-circuit current density J_{thz}, respectively the maximal permissible short-circuit current I_{thz}, for the given cross-section q_n is calculated by

$$J_{thz} = \sqrt{\frac{Q_c * (\beta + 20°C)}{\rho_{20}} * \ln\left(\frac{\delta_1 + \beta}{\delta_0 + \beta}\right)} * \frac{1}{\sqrt{T_{kr}}} \tag{10.7a}$$

$$I_{thz} = \sqrt{\frac{Q_c * (\beta + 20°C)}{\rho_{20}} * \ln\left(\frac{\delta_1 + \beta}{\delta_0 + \beta}\right)} * \frac{q_n}{\sqrt{T_{kr}}} \tag{10.7b}$$

where Q_c is the specific caloric heat, α_0 is the thermal coefficient of resistance, β is the parameter: $\beta = 1/\alpha_0 - 20°C$, δ_1 is the maximal permissible temperature at end of short-circuit, δ_0 is the maximal permissible temperature at beginning of short-circuit, ρ_{20} is the specific resistance at 20°C, q_n the nominal cross-section of screening or sheath and T_{kr} is the rated short-circuit duration.

The calculation for short-circuit duration different from the rated short-circuit duration is performed using

$$I_{th} \leq I_{thr} * \sqrt{\frac{T_{kr}}{T_k}} \quad \text{for } T_k \geq T_{kr} \tag{10.4b}$$

10.2.2 Mechanical short-circuit strength of rigid conductors

10.2.2.1 General

Currents in conductors induce electromagnetic forces into other conductors. The arrangement of parallel conductors, such as busbars and conductors of overhead lines, is of special interest as the electromagnetic forces will be maximal as compared with transversal arrangements. Three-phase and double-phase short-circuits without earth connection normally cause the highest forces. The currents inducing the electromagnetic forces are a function of time; therefore the forces are also a function of time. Electromagnetic forces lead to stresses in rigid conductors, to forces (bending, compression and tensile stress) on support structures and to tensile forces in slack conductors. Within this section, only stresses on rigid conductors are explained.

10.2.2.2 Electromagnetic forces

Figure 10.5 shows the arrangement of parallel conductors as can be found in busbar arrangements. In the case of a double-phase short-circuit without earth connection the forces on the conductors Y and B are

$$F_{k2} = \frac{\mu_0}{2\pi} * \frac{l}{a} * i_Y * i_B \tag{10.8}$$

where μ_0 is the permeability, a is the spacing of conductors, l is the length of conductors and i_Y; i_B are the peak values of short-circuit current in phases Y and B.

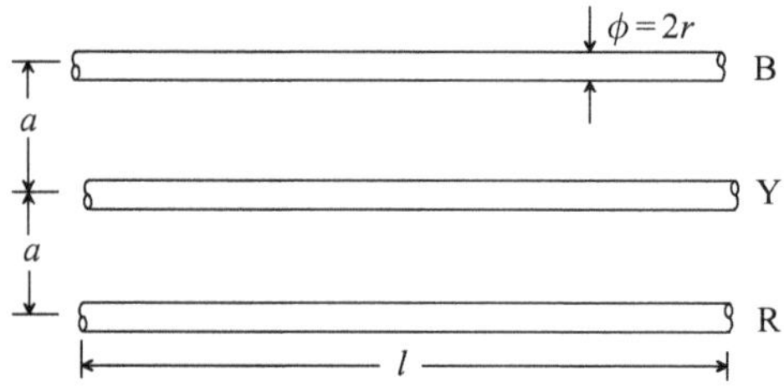

Figure 10.5 Arrangement of parallel conductors

Assuming that the distance between the support structures is large as compared with the spacing between the conductors [13], which on the other hand is assumed to be large as compared with the conductor radius r ($l/a > 10$; $a/r > 10$), typical in high-voltage installations, the Skin- and Proximity-effects can be neglected. The amplitude of the force F_{k2} is acting always in the same direction.

In the case of a three-phase short-circuit, the central conductor is exposed to the maximal force, as the magnetic fields caused by the outer conductors R and B are superimposed. The force F_{k3} on the central conductor acting in the opposite direction is

$$F_{k3} = \frac{\mu_0}{2\pi} * \frac{l}{a} * i_Y * (i_R - i_B) \tag{10.9}$$

Due to the decaying d.c. component, the maximal forces act immediately after the initiation of the short-circuit. With the peak short-circuit current i_{p2}, respectively i_{p3}, being the maximal value of the short-circuit current the maximal forces are

$$F_{k2\,max} = \frac{\mu_0}{2\pi} * \frac{l}{a_m} * i_{p2}^2 \tag{10.10}$$

$$F_{k3\,max} = \frac{\mu_0}{2\pi} * \frac{l}{a_m} * \frac{\sqrt{3}}{2} * i_{p3}^2 \tag{10.11}$$

In the case the conductors are arranged at the edges of an isosceles triangle, the calculation of the forces is identical to that mentioned above.

The assumptions made above for the spacing and lengths ($l/a > 10$; $a/r > 10$), are normally not fulfilled in low-voltage and medium-voltage installations. The Skin- and Proximity-effects cannot be neglected and the influence on the electromagnetic forces is considered by the effective distance a_m using a correction factor k_{12}

$$a_m = \frac{a}{k_{12}} \tag{10.12}$$

A suitable range of values of the correction factor k_{12} for conductors with rectangular cross-section is given in Figure 10.6. The electromagnetic force is increased in the case of flat conductor arrangement and reduced in the case of standing arrangement as compared with circular cross-section.

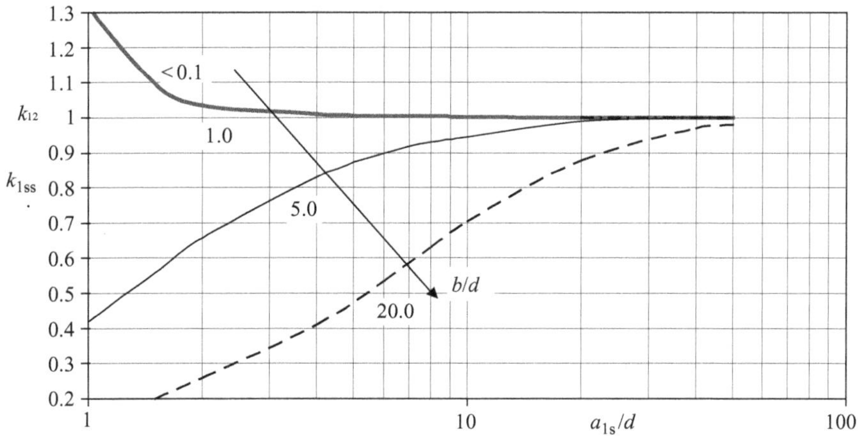

Figure 10.6 Correction factor k_{12} for the calculation of effective distance (according to Figure 1 of IEC 61660-2:1997)

If the main conductor is composed of several sub-conductors (number n), each sub-conductor carries only the nth part of total current and the maximal force $F_{s\,max}$ on the sub-conductors is

$$F_{s\,max} = \frac{\mu_0}{2\pi} * \frac{l_s}{a_s} * \left(\frac{i_p}{n}\right)^2 \tag{10.13}$$

where l_s is the length of sub-conductor, n is the number of sub-conductors, a_s is the spacing between sub-conductors and i_p is the peak short-circuit current.

The three-phase short-circuit in three-phase systems, respectively the two-phase short-circuit in two-phase systems, will cause the maximal force. The effective distance of the sub-conductors a_s in the case of circular cross-section is

$$\frac{1}{a_s} = \sum_{i=2}^{n} a_{1i} \tag{10.14a}$$

and for rectangular cross-section

$$\frac{1}{a_\mathrm{s}} = \sum_{i=2}^{n} \frac{k_{1\mathrm{i}}}{a_{1\mathrm{i}}} \tag{10.14b}$$

whereas the factor $k_{1\mathrm{i}}$ (named $k_{1\mathrm{ss}}$) can be taken from Figure 10.6.

10.2.2.3 Calculation of stresses in rigid conductors

The forces on the support structures of conductors and the stresses in the conductors themselves depend on the type of mechanical fixing and the elasticity. The mechanical system composed of conductor, fixing and supporting structure has a mechanical natural frequency, which can be actuated by the frequency of the current (50 Hz or 60 Hz), thus increasing the mechanical forces. As axial forces in rigid conductors can be neglected, the bending stress σ_m for the main conductor, respectively σ_s for the sub-conductors, are calculated using

$$\sigma_\mathrm{m} = V_\sigma * V_\mathrm{r} * \beta * \frac{F_\mathrm{m} * l}{8 * Z} \tag{10.15a}$$

$$\sigma_\mathrm{s} = V_{\sigma\mathrm{s}} * V_\mathrm{rs} * \frac{F_\mathrm{s} * l_\mathrm{s}}{8 * Z_\mathrm{s}} \tag{10.15b}$$

where l; l_s are the lengths of conductor, respectively sub-conductor, F_m; F_s are the forces on main conductor respectively on sub-conductor, V is the factor as explained below and Z_s is the section moduli depending on the shape of conductor. Whereas F_m respectively F_s is the force F_m3 in the case of three-phase short-circuits in three-phase systems. In two-phase systems the force F_m2 in the case of two-phase short-circuits has to be used. Typical ranges of values of the factors V_σ, V_r (respectively $V_{\sigma\mathrm{s}}$, V_rs) are shown in Figures 10.7 and 10.8. The factor β takes account of type and number of supports and can be obtained from Table 10.3.

Table 10.3 Factors α, β and γ for different arrangement of supports (according to Table 3 of IEC 60865-1:1993)

Type of bar and fixing of support		Factor α		Factor β	Factor γ
		Support A	Support B		
Single-span bar	Both simple	0.5	0.5	1	1.57
	A: fixed B: simple	0.625	0.375	0.73	2.45
	Both fixed	0.5	0.5	0.5	3.56
Multiple-span bar with equidistant simple supports	2 spans A:fixed B: simple	0.375	1.25	0.73	2.45
	3 and more bars A: fixed B: simple	0.4	1.1	0.73	3.56

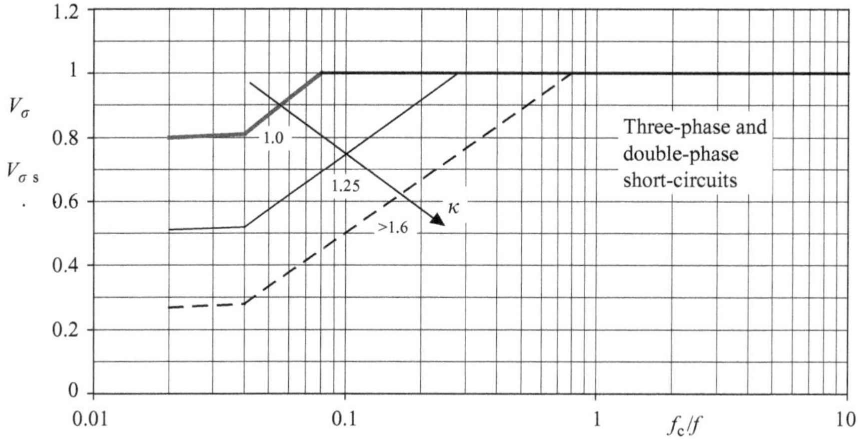

Figure 10.7 Factors V_σ and $V_{\sigma s}$ for the calculation of bending stress (according to Figure 4 of IEC 60865-1:1993)

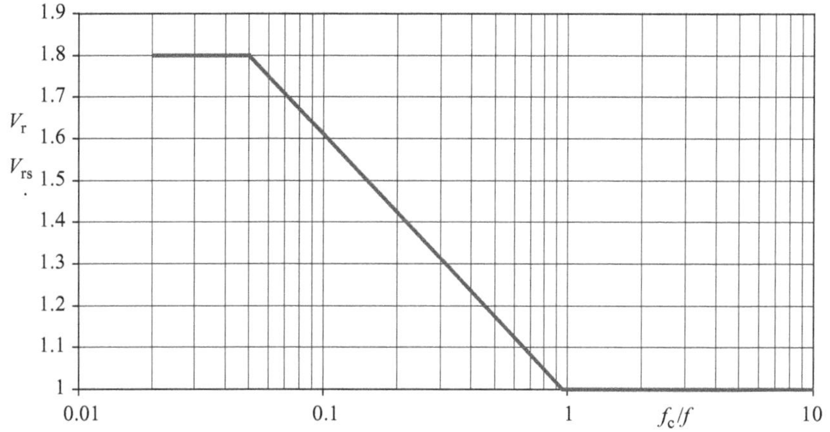

Figure 10.8 Factors V_r and V_{rs} for the calculation of bending stress (according to Figure 5 of IEC 60865-1:1993)

If the busbar consists of multiple bars supported in unequal distances, the maximal supporting distance is regarded. If the supporting distance is less than 20 per cent of the distance of neighbouring bars, the bars have to be coupled by joints. Joints between two supports are permitted, if the distance between the supporting points is less than 70 per cent of the supporting distances of neighbouring bars.

The values of section moduli Z, respectively Z_s, for typical arrangements of rectangular cross-sections with stiffening elements are outlined in Table 5 of IEC 60865-1:1993. Values for Z are between 0.867 d^2b and 3.48 d^2b, with b being the height of the conductor and d the thickness of the conductor, respectively the stiffening element. Furthermore, the factor of plasticity q has to be considered. Values of q are given for typical arrangements in Table 4 of IEC EN 60865-1:1993. Values for q are between 1.19 and 1.83 for rectangular cross-section and for U-, H- and I-shape profiles. In the case of circular ring section type conductors, the factor of plasticity depends on the diameter and the thickness of the wall. Reference is made to IEC 61865-1:1993.

The conductor has sufficient electromagnetic strength with respect to bending stress, if the bending stress value σ_m is below the product of factor of plasticity and the stress corresponding to the yield point.

$$\sigma_m \leq q * R_{p0.2} \tag{10.16}$$

When the distance between the conductors is affected significantly by the short-circuit, the value of the plasticity factor shall be set to $q = 1$. If the conductor is composed of sub-conductors the total bending stress, i.e., the sum of σ_m and σ_s as per above has to be considered. The short-circuit strength is given, if

$$\sigma_{tot} = \sigma_m + \sigma_s \leq q * R_{p0.2} \tag{10.17}$$

If only limit values of the stress corresponding to the yield point rather than readings are available, the minimal values should be used.

Quantities as per Equations (10.16) and (10.17) are

q	Factor of plasticity
$R_{p0.2}$	Stress corresponding to the yield point

More information on mechanical short-circuit stress on rigid conductors is included in [47].

10.2.2.4 Forces on supports

The relevant force to be considered for short-circuit strength is the dynamic force F_d to be calculated by

$$F_d = V_F * V_r * \alpha * F_m \tag{10.18}$$

The three-phase short-circuit in three-phase systems, respectively the two-phase short-circuit in two-phase systems, will cause the maximal force to be used for the force F_m. Typical values for the factors V_r and V_F are shown in Figures 10.8 and 10.9. The factor α depends on the type and the number of supports and can be obtained from Table 10.3.

The short-circuit strength of supports and fixing material is sufficient, if the dynamic force F_d is below the rated force F_{rB} as per manufacturer's data.

$$F_d \leq F_{rB} \tag{10.19}$$

Standards for the short-circuit stress on foundations are actually under discussion.

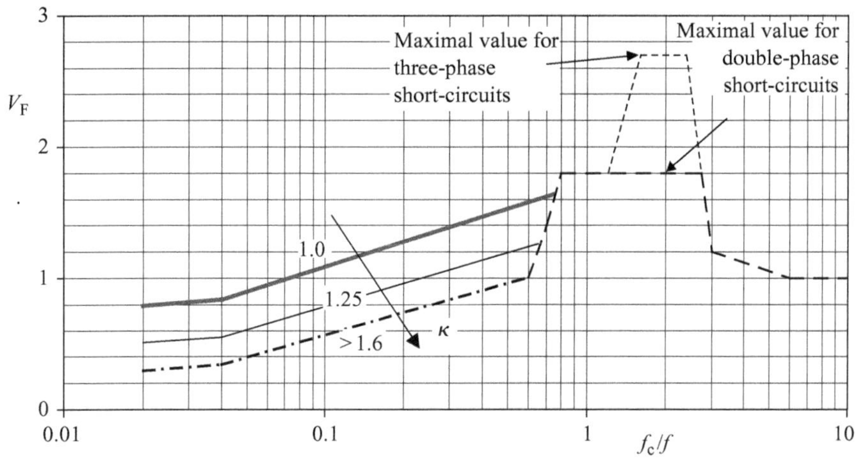

Figure 10.9 *Factor V_F for the calculation of bending stress (according to Figure 4 of IEC 60865-1:1993)*

10.2.2.5 Influence of conductor oscillation

In Sections 10.2.2.3 and 10.2.2.4, factors are explained which take account of the function of time of the bending stress and forces. Deviations from values for V_σ, V_r, $V_{\sigma s}$, V_{rs} and V_F are permitted, if the mechanical natural frequency of the arrangement is taken into account. It should be noted in this respect that the required data are difficult to obtain.

The mechanical natural frequency f_c of a conductor, either main or sub-conductor is calculated by

$$f_c = \frac{\gamma}{l^2} * \sqrt{\frac{E * J}{m'}} \tag{10.20}$$

The mechanical natural frequency of main conductors, composed of several sub-conductors with rectangular cross-sections, can be calculated by

$$f_c = c * \frac{\gamma}{l^2} * \sqrt{\frac{E * J_s}{m'_s}} \tag{10.21a}$$

and for sub-conductors by

$$f_{cs} = \frac{3.56}{l^2} * \sqrt{\frac{E * J_s}{m'_s}} \tag{10.21b}$$

Quantities as per Equations (10.20) and (10.21) are:

γ	Factor taking account of type and number of supports (Table 10.3)
E	Young's modulus
J; J_s	Second mechanical moment of the conductor, respectively sub-conductor
m'; m'_s	Specific mass (mass per length), respectively sub-conductor
l; l_s	Conductor length, respectively sub-conductor length
c	Factor as per Equation (10.21c) taking account of stiffening elements (Figure 10.10).

$$c = \frac{c_c}{\sqrt{1 + \xi_m (m_z/(n * l * m'_s))}} \tag{10.21c}$$

where c_c is the factor as per Table 10.4, ξ_m is the factor as per Table 10.4, m_z the specific mass of stiffening elements and n is the number of sub-conductors.

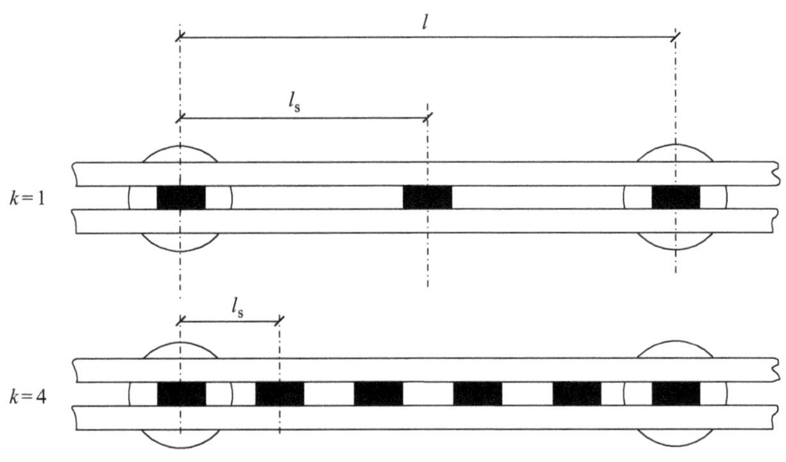

Figure 10.10 Calculation of mechanical natural frequency (Factor c). Arrangement of distance elements and calculation equation (according to Figure 3 of IEC 60865-1:1993)

Number k and type of distance of stiffening elements, the ratio of the mass of stiffening (distance) elements m_z to the specific mass of the conductor $m'_s l$ separately for the two swing directions in parallel or rectangular to the side with largest cross-section and the second moments J and J_s of the conductor, respectively the sub-conductor area, are duly considered. The Young's modulus E and the specific mass m' depend on the construction and on the type of material. The remarks as per Annex A.3 of IEC 60865:1993 shall be observed.

The bending stress and the dynamic force on supports are calculated taking account of the mechanical natural frequency f_c and the factors V_σ, V_r , $V_{\sigma s}$, V_{rs} and V_F

Table 10.4 Factors for the calculation of mechanical natural frequency. Swing is at right angle to the area of sub-conductor

Stiffening or distance elements			l_s/l	ξ_m
Number k	Stiffening element c_c	Distance element c_c		
0	1.0	1.0	—	0.0
1	1.0	1.0	0.5	2.5
2	1.48	1.0	0.33	3.0
3	1.75	1.0	0.25	4.0
4	2.14	1.0	0.2	5.0

obtained from Figures 10.7 to 10.9. Parameters are the factor κ as per Chapter 4 and the ratio of mechanical natural frequency f_c to power frequency f (50 or 60 Hz). Calculation equations are included in the Annex of IEC 60865-1:1993.

More information on the electromechanical effects and conductor oscillations can be found in [46,49,51,53,55].

10.3 d.c. auxiliary installations

10.3.1 Substitute rectangular function

The calculation of thermal and electromagnetic effects of short-circuit currents in d.c. auxiliary installations is carried out similar to the analysis in a.c. systems. In general, two alternatives for the calculation are possible. The simplified approach (first alternative) is based on the standard approximation function [41] of the short-circuit current

$$i_1(t) = i_p * \frac{1 - e^{-t/\tau_1}}{1 - e^{-t_p/\tau_1}} \quad \text{for } 0 \le t \le t_p \tag{10.22a}$$

$$i_2(t) = i_p \left(\left(1 - \frac{I_k}{i_p} \right) * e^{-(t-t_p)/\tau_2} + \frac{I_k}{i_p} \right) \quad \text{for } t_p \le t \le T_k \tag{10.22b}$$

The electromagnetic effects are calculated using the peak short-circuit current i_p of the standard approximation function. The second alternative is based on the substitute rectangular function, which achieves the same effects as the standard approximation function [40,41]. Quantities as per Equations (10.22) are explained in Figure 10.11. The substitute rectangular function is defined by I_R^2 and t_R according to Figure 10.11

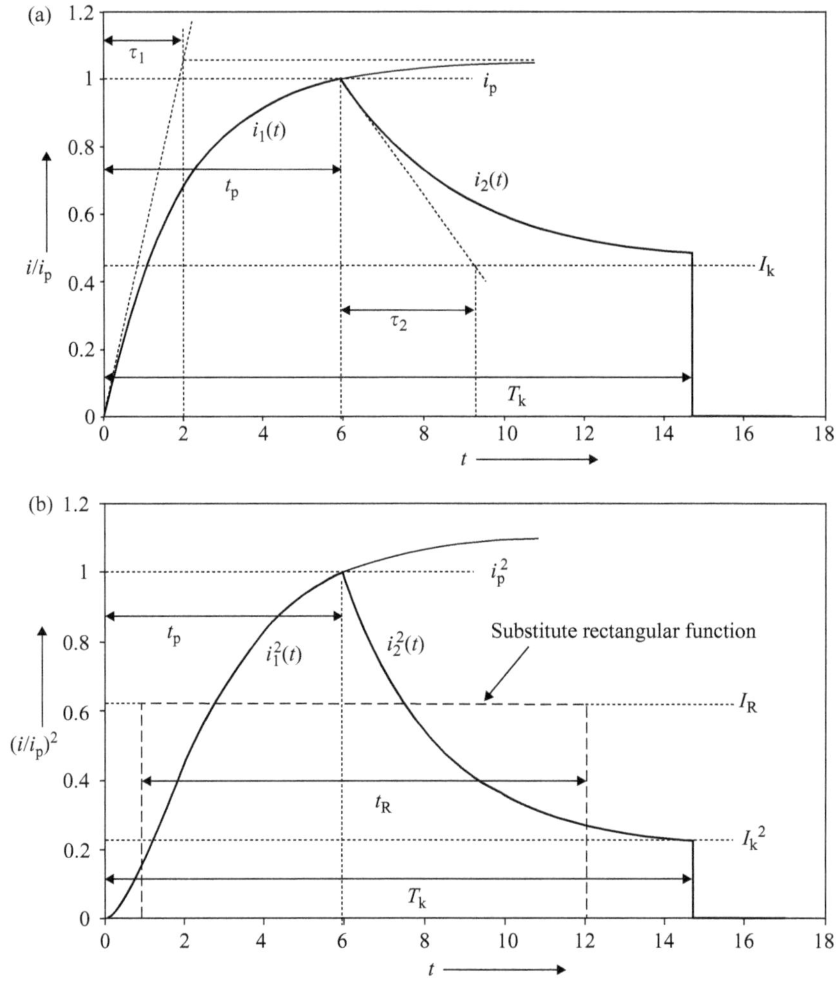

Figure 10.11 Standard approximation function (a) and substitute rectangular function (b) (according to Figure 4 of IEC 60660-2:1997). Not to scale

and calculated by

$$I_R^2 = 0.2887 * \sqrt{\frac{A_i^3}{I_g}} \tag{10.23a}$$

$$t_R = 3.464 * \sqrt{\frac{I_g}{A_i}} \tag{10.23b}$$

whereas A_i and I_g are calculated by factors $m_{\theta 1}$, $m_{\theta 2}$, m_{g1}, m_{g2}, m_{Ig1} and m_{Ig2}, depending on the peak short-circuit current i_p, the short-circuit duration T_k and the time to peak t_p

$$A_i = i_p^2 * (t_p * m_{\theta 1} + (T_k - t_p) * m_{\theta 2}) \tag{10.24a}$$

$$I_g = i_p^2 \left[\frac{t_p^3}{12} m_{Ig1} + \frac{(T_k - t_p)^3}{12} m_{Ig2} + m_{\theta 1} * t_p * (t_g - m_{g1} * t_p)^2 \right.$$
$$\left. + m_{\theta 2} * (T_k - t_p) * (t_p + m_{g2} * (T_k - t_p) - t_g)^2 \right] \tag{10.24b}$$

where

$$t_g = \frac{m_{\theta 1} * m_{g1} * t_p^2 + m_{\theta 2} * (T_k - t_p) * (t_p + m_{g2} * (T_k - t_p))}{m_{\theta 1} * t_p + m_{\theta 2} * (T_k - t_p)} \tag{10.24c}$$

or can be taken from diagrams in IEC 60865, which include calculation equations for the factors $m_{\theta 1}, m_{\theta 2}, m_{g1}, m_{g2}, m_{Ig1}$ and m_{Ig2}.

The mechanical natural frequency of either main or sub-conductors are to be taken into account for the calculation of the substitute rectangular function. The mechanical natural frequency is calculated by

$$f_c = \frac{\gamma}{l^2} * \sqrt{\frac{E * J}{m'}} \tag{10.25a}$$

The factor γ takes account of the type and number of supports and can be obtained from Table 10.3. For main conductors composed of several sub-conductors, the mechanical natural frequency is to be calculated by

$$f_c = c * \frac{\gamma}{l^2} * \sqrt{\frac{E * J_s}{m_s'}} \tag{10.25b}$$

For the calculation of the bending stress of sub-conductors, the mechanical natural frequency [43,44] is calculated by

$$f_{cs} = \frac{3.56}{l^2} * \sqrt{\frac{E * J_s}{m_s'}} \tag{10.25c}$$

Quantities as per Equations (10.25) are:

γ	Factor taking account of the type and number of supports (Table 10.3)
E	Young's modulus
$J; J_s$	Second mechanical moment of the conductor, respectively sub-conductor
$m'; m_s'$	Specific mass (mass per length), respectively sub-conductor
$l; l_s$	Conductor length, respectively sub-conductor length
c	Factor taking account of stiffening elements (Equation 10.21c).

The vibration period T_{me} of the main conductor, respectively T_{mes} of the sub-conductor is calculated by

$$T_{me} = \frac{1}{f_c} \tag{10.26a}$$

$$T_{mes} = \frac{1}{f_{cs}} \tag{10.26b}$$

where f_c; f_{cs} are the natural mechanical frequency of the conductor, respectively sub-conductors.

The parameters t_R and I_R^2 of the substitute rectangular function are calculated for the short-circuit duration T_k, in the case $T_k \leq 0.5T_{me}$. In the case $T_k > 0.5T_{me}$ the parameters are calculated for the equivalent short-circuit duration being $T_{ke} = \text{MAX}\{0.5T_{me}; 1.5t_p\}$. In the case $T_k = t_p$, i.e., the short-circuit current has a decreasing function only, the substitute rectangular function is calculated, independently from the vibration period T_{me} for the total short-circuit duration T_k.

The calculation of the force for the main conductor is done by

$$F_{R\,max} = \frac{\mu_0}{2\pi} * \frac{l_s}{a_m} * I_R^2 \tag{10.27a}$$

respectively, in the case of sub-conductors (number n) by

$$F_{Rs\,max} = \frac{\mu_0}{2\pi} * \frac{l_s}{a_s} * \left(\frac{I_{Rs}}{n}\right)^2 \tag{10.27b}$$

where I_R; I_{Rs} are the current of the substitute rectangular function, l; l_s is the length of conductor, respectively sub-conductor, a_m; a_s is the effective distance between conductor, respectively sub-conductor, and n is the number of sub-conductors.

10.3.2 Mechanical short-circuit strength of rigid conductors

10.3.2.1 Forces

The calculation of electromagnetic effects on rigid conductors is based on the substitute rectangular function as per Figure 10.11 and described in Section 10.3.1. The substitute rectangular function leads to identical bending stress and forces as the standard approximation function. The forces between main conductors are calculated by

$$F_R = \frac{\mu_0}{2\pi} * \frac{l}{a_m} * I_R^2 \tag{10.28a}$$

and between sub-conductors (number n) by using

$$F_{Rs} = \frac{\mu_0}{2\pi} * \frac{l_s}{a_s} * \left(\frac{I_{Rs}}{n}\right)^2 \tag{10.28b}$$

Quantities as per Equations (10.28) are:

I_R; I_{Rs}	Current of the substitute rectangular function
l; l_s	Length of conductor, respectively sub-conductor
a_m; a_s	Effective distance between conductor, respectively sub-conductor
n	Number of sub-conductors

The bending stress and forces on supports for both main and sub-conductors are calculated similar to a.c. installations as described in Sections 10.2.2.2 to 10.2.2.4.

The effective distance of main conductors a_m is calculated from the distance a by

$$a_m = \frac{a}{k_{12}} \tag{10.29}$$

with the correction factor k_{12} according to Figure 10.6.

The effective distance of sub-conductors a_s in the case of circular conductors is calculated by

$$\frac{1}{a_s} = \sum_{i=2}^{n} a_{1i} \tag{10.30a}$$

respectively, in the case of rectangular cross-sections

$$\frac{1}{a_s} = \sum_{i=2}^{n} \frac{k_{1i}}{a_{1i}} \tag{10.30b}$$

with the factor k_{1i} according to Figure 10.6.

10.3.2.2 Bending stress

The bending stress on main conductors σ_m, respectively on sub-conductors σ_s, resulting from the bending forces is calculated using

$$\sigma_m = V_\sigma * \beta * \frac{F_R * l}{8 * Z} \tag{10.31a}$$

$$\sigma_s = V_{\sigma s} * \frac{F_{Rs} * l_s}{16 * Z_s} \tag{10.31b}$$

where Z; Z_s are the section moduli of the main, respectively sub-conductors, V is the factor as explained below, l; l_s is the length of conductor, respectively sub-conductor, and β is the factor taking account of the type and number of supports (Table 10.3).

The forces F_R and F_{Rs} are the forces as calculated for the substitute rectangular function. Typical values of V_σ and $V_{\sigma s}$ are outlined in Figure 10.12. IEC 61660-2:1997 indicates in Table 2 that V_σ and $V_{\sigma s}$ should be lower than 1.

If the busbar consists of multiple bars supported in unequal distances, the maximal supporting distance is to be considered. If a supporting distance is less than 20 per cent of the distance of neighbouring bars, the bars have to be coupled by joints. Joints between two supports are permitted, if the distance between the supporting points is less than 70 per cent of the supporting distances of neighbouring bars.

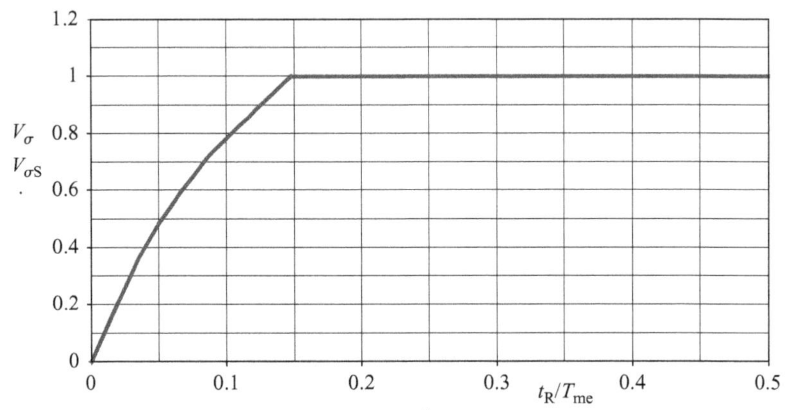

Figure 10.12 Factors V_σ and $V_{\sigma s}$ for the calculation of bending stress on conductors (according to Figure 9 of IEC 61660-2:1997)

The conductor has sufficient electromagnetic strength with respect to bending stress, if the bending stress σ_m is below the product of factor of plasticity and stress corresponding to the yield point

$$\sigma_m \leq q * R_{p0.2} \tag{10.32a}$$

In the case of several sub-conductors the total bending stress, i.e., the sum of the stresses σ_m and σ_s is to be considered

$$\sigma_{tot} = \sigma_m + \sigma_s \leq q * R_{p0.2} \tag{10.32b}$$

where q is the factor of plasticity and $R_{p0.2}$ is the stress corresponding to the yield point.

Values for the factor of plasticity q for typical arrangements are included in Table 4 of IEC 60660-2:1997. Values are between $q = 1.19$–1.83 for rectangular cross-sections, U-, H- and I-shape profiles. For circular ring section type conductors the factor of plasticity depends on the diameter and wall thickness. Reference is made to IEC 61660-2:1997.

10.3.2.3 Forces on supports

The dynamical force F_d is the dominating parameter for the short-circuit strength of supports to be calculated by

$$F_d = V_F * \alpha * F_R \tag{10.33}$$

whereas the force F_R is the force calculated on the basis of the substitute rectangular function. Typical values of V_F are outlined in Figure 10.13. The factor α takes account of the type and number of supports and can be obtained from Table 10.3.

The requirements regarding the short-circuit strength of supports and fixing material are fulfilled if the dynamical force F_d remains below the rated value F_{rB}

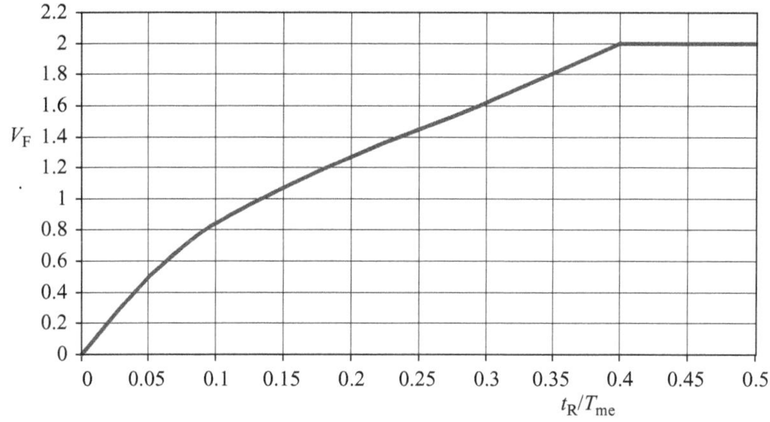

Figure 10.13 Factor V_F for the calculation of forces on supports (according to Figure 9 of IEC 61660-2:1997)

as per manufacturer's data

$$F_d \leq F_{rB} \tag{10.34}$$

Standardisation of stresses on foundations is actually under consideration. More information on the theoretical background can be found in [43–45,48].

10.3.3 Thermal short-circuit strength

The thermal short-circuit strength is analysed using the thermal equivalent short-time current I_{th} and the short-circuit duration T_k.

$$I_{th} = \sqrt{\frac{A_i}{T_k}} \tag{10.35}$$

with A_i as per Equation (10.24a). The upper limit of the thermal equivalent short-time current is defined as $I_{th} = i_p$. IEC 61660-2 indicates that separate considerations for equipment (e.g., transformers, transducers) and conductors (e.g., busbars, cables) have to be taken.

The thermal short-circuit strength of equipment is fulfilled if the thermal equivalent short-time current I_{th} is below the rated thermal equivalent short-time current I_{thr}

$$I_{th} \leq I_{thr} \quad \text{for } T_k \leq T_{kr} \tag{10.36a}$$

with the rated short-circuit duration T_{kr} assumed equal to 1 s. If the short-circuit duration T_k is longer than the rated short-circuit duration T_{kr}, the thermal short-circuit strength is given if

$$I_{th} \leq I_{thr} * \sqrt{\frac{T_{kr}}{T_k}} \quad \text{for } T_k \geq T_{kr} \tag{10.36b}$$

Permissible values for conductor material of cables are outlined in Table 10.1 and in Figure 10.3.

In order to analyse the short-circuit strength of bare conductors, the thermal equivalent short-time current density J_{th} is calculated using the thermal equivalent current and the cross-section q_n

$$J_{th} = \frac{I_{th}}{q_n} \tag{10.37}$$

where I_{th} is the thermal equivalent current and q_n the cross-section of the conductor. The steel part of Al/St-conductors is not taken into account for the cross-section.

Conductors have sufficient thermal strength if the thermal equivalent short-time density J_{th} is below the rated thermal equivalent short-time density J_{thr}, whereas short-circuit durations different from the rated short-circuit duration have to be considered.

$$J_{th} \leq J_{thr} * \sqrt{\frac{T_{kr}}{T_k}} \tag{10.38}$$

Data on the rated short-time current density of conductor material are given in Table 10.1 and Figure 10.3. In non-insulated conductors, i.e., bare conductors and busbars, the thermal equivalent short-time current density is allowed to be above the rated short-time current density in the case $T_k < T_{kr}$.

10.4 Calculation examples (a.c. system)

10.4.1 Calculation of thermal effects

The thermal short-circuit strength of a cable N2XS2Y 240 6/10 kV is to be analysed. Figure 10.14 indicates the equivalent circuit diagram of a power system. In the case of near-to-generator short-circuit (three-phase) the maximal short-circuit current at the busbars Q, respectively A, and the thermal equivalent short-circuit current I_{th} for the short-circuit duration 0.1 s; 1.2 s; 2 s and 4 s are to be calculated as well as the

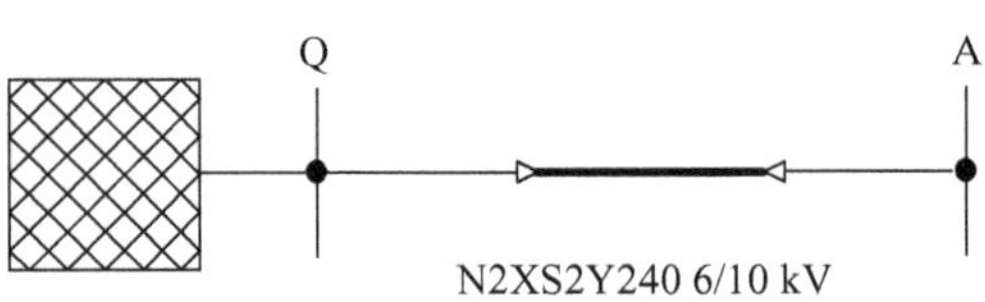

Figure 10.14 Equivalent circuit diagram, data of equipment, resistance at 20°C

short-circuit current density J_{thz} permitted for a cable sheath of copper (initial temperature 80°C; temperature at the end of short-circuit 350°C; short-circuit duration 0.2 s; 1.2 s; 2 s).

Data of equipment are given below:

$$U_{nQ} = 10\ \text{kV};\ S''_{kQ} = 520\ \text{MVA}$$

$$R'_K = 0.0754\ \Omega/\text{km};\ X'_K = 0.11\ \Omega/\text{km};\ l = 2\ \text{km}$$

The maximal short-circuit current is given for short-circuits at the sending-end of the cable (approximately identical to location Q) $I''_{k3Q} = 30.03$ kA. The maximal short-circuit current at location A is $I''_{k3A} = 13.28$ kA. Table 10.1 indicates a value of $J_{thr} = 143$ A/mm^2. The thermal equivalent rated current is $I_{thr} = 34.32$ kA. The calculation results are outlined in Table 10.5.

Table 10.5 Results of calculation of thermal equivalent currents

t_k (s)	I_{thz} (kA)	I_{thQ} (kA)	I_{thA} (kA)	J_{thz} (A/mm^2)
0.2	76.26	14.12	31.92	689
1.2	31.33	11.73	26.52	285
2	24.27	9.62	21.86	221
4	17.16	8.81	19.92	—

The short-circuit strength is sufficient if $I_{th} \leq I_{thz}$. This is fulfilled if $t_k < 2$ s for short-circuits near busbar A and $t_k < 4$ s for short-circuits near busbar Q.

10.4.2 Electromagnetic effect

A wind power plant is connected to the public supply system by four cables in parallel laid on racks as indicated in Figure 10.15. The relevant parameters of the short-circuit current and the maximal forces on the cables, respectively on the fixing material, are to be calculated. The permissible distance of the fixing clamps is to be determined. The maximal permissible force on the clamps is $F_{zul} = 40$ kN.

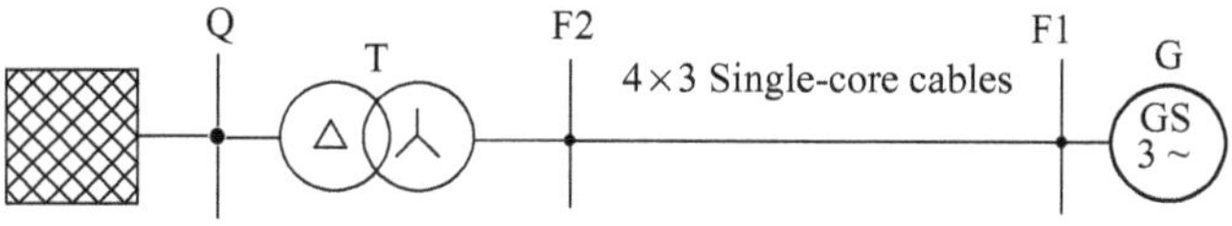

Figure 10.15 Equivalent circuit diagram of a power system with wind power plant

The data of equipment are given below:

$$S''_{kQ} = 1000\ \text{MVA};\ U_{nQ} = 110\ \text{kV}$$

$$S_{rT} = 2.5\ \text{MVA};\ u_{krT} = 6\%;\ u_{RrT} = 0.8\%;\ t_{rT} = 20\ \text{kV}/0.66\ \text{kV}$$

$S_{rG} = 2.5$ MVA; $U_{rG} = 0.66$ kV; $\cos\varphi_{rG} = 0.85$; $x''_{dG} = 18$ per cent

Each cable: $X_L = 9.84\ \text{m}\Omega$; $R_L = 10.82\ \text{m}\Omega$

Short-circuits are fed from the generator as well as from the power system feeder. Short-circuit currents are given in Table 10.6.

Table 10.6 Results of short-circuit current calculation

Parameter	Short-circuit location	
	F1 (kA)	F2 (kA)
Generator	$I''_{k3} = 13.3$	$I''_{k3} = 12.17$
	$i_p = 32.05$	$i_p = 27.07$
Power system feeder	$I''_{k3} = 34.48$	$I''_{k3} = 36.43$
	$i_p = 65.59$	$i_p = 86.66$

The maximal short-circuit current (peak short-circuit current) $i_p = 86.66$ kA is given for a short-circuit at location F2 and is taken as the basis for the analysis. The force on the fixing material is $F'_s = 51.79$ N/m. The distance of the fixing clamps shall be below $d \leq 0.77$ m.

10.5 Calculation examples (d.c. system)

10.5.1 Thermal effect

The thermal effect of the short-circuit current is to be calculated. The conductor of the main busbar has a cross section $q_n = 2 \times 400\ \text{mm}^2 (d \times b = 10\ \text{mm} \times 40\ \text{mm}$; Cu), $\rho = 0.0173\ \Omega\text{mm}^2/\text{m}$, see Figure 10.16. The short-circuit duration is $T_k = 100$ ms. Temperature at beginning of short-circuit is $\delta_0 = 20°\text{C}$ and at the end $\delta_1 = 250°\text{C}$.

The short-circuit parameters are summarised below:

Peak short-circuit current	$i_p = 50.5$ kA
Quasi steady-state short-circuit current	$I_k = 46.6$ kA
Time-to-peak	$t_p = 12.1$ ms
Rise-time constant	$\tau_1 = 4.03$ ms
Decay-time constant	$\tau_2 = 17.3$ ms

The time course of the short-circuit current is given by

$$i_1(t) = 47.99\ \text{kA} * (1 - \text{e}^{-t/4.03}) \qquad \text{for } 0 \leq t \leq 12.1\ \text{ms}$$
$$i_2(t) = 50.5\ \text{kA} * (0.077 * \text{e}^{-(t-12.1)/17.3} + 0.923) \qquad \text{for } 12.1\ \text{ms} \leq t \leq T_k$$

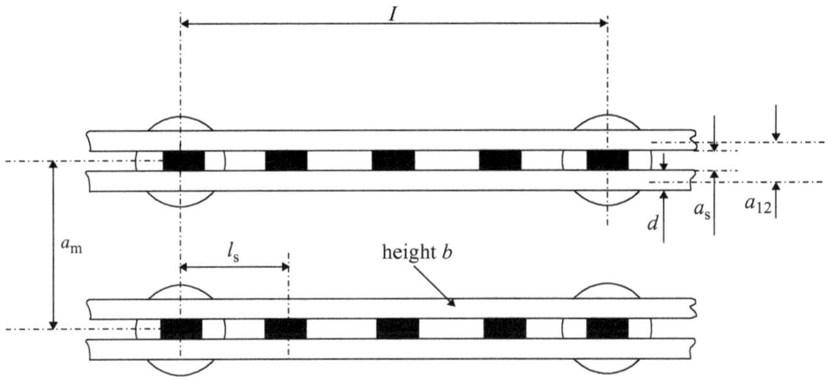

Figure 10.16 Arrangement of busbar conductor (data, see text)

The thermal equivalent short-time current I_{th} as per Equation (10.35)

$$I_{th} = \sqrt{\frac{A_i}{T_k}} = \sqrt{\frac{206.11\ kA^2}{0.1\ s}} = 45.4\ kA$$

with A_i as per Equation (10.24a)

$$A_i = i_p^2 * (t_p * m_{\theta 1} + (T_k - t_p) * m_{\theta 2}) = 50.5^2\ kA^2$$
$$* (0.0121\ s * 0.65 + (0.1\ s - 0.0121\ s) * 0.83) = 206.11\ kA^2$$

Factors $m_{\theta 1}$ and $m_{\theta 2}$ are calculated as per IEC 61660-2

$$m_{\theta 1} = 0.65$$
$$m_{\theta 2} = 0.83$$

The thermal equivalent short-time current density J_{th} according to Equation (10.37) is calculated as

$$J_{th} = \frac{I_{th}}{q_n} = \frac{45(4)\ kA}{2(400)\ mm^2} = 56.75\ A/mm^2$$

Conductors have sufficient thermal strength if the thermal equivalent short-time density J_{th} is below the rated thermal equivalent short-time density J_{thr} taken from Figure 10.3 ($J_{thr} = 190\ A/mm^2$).

$$J_{th} \leq J_{thr} * \sqrt{\frac{T_{kr}}{T_k}}$$

$$56.75\ A/mm^2 \leq 190\ A/mm^2 * \sqrt{\frac{1\ s}{0.1\ s}} = 600\ A/mm^2$$

The busbar-conductor has sufficient thermal short-circuit strength.

10.5.2 Electromagnetic effect

The configuration as per Sections 9.7 and 10.5.1 is regarded. The conductor arrangement is outlined in Figure 10.16. The distances are: $a_m = 75$ mm; $b = 40$ mm; $a_s = 10$ mm; $l = 1050$ mm; $l_s = 35$ mm. The mechanical constants of the busbar are:

Stress corresponding to the yield point	$R_{p0.2} = 340\ \mathrm{N/mm^2}$
Specific mass of subconductor	$m'_s = 3.5\ \mathrm{kg/m}$
Young's modulus	$E = 106\ \mathrm{kN/mm^2}$

Stiffening elements have dimension 40/40/10 mm.

10.5.2.1 Calculation of forces with simplified approach

Peak force between main conductors

$$F_{Rm} = \frac{\mu_0}{2\pi} * \frac{l}{a_m} i_p^2 = \frac{4\pi * 10^{-7} * (\mathrm{Vs/Am})}{2\pi} * \frac{1.05\ \mathrm{m}}{0.0765\ \mathrm{m}} * 50.5^2 * \mathrm{kA}^2$$
$$= 7000.7\ \mathrm{N}$$

Effective distance of main conductors a_m according to Equation (10.29)

$$a_m = \frac{a}{k_{12}} = \frac{0.075\ \mathrm{m}}{0.98} = 0.0765\ \mathrm{m}$$

with the correction factor $k_{12} = 0.98$ according to Figure 10.6.

Peak force between sub-conductors

$$F_{Rs} = \frac{\mu_0}{2\pi} * \frac{l_s}{a_s} \left(\frac{i_p}{n}\right)^2 = \frac{4\pi * 10^{-7} * (\mathrm{Vs/Am})}{2\pi} * \frac{0.35\ \mathrm{m}}{0.028\ \mathrm{m}} * \left(\frac{50.5 * \mathrm{kA}}{2}\right)^2$$
$$= 1593.9\ \mathrm{N}$$

Effective distance of sub-conductors a_s

$$a_s = \frac{a_{12}}{k_{12}} = \frac{0.02\ \mathrm{m}}{0.72} = 0.028\ \mathrm{m}$$

with the correction factor $k_{12} = 0.72$ according to Figure 10.6.

10.5.2.2 Calculation of forces with substitute rectangular function

Parameters of short-circuit current remain identical to those mentioned above.

$$i_p = 50.5\ \mathrm{kA}; \qquad t_p = 12.1\ \mathrm{ms}; \qquad T_k = 100\ \mathrm{ms}$$

Mechanical natural frequency of main conductor as per Equation (10.25)

$$f_c = c * \frac{\gamma}{l^2} * \sqrt{\frac{E * J_s}{m'_s}}$$

with $c \approx 1.44$ as per Figure 10.10, factor γ as per Table 10.3 and the second mechanical moment J of the conductor, respectively sub-conductor

$$J = J_s = \frac{d^3 * b}{12} = \frac{0.01^3 * 0.04}{12}\,\text{m}^4 = 3.33 * 10^{-9}\,\text{m}^4$$

$$f_c = 1.44 * \frac{3.56}{1.05^2\,\text{m}^2} * \sqrt{\frac{106\,\text{kN/mm}^2 * 3.33 * 10^{-9}\,\text{m}^4}{3.5\,\text{kg/m}}} = 46.7\,\text{Hz}$$

Vibration period T_{me} as per Equation (10.26)

$$T_{me} = \frac{1}{f_c} = \frac{1}{46.7}\,\text{s} = 21.4\,\text{ms}$$

Mechanical natural frequency of sub-conductors

$$f_{cs} = \frac{3.56}{l_s^2} * \sqrt{\frac{E * J_s}{m_s'}} = \frac{3.56}{0.35^2\,\text{m}^2} * \sqrt{\frac{106\,\text{kN/mm}^2 * 3.33 * 10^{-9}\,\text{m}^4}{3.5\,\text{kg/m}}}$$
$$= 291.8\,\text{Hz}$$

Vibration period T_{mes}

$$T_{mes} = \frac{1}{f_c} = \frac{1}{291.8}\,\text{s} = 3.43\,\text{ms}$$

Substitute rectangular function to be calculated for the equivalent short-circuit duration T_{ke}, as $T_k > 0.5 T_{me}$.

$$T_{ke} = \text{MAX}\{0.5 * T_{me};\ 1.5 * t_p\} = \text{MAX}\{0.5 * 21.4\,\text{ms};\ 1.5 * 12.1\,\text{ms}\}$$
$$= 18.2\,\text{ms}$$

Parameters of the substitute rectangular function (in this case identical for the main conductor and for the sub-conductors) as per Equations (10.23) and (10.24)

$$I_R^2 = 0.2887 * \sqrt{\frac{A_i^3}{I_g}}$$

$$t_R = 3.464 * \sqrt{\frac{I_g}{A_i}}$$

$$I_g = i_p^2 \left[\frac{t_p^3}{12} m_{Ig1} + \frac{(T_k - t_p)^3}{12} m_{Ig2} + m_{\theta 1} * t_p * (t_g - m_{g1} * t_p)^2 \right.$$
$$\left. + m_{\theta 2} * (T_k - t_p) * (t_p + m_{g2} * (T_k - t_p) - t_g)^2 \right]$$

$$A_i = i_p^2 * (t_p * m_{\theta 1} + (T_k - t_p) * m_{\theta 2})$$

$$t_g = \frac{m_{\theta 1} * m_{g1} * t_p^2 + m_{\theta 2} * (T_k - t_p) * (t_p + m_{g2} * (T_k - t_p))}{m_{\theta 1} * t_p + m_{\theta 2} * (T_k - t_p)}$$

Factors $m_{\theta 1}, m_{\theta 2}, m_{g1}, m_{g2}, m_{Ig1}$ and m_{Ig2} are calculated according to IEC 61660-2

$$m_{\theta 1} = 0.65; \qquad m_{\theta 2} = 0.83$$
$$m_{g1} = 0.63; \qquad m_{g2} = 0.47$$
$$m_{Ig1} = 0.42; \qquad m_{Ig2} = 0.86$$

$$A_i = (50.5\ \text{kA})^2 * (0.0121\ \text{s} * 0.65 + (0.0182 - 0.0121\ \text{s}) * 0.83)$$
$$= 32.97\ \text{kA}^2\text{s}$$

$$t_g = [0.65 * 0.63 * (0.0121\ \text{s})^2 + 0.83 * (0.0182 - 0.0121\ \text{s}) * (0.0121 + 0.47 * (0.0182 - 0.0121\ \text{s}))] * [0.65 * 0.0121\ \text{s} + 0.83 * (0.0182 - 0.0121\ \text{s})]^{-1}$$

$$t_g = 0.0105\ \text{s}$$

$$I_g = (50.5\ \text{kA})\left[\frac{t_p^3}{12} m_{Ig1} + \frac{(T_k - t_p)^3}{12} m_{Ig2} + m_{\theta 1} * t_p * (t_g - m_{g1} * t_p)^2 + m_{\theta 2} * (T_k - t_p) * (t_p + m_{g2} * (T_k - t_p) - t_g)^2\right]$$

$$I_g = (50.5\ \text{kA})^2 \left[\begin{array}{l} \frac{(0.0121\ \text{s})^3}{12} * 0.42 + \frac{(0.0182 - 0.0121\ \text{s})^3}{12} * 0.86 + \cdots \\ \cdots 0.65 * 0.0121\ \text{s} * (0.0105\ \text{s} - 0.63 * 0.0121\ \text{s})^3 + \cdots \\ \cdots 0.83 * (0.0182 - 0.0121\ \text{s}) * (0.0121\ \text{s} + 0.47 \\ \quad *(0.0182 - 0.0121\ \text{s}) - 0.0105\ \text{s})^2 \end{array}\right]$$

$$I_g = 457.2\ \text{A}^2\text{s}^3$$

The parameters of the substitute rectangular function

$$I_R^2 = 0.2887 * \sqrt{\frac{(32.97 * 10^6\ \text{A}^2\text{s})^3}{457.2\ \text{A}^2\text{s}^3}} = 2556 * 10^6\ \text{A}^2$$

$$t_R = 3.464 * \sqrt{\frac{457.2\ \text{A}^2\text{s}^3}{32.97 * 10^6\ \text{A}^2\text{s}}} = 12.9\ \text{ms}$$

The standardised rectangular function and the approximated total short-circuit current (see e.g., Section 9.7) are outlined in Figure 10.17.

Peak force between main conductors as per Equation (10.27)

$$F_{Rm} = \frac{\mu_0}{2\pi} * \frac{l}{a_m} * I_R^2 = \frac{4\pi * 10^{-7} * (\text{Vs/Am})}{2\pi} * \frac{1.05\ \text{m}}{0.0765\ \text{m}} * 2556 * 10^6\ \text{A}^2$$
$$= 7016.5\ \text{N}$$

Effective distance of main conductors $a_m = 0.0765$ m as per above.

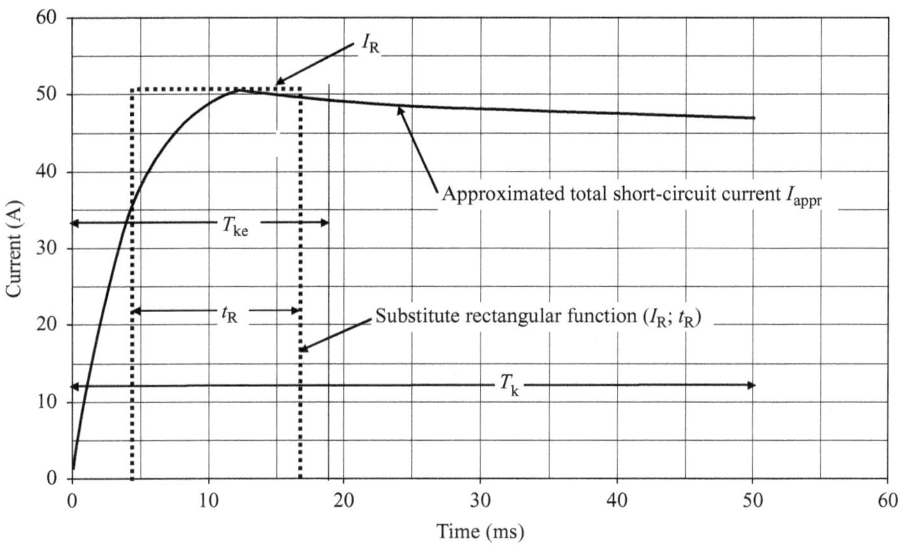

Figure 10.17 Standardised rectangular function and approximated total short-circuit current

Peak force between sub-conductors

$$F_{Rs} = \frac{\mu_0}{2\pi} * \frac{l_s}{a_s}\left(\frac{I_{Rs}}{n}\right)^2 = \frac{4\pi * 10^{-7} * (\mathrm{Vs/Am})}{2\pi} * \frac{0.35\ \mathrm{m}}{0.028\ \mathrm{m}}$$

$$* \left(\frac{2556 * 10^6\ \mathrm{A}^2}{4}\right) = 1597.5\ \mathrm{N}$$

10.5.2.3 Bending stress

Bending stresses σ_m; σ_s on main, respectively, sub-conductors (Equation (10.31)) are calculated for the forces obtained by the substitute rectangular function only.

$$\sigma_m = V_\sigma * \beta * \frac{F_{Rm} * l}{8 * Z} = 1.0 * 0.73 * \frac{7016.5\ \mathrm{N} * 1.05\ \mathrm{m}}{8 * 3.47 * 10^{-6}\ \mathrm{N/mm^2}}$$

$$= 193.7\ \mathrm{N/mm^2}$$

$$\sigma_s = V_{\sigma s} * \frac{F_{Rs} * l_s}{16 * Z_s} = 1.0 * \frac{1597.5\ \mathrm{N} * 0.35\ \mathrm{m}}{16 * 0.667 * 10^{-6}\ \mathrm{N/mm^2}} = 49.9\ \mathrm{N/mm^2}$$

with V_σ and $V_{\sigma s}$ equal One as per Figure 10.12 and the factor $\beta = 0.73$ as per Table 10.3.

Total bending stress

$$\sigma_{tot} = \sigma_m + \sigma_s = (193.7 + 49.9)\ \mathrm{N/mm^2} = 243.6\ \mathrm{N/mm^2} \le q * R_{p0.2}$$

$$= 1.5 * 340\ \mathrm{N/mm^2}$$

Conductors have sufficient electromagnetic strength as

$$\sigma_{\text{tot}} \leq q * R_{\text{p0.2}} = 1.5 * 340\ \text{N/mm}^2 = 510\ \text{N/mm}^2$$

and if

$$\sigma_{\text{s}} \leq R_{\text{p0.2}} = 340\ \text{N/mm}^2$$

10.5.2.4 Forces on supports

Forces on outer supports

$$F_{\text{d}} = V_{\text{F}} * \alpha * F_{\text{Rm}} = 2.0 * 0.4 * 7016.5\ \text{N} = 5.61\ \text{kN}$$

Forces on inner supports

$$F_{\text{d}} = V_{\text{F}} * \alpha * F_{\text{Rm}} = 2.0 * 1.1 * 7016.5\ \text{N} = 15.4\ \text{kN}$$

Chapter 11
Limitation of short-circuit currents

11.1 General

The expansion of electrical power systems by new power stations and new lines (overhead transmission lines and cable circuits) results in an increase of short-circuit currents due to an increase in sources feeding the short-circuit and due to a reduction of system impedance. The improvement of existing installations and the replacement of equipment are necessary, in case the permissible short-circuit current will be exceeded. Measures to limit the short-circuit currents can also be realised which might be more economic than the replacement of equipment and installations. Different measures have to be taken into account such as measures affecting the whole system (higher voltage level), measures concerning installations and substations (separate operation of busbars) and measures related to equipment (Ip-limiter).

All measures have an influence on the system reliability as well, which must be guaranteed under outage conditions of equipment after the measures for limitation of short-circuit currents are in operation. Measures for short-circuit current limitation decrease the voltage stability, increase the reactive power requirement, reduce the dynamic stability and increase the complexity of operation. Furthermore some measures to limit short-circuit currents will contradict requirements for a high short-circuit level, e.g., to reduce flicker in the case of connection of arc-furnace.

The decision on location of power stations is determined beside other criteria by the availability of primary energy (lignite coal fired power stations are build nearby the coal mine), requirement of cooling water (thermal power stations are placed near the sea or at large rivers), geological conditions (hydro power stations can only be build if water reservoirs are available), requirements of the power system (each power station requires a system connection at suitable voltage level) and the vicinity to consumers (combined heat and energy stations need heat consumers nearby).

The connection of large power stations is determined by the branch short-circuit current from the generators. Figure 11.1 outlines considerations to select the suitable

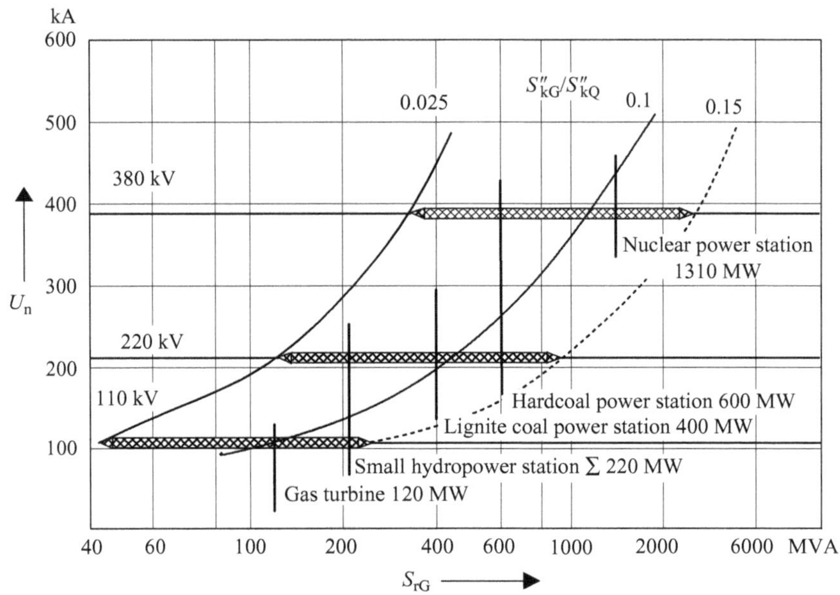

Figure 11.1 Selection of suitable voltage level for the connection of power stations

voltage level for the connection of power stations to the power system. It is assumed that more than one power station is connected to the system.

As generation of electrical energy without consumers is without any sense, a suitable power system has to be planned and constructed accordingly. An increased amount of small combined heat and power stations (distributed generation) with connection to the medium-voltage and even to the low-voltage system requires additional considerations with respect to protection, operation and short-circuit level in the different voltage levels [25].

11.2 Measures

11.2.1 Measures in power systems

11.2.1.1 Selection of nominal system voltage

A higher nominal system voltage by constant rated power of feeding transformers will reduce the short-circuit level proportionally. The selection of nominal system voltage must take into account the recommended voltages as per IEC 60038:1987 and the common practice in the utility itself and maybe in the whole country. Table 11.1 lists a selection of recommended voltages. The table also includes information on typical applications in Europe.

The short-circuit current is directly proportional to the voltage level, respectively to the voltage ratio of feeding transformers, if all other parameters are constant.

Table 11.1 Selection of recommended voltage as per IEC 60038:1987

	Nominal voltage	Application	Remarks
Low voltage (V)	400–230	Private consumers Small industrial consumers	According to IEC Table I
	500	Motor connection in industry	Not listed in IEC
Medium voltage (kV)	6	HV-motors in industry, auxiliary supply in power stations	According to IEC Table III
	10	Urban distribution systems, industrial systems	According to IEC Table III
	20	Industrial systems, rural distribution systems	According to IEC Table III
	30	Electrolysis, arc furnace, rectifiers	Not listed in IEC
High voltage (kV)	110	Urban transport systems	According to IEC Table IV
	220	Transport system with regional task	According to IEC Table IV
	380	Transmission system country-wide	According to IEC Table V the highest voltage of equipment $U_{bmax} = 420$ kV is defined

The selection of a new nominal system voltage normally is only possible when new electrification projects are considered. As the impedance voltage of transformers increases with increasing voltage an additional positive effect on the reduction of short-circuit currents is seen. As a by-effect it should be noted that the transmittable power of overhead lines and cables is increased with increasing voltage without increasing the cross-section of the conductor. On the other hand, the voltage drop of the transformer is increased in the case of increase of impedance voltage.

11.2.1.2 Operation as separate subsystems

The power system is operated as several subsystems, which are connected at higher voltage level. Figure 11.2 outlines the general structure of a 132-kV-cable system (total load approximately 1500 MW). The system is supplied from the 400-kV-system and by a power station connected to the 132-kV-level. Assuming a meshed system operation, i.e., the 132-kV-system is operated as one system with all breakers closed, the short-circuit currents in the case of three-phase and single-phase short-circuits are $I''_{k3} = 26.0$–37.4 kA and $I''_{k1} = 37.3$–45.7 kA. Operating the 132-kV-system as two separate subsystems coupled only on the 400 kV-level, the short-circuit currents will be reduced to the values as outlined in Figure 11.2.

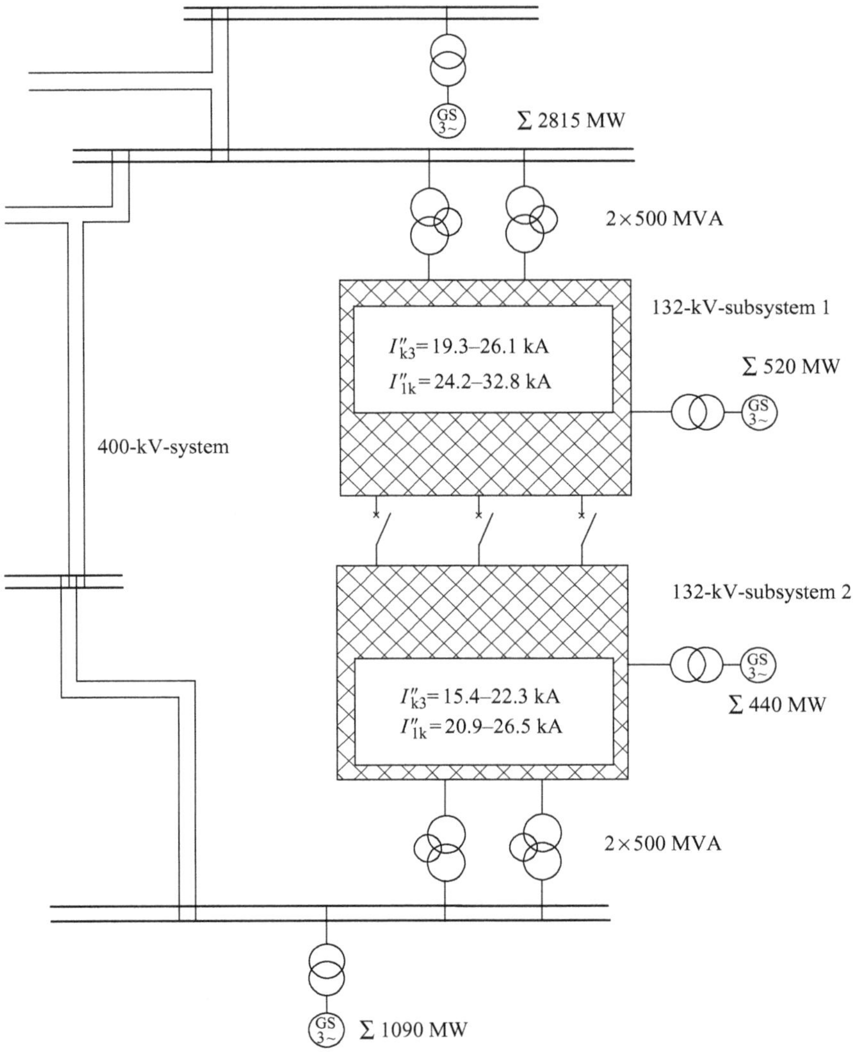

Figure 11.2 Schematic diagram of a 400/132-kV-system for urban load; values of short-circuit currents in case of operation as two subsystems

Operating the 132-kV-system as two separate subsystems will require additional cable circuits and an extension of the switchgear to fulfil the $(n-1)$-criteria for a reliable power supply.

11.2.1.3 Distribution of feeding locations

Power stations and system feeders from higher voltage levels are to be connected to several busbars in the system. This measure was realised in the power system as per Figure 11.2, which is a by-effect to the system separation. A further example is outlined in Figure 11.3. A power station of 395 MW is connected to a 132-kV-system, which has a second supply from the 220-kV-system. The 132-kV-system is a pure cable network and the shortest cable length between any two substations is 11.2 km. In the case when the busbar-coupler K in the power station is closed, the three-phase short-circuit current at the busbar is $I''_{k3} = 37.6$ kA; the short-circuit currents at the busbars in the 132-kV-system remain below $I''_{k3} = 33.5$ kA. If the busbar-coupler K is operated opened, the short-circuit currents at the busbar in the power station are $I''_{k3} = 28.0$ kA and $I''_{k3} = 29.3$ kA. For short-circuits at the busbars in the system itself the short-circuit currents are reduced up to 4.1 kA.

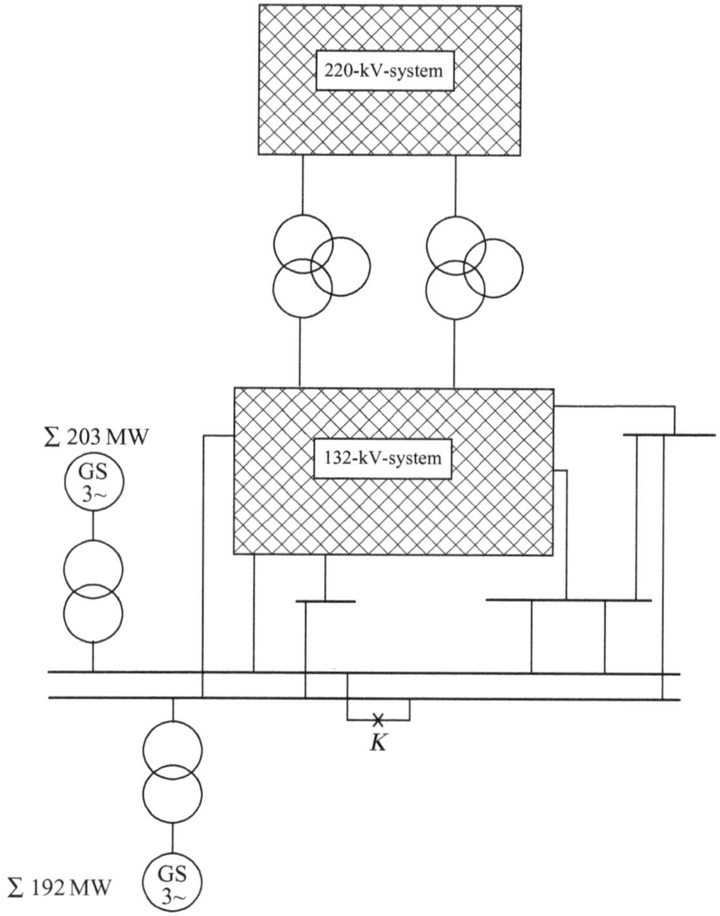

Figure 11.3 Schematic diagram of a 132-kV-system with power station

The generators and the 132-kV-cables in the power stations need to be switched-on to the busbars in such a way that the generated power can be transferred to the power system without overloading any of the cable even under outage conditions.

11.2.1.4 Coupling of power system at busbars with low short-circuit level

Different parts of the power system shall be connected only at busbars with low short-circuit level. Figure 11.4 outlines a 30-kV-system with overhead lines, which is fed from the 110-kV-system by two transformers operated in parallel. The three-phase short-circuit current is $I''_{k3} = 10.09$ kA. If the transformers are not operated in parallel and the system is coupled at busbar K7 the short-circuit current at the feeding busbar is $I''_{k3} = 5.94$ kA.

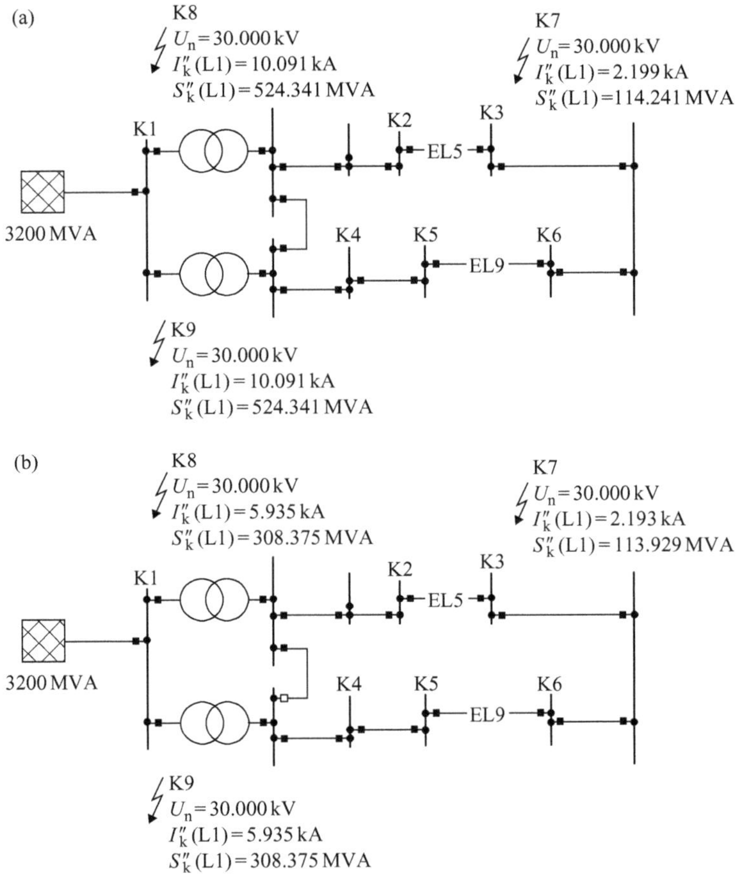

Figure 11.4 Equivalent circuit diagram of a 30-kV-system with feeding 132-kV-system: (a) Operation with transformers in parallel and (b) limitation of short-circuit current. Result of three-phase short-circuit current: $S''_{kQ} = 3.2$ GVA; $S_{rT} = 40$ MVA; $u_{krT} = 12\%$; $t_{rT} = 110/32$; OHTL 95Al; $l_{tot} = 56$ km

It should be noted that the short-circuit level at busbar K7 is affected only to a minor extent. If the transformers are loaded only up to 50 per cent of their rated power and if the lines have sufficient thermal rating, both system configurations have the same supply reliability.

11.2.1.5 Restructuring of the power system

Restructuring of power systems is comparatively costly and complicated. In medium voltage systems restructuring is in most cases only possible together with the commissioning of new primaries, loop-in and loop-out of cable (overhead line) circuits and the operation of the system as a radial system. In a high voltage system, restructuring requires a total different system topology. Figure 11.5 outlines the comparison of two system topologies, i.e., ring fed system and radial fed system.

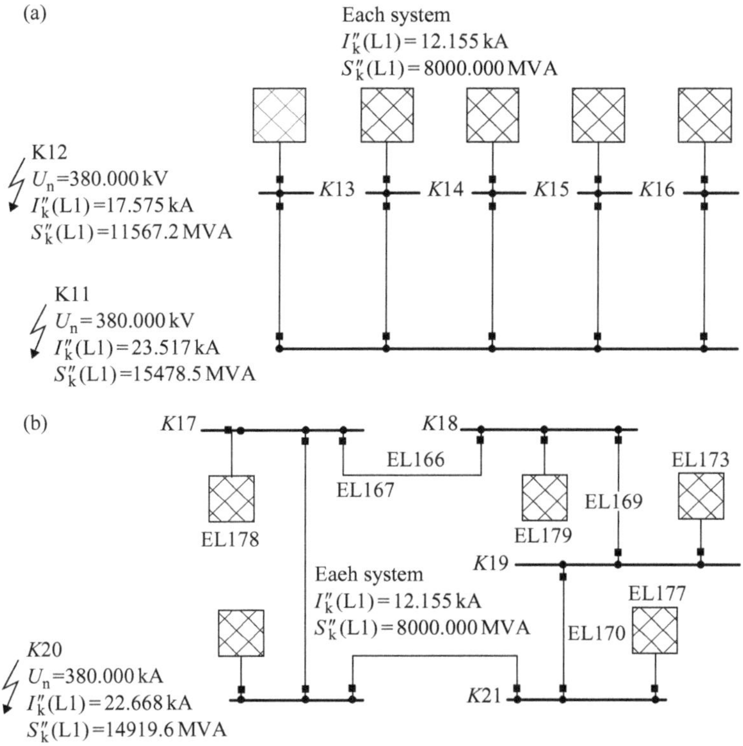

Figure 11.5 Equivalent circuit diagram of a 380-kV-system and results of three-phase short-circuit current calculation: (a) Radial fed system and (b) ring fed system. $S''_{kQ} = 8$ *GVA; OHTL ACSR/AW 4 × 282/46;* $l_i = 120$ *km*

As can be seen from Figure 11.5 the short-circuit currents are reduced from $I''_{k3} =$ 23.6 kA to $I''_{k3} = 22.7$ kA (3.8 per cent) with the new topology. The reduction of the short-circuit currents is comparatively small, but will be more significant, if an increased number of feeders (or generators) shall be connected [2].

11.2.2 Measures in installations and switchgear arrangement

11.2.2.1 Multiple busbar operation

The connection of lines and feeders to more than one busbar per substation is advantageous as compared with the operation of the substation with single busbar or with busbar coupler closed. Figure 11.6 outlines the schematic diagram of a 110-kV

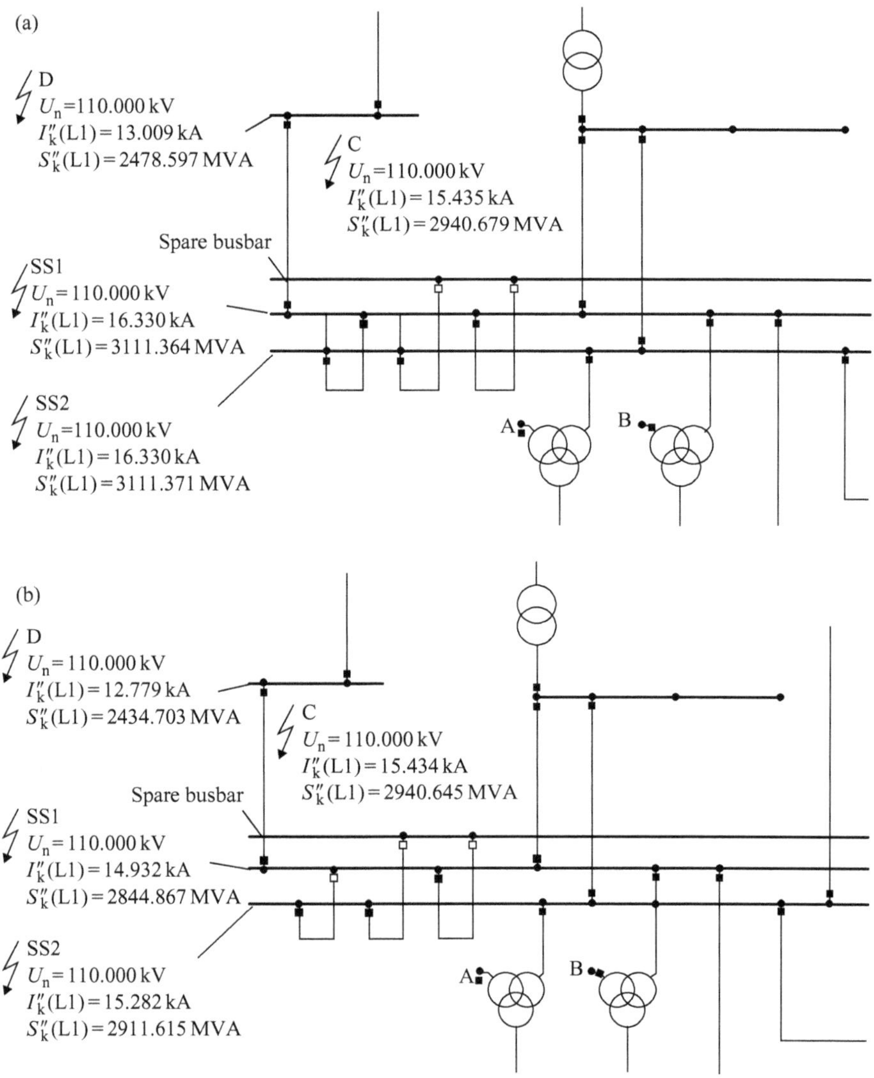

Figure 11.6 Schematic diagram of a 110-kV-substation fed from the 220-kV-system: (a) Operation with buscoupler closed and (b) operation with buscoupler open. Result of three-phase short-circuit current calculation

system. The 110-kV-substation is equipped with a double busbar and one additional spare busbar. The substation is fed from the 220-kV-system; outgoing 110-kV-cables are connected to each of the two busbars in operation.

The operation with two busbars reduces the three-phase short-circuit current from $I''_{k3} = 16.3$ kA to $I''_{k3} = 14.9$ kA (8.6 per cent) at SS1 and $I''_{k3} = 15.3$ kA (6.1 per cent) at SS2. Each of the two busbars SS1 and SS2 can be switched-on to the spare busbar without coupling the busbars.

11.2.2.2 Busbar sectionaliser in single busbar switchgear

Single busbars can be equipped with busbar sectionaliser, so that an operation mode similar to double busbar operation is possible. The outgoing cables and the feeding transformers need to be connected to the busbar section in such a way that the loading of feeders is approximately equal. Figure 11.7 indicates an industrial system with nominal voltage of 6 kV, which is fed from the 30-kV-system.

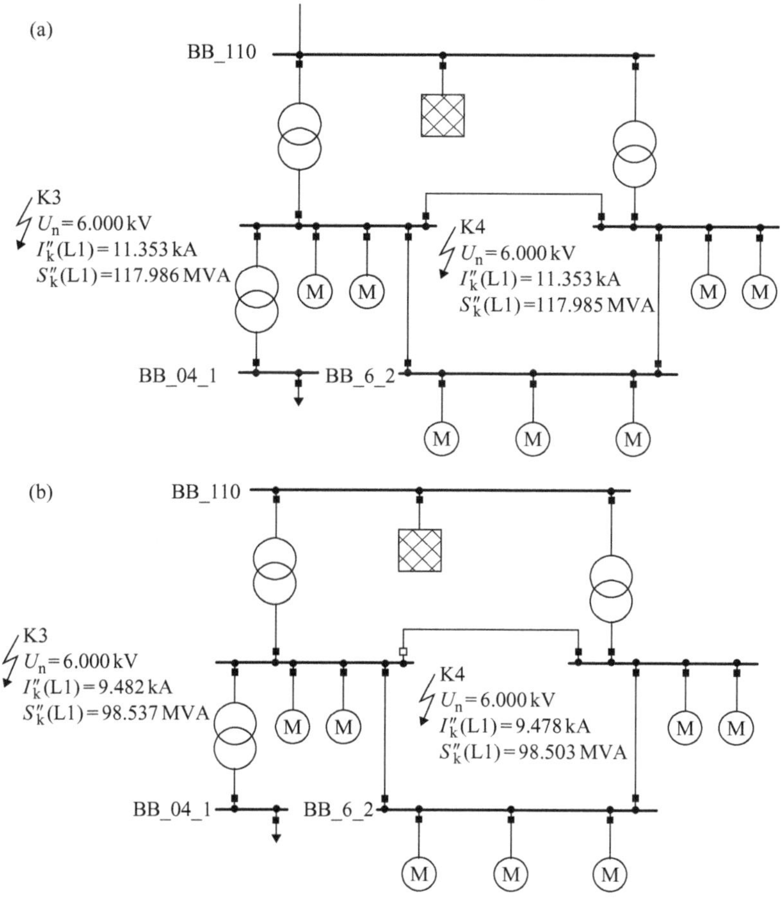

Figure 11.7 Equivalent circuit diagram of a 6-kV-industrial system. Results of three-phase short-circuit current calculation: (a) Busbar sectionaliser closed and (b) busbar sectionaliser open

The short-circuit current at the feeding busbar is reduced by 16.8 per cent from $I''_{k3} = 11.4$ kA to $I''_{k3} = 9.48$ kA in the case when the busbar sectionaliser is kept open. The outgoing feeders have to be arranged in such a way that the loading will be approximately equal for both busbar sections K3 and K4.

11.2.2.3 Short-circuit current limiting equipment

Short-circuit current limiting equipment and fuses (medium voltage and low voltage systems) can be installed to reduce the short-circuit level in parts of the installations. In medium voltage installations, Ip-limiter can be installed. Figure 11.8 outlines the schematic diagram of an industrial system. The existing switchgear A with low short-circuit rating shall be extended with the busbar section B, which is fed by an additional system feeder Q2. The maximal permissible short-circuit current I''_{kAmax} of busbar section A is exceeded by this extension.

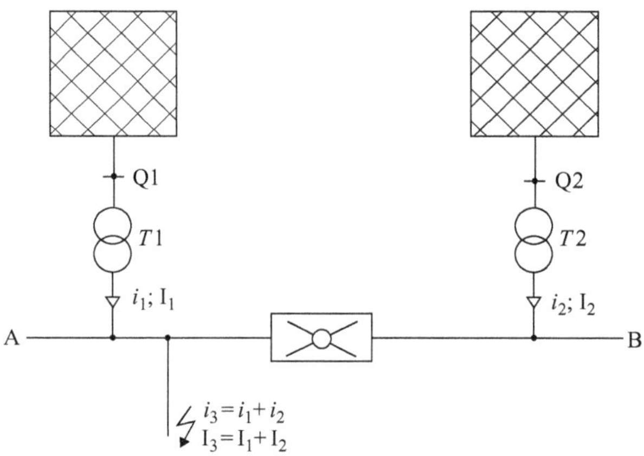

Figure 11.8 Equivalent circuit diagram of switchgear with single busbar

The total short-circuit current from both system feeders shall be limited to the permissible short-circuit current I''_{kAmax} of busbar section A in the case of a short-circuit at busbar A. If the relation I''_{kQ1}/I''_{kQ2} depends on the ratio Z_{Q1}/Z_{Q2} of the feeders Q1 and Q2 it is sufficient to measure the partial short-circuit current through the Ip-limiter. The current ratio is

$$\frac{I_1}{I_2} = \frac{I''_{kQ1}}{I''_{kQ2}} \tag{11.1}$$

and the total short-circuit current

$$I_3 = I_2 * \left(1 + \frac{I''_{kQ1}}{I''_{kQ2}}\right) \le I''_{kAmax} \tag{11.2}$$

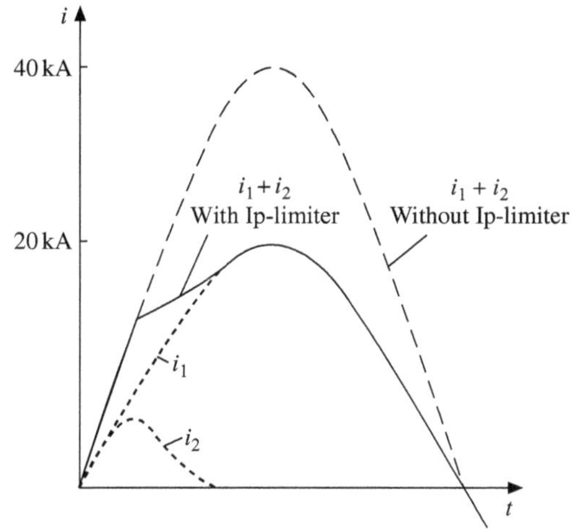

Figure 11.9 *Time course of short-circuit current in installations with and without Ip-limiter*

The threshold value $I_{2\text{an}}$ of the Ip-limiter is

$$I_{2\text{an}} = I''_{\text{kA max}} * \frac{I''_{\text{kQ2}}}{I''_{\text{kQ1}} + I''_{\text{kQ2}}} \tag{11.3}$$

When the permissible short-circuit currents I''_{kAmax} and I''_{kBmax} of both busbar sections A and B are exceeded, the threshold value $I_{1\text{an}}$ of the Ip-limiter for short-circuits at busbar section B is needed as well

$$I_{1\text{an}} = I''_{\text{kB max}} * \frac{I''_{\text{kQ1}}}{I''_{\text{kQ2}} + I''_{\text{kQ1}}} \tag{11.4}$$

The threshold value I_{an} of the Ip-limiter is set to the minimum of both values

$$I_{\text{an}} = \text{MIN}\{I_{1\text{an}}; I_{2\text{an}}\} \tag{11.5}$$

The detailed design and determination of the settings are determined, besides other factors, by different topologies of the power system, different phase angles of the branch short-circuit currents and different rating of the switchgear in the system.

Figure 11.9 outlines the time curves of short-circuit currents at section A as per Figure 11.8. The branch short-circuit current i_2 from system feeder Q2 is switched off by the Ip-limiter within 7 ms, thus reducing the peak short-circuit current significantly.

The technical layout of one phase of an Ip-limiter is shown in Figure 11.10. Inside an insulating tube (1) the main current conductor (3) with a breaking element, blown by a triggerable explosive loading, (2) is located. When the threshold value is exceeded, the tripping circuit triggers the explosive loading; the arc inside the insulating tube cannot be quenched and is commutated to the fuse element (4), which

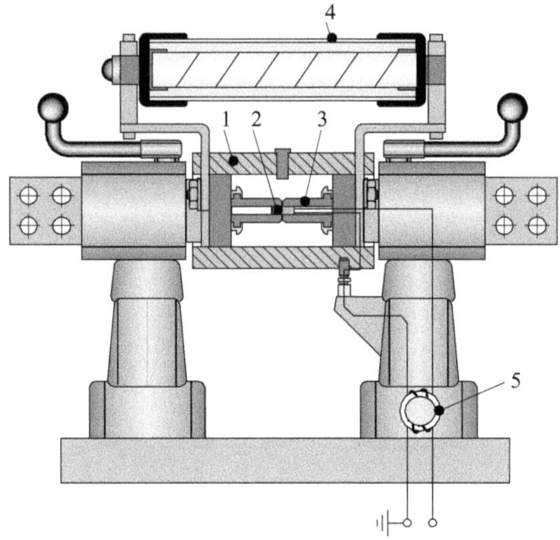

Figure 11.10 *Cutaway view of an Ip-limiter support: (1) insulating tube, (2) explosive loading, (3) main conductor, (4) fuse element and (5) transducer*

Source: ABB Calor Emag Schaltanlagen AG

is able to quench the short-circuit according to the fuse It-characteristic. The main elements, i.e., the isolating tube with main conductor and fuse element, need to be replaced after operation of the Ip-limiter. A measuring unit with tripping circuit is needed to compare the actual current value with the threshold value.

Ip-limiters are nowadays available with thyristor technique. The short-circuit current can be limited within 1–2 ms after initiation of the fault. The Ip-limiter is back in operation after fault clearing; an exchange of main conductor and fuse is not necessary. Additional operational functions, such as limitation of start-up current of large motors can also be realised. Superconducting Ip-limiters are actually in laboratory tests [7].

11.2.3 Measures concerning equipment

11.2.3.1 Impedance voltage of transformers

Transformers with high impedance voltage are reducing the short-circuit level, however the reactive power losses are increased and the tap-changer needs to be designed for higher voltage drops. Figure 11.11 indicates the equivalent circuit diagram of a 10-kV-system fed from a 110-kV-system by three transformers $S_{rT} = 40$ MVA. The system load is $S_L = 72$ MVA, $\cos\varphi = 0.8$. The short-circuit power of the 110-kV-system is $S''_{kQ} = 2.2$ GVA; the voltage at the 10-kV-busbar shall be controlled within a bandwidth of ±0.125 kV around $U = 10.6$ kV.

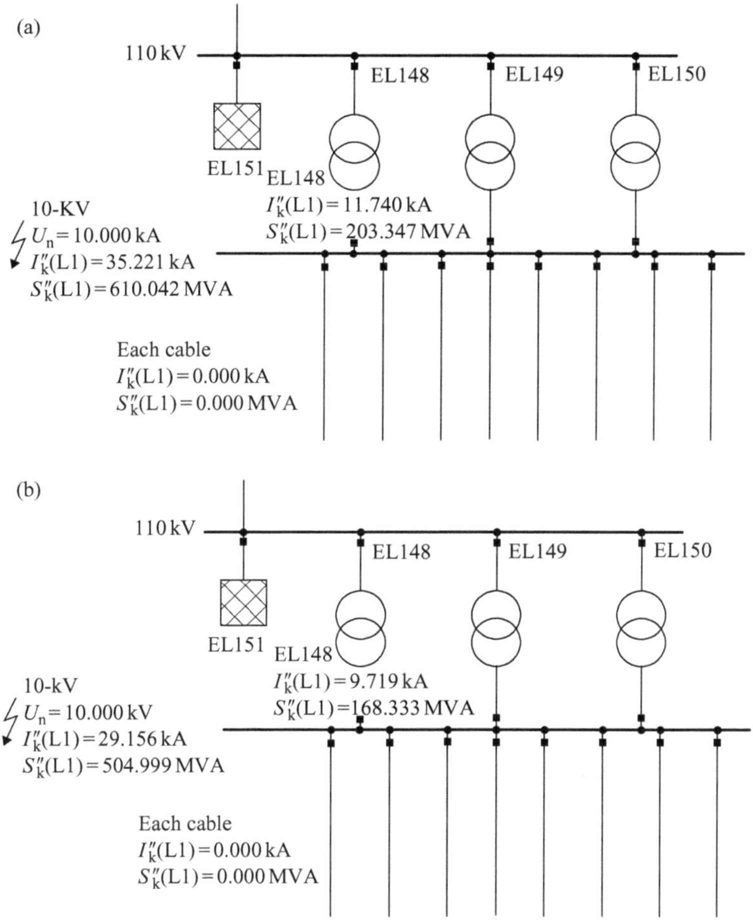

Figure 11.11 Equivalent circuit diagram of a 10-kV-system with incoming feeder. Results of three-phase short-circuit current calculation: (a) Impedance voltage 13% and (b) impedance voltage 17.5%

Table 11.2 Result of loadflow and short-circuit analysis as per Figure 11.11

U_{krT} (%)	I''_{k3} (kA)	I''_{k1} (kA)	Tap-changer position	Reactive power losses of one transformer (Mvar)
13	35.2	22.5	+6	2.61
17.5	29.2	20.7	+8	3.58

The relevant results of loadflow and short-circuit analysis are outlined in Table 11.2. As can be seen the increase of the impedance voltage from 13 to 17.5 per cent reduces the short-circuit current, but increases the reactive power losses and increases the number of steps at the tap-changer to control the voltage.

11.2.3.2 Short-circuit limiting reactor

The application of short-circuit limiting reactors can be defined as a measure related to switchyards or a measure related to equipment. Figure 11.12 outlines the equivalent circuit diagram of a 10-kV-system in industry with direct connection to an urban 10-kV-system. Two reactors are installed to limit the short-circuit currents. The three-phase short-circuit current without local generation in the industrial system at the coupling busbar between industry and utility is $I''_{k3} = 20.43$ kA.

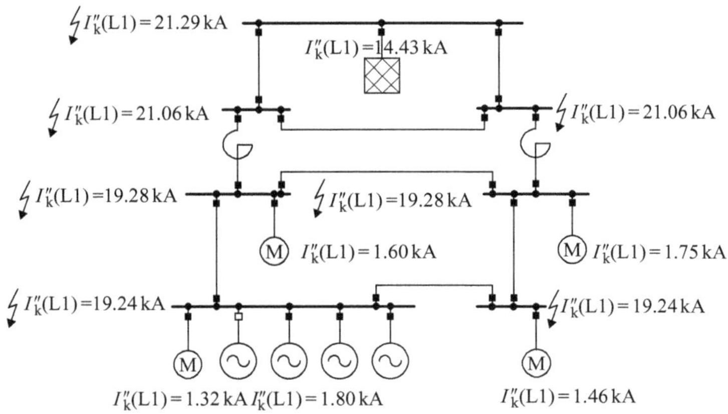

Figure 11.12 Equivalent circuit diagram of a 10-kV-system with short-circuit limiting reactors. Results of three-phase short-circuit current calculation

The industrial system is connected to a heat and power plant with four generators 6.25 MVA each, three out of four are allowed to be in operation at the same time. The short-circuit current is increased by this to 25.6 kA. To limit the short-circuit current to $I''_{k3} \leq 21.5$ kA reactors with $I_n = 1600$ A; $u_k = 20\%$ were installed. The short-circuit current is reduced to $I''_{k3} = 21.3$ kA.

11.2.3.3 Earthing impedances

Single-phase short-circuit currents can be reduced significantly by the installation of earthing impedances in the neutral of transformers or at artificial neutrals without affecting the three-phase short-circuit currents. Figure 11.13 represents an 11.5-kV-system fed from the 132-kV-system. Each substation is equipped with four transformers ($S_r = 40$ MVA, $u_k = 14\%$). The 132-kV-system has direct neutral earthing, the short-circuit currents are $I''_{k3} \approx 29.3$ kA and $I''_{k1} \approx 37.3$ kA.

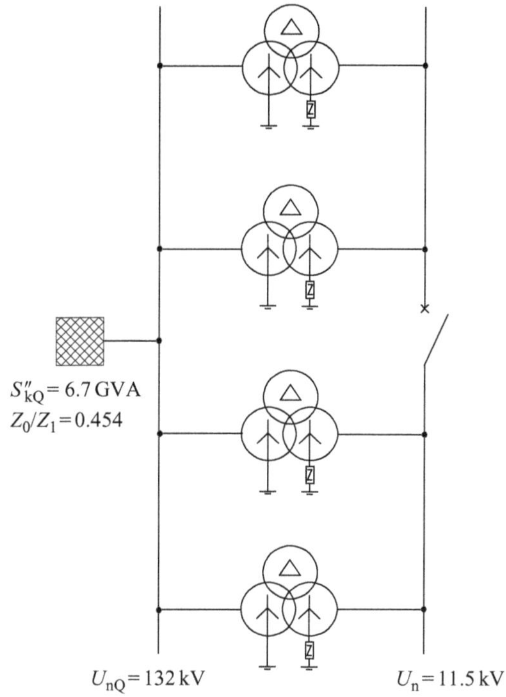

Figure 11.13 Equivalent circuit diagram of 11.5-kV-system fed from the 132-kV-system

The permissible short-circuit current in the 11.5-kV-system is 25 kA. The single-phase short-circuit currents at 11.5-kV-busbar are $I''_{k1} = 15.04$ kA when one transformer is in operation and $I''_{k1} = 29.27$ kA when two transformers are operated in parallel.

In order to limit the single-phase short-circuit current on the 11.5-kV-side to $I''_{k1} < 25$ kA (two transformers in parallel), an earthing resistance of $R_E = 0.31\ \Omega$ or an earthing reactor of $X_E = 0.1\ \Omega$ need to be installed in the 11.5-kV-neutral of each of the transformers [3].

11.2.3.4 Increased subtransient reactance of generators

Generators are the direct sources for short-circuit currents; the contribution of one generator to the short-circuit current is inversely proportional to the subtransient reactance X''_d when the voltage is not changed, see Chapters 3 and 4. An increased subtransient reactance reduces the branch short-circuit current and by this the total short-circuit current. Figure 11.14 indicates the results of short-circuit current calculation of a power station. Generators of different make but identical rating $S_{rG} = 150$ MVA are installed. The three-phase branch short-circuit currents are in the range of $I''_{k3} = 2.32$–2.75 kA depending on the subtransient reactance.

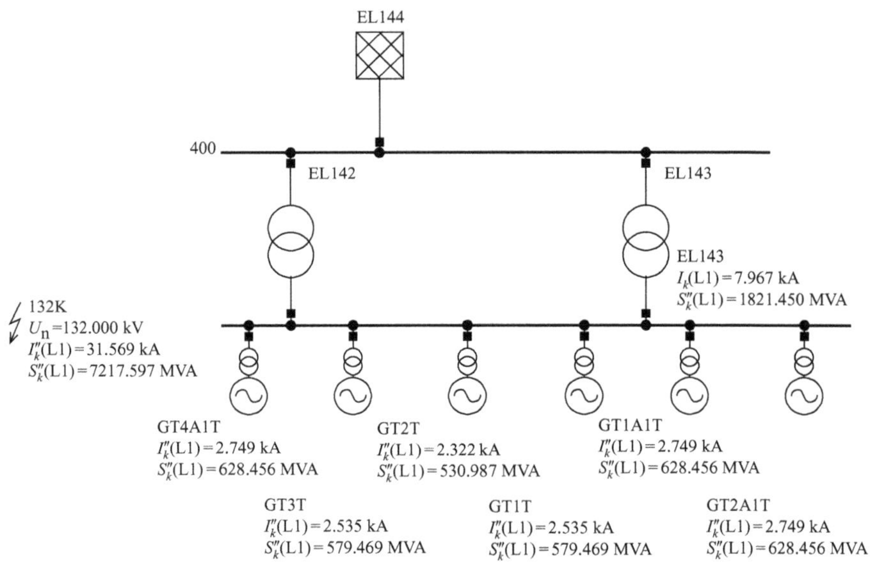

Figure 11.14 *Equivalent circuit diagram of a power station with 132-kV-busbar. Results of three-phase short-circuit current calculation:* $S_{rG} = 150$ *MVA;* $x''_d = 12$–*17.8%*

High subtransient reactance of generators has a negative impact on the dynamic stability performance of the generators. In the case of short-circuits on the transmission line with subsequent fault clearing the transmittable power from a power station is reduced if the fault clearing time of the protection is kept constant, respectively, the fault clearing time must be reduced to keep the transmittable power constant. Details can be obtained from [2, 26].

11.3 Structures of power systems

11.3.1 General

It should be noted that some of the measures as per Section 11.2 to reduce short-circuit currents can only be applied in certain power systems. Ip-limiters are only available in low voltage and medium voltage systems. When only a single busbar is installed, the operation with two busbars is not possible and in a radial fed system, no additional feeding point is normally available. Within this section, the main structures of power systems are introduced:

- Radial system
- Ring-main system
- Meshed system

More details can be obtained from [26].

11.3.2 Radial system

Radial systems represent the simplest topology of a power system and can usually be found in low voltage systems. Figure 11.15 outlines the general structure, with one feeding point and distributing of the lines in several branches. This structure is suitable in the case of low load density but also for the connection of high bulk-supply loads. The calculation of short-circuit currents is comparatively easy, as there are no meshed lines in the system and only one topology has to be analysed.

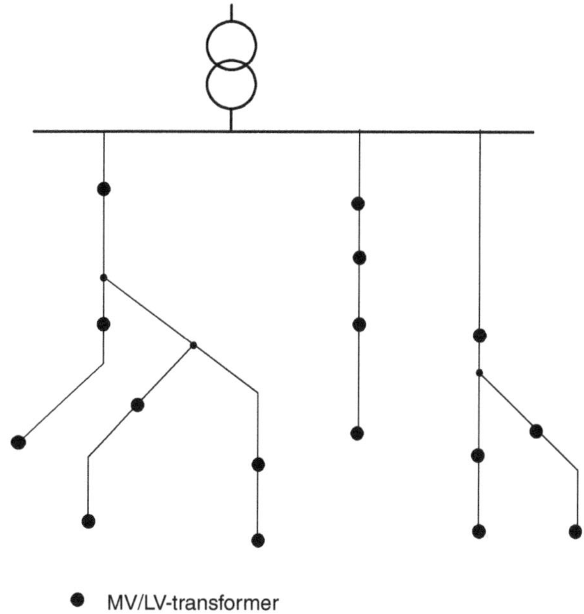

Figure 11.15 General structure of a radial system with one incoming feeder

11.3.3 Ring-main system

In a ring-main system the receiving end of each line of a radial system is to be connected either back to the feeding busbar or to an additional feeding busbar. Ring-main systems are most often planned for medium voltage systems, and in rare cases for low voltage systems as well. Normally ring-main systems are operated with open breaker or isolator in one primary as indicated in Figure 11.16(a). This enables an operation similar to radial systems but with a switchable reserve for all consumers. Feeding busbars can be planned at several locations of the system as indicated in Figure 11.16(b). Short-circuit current calculation is more complicated than in radial systems as several operating conditions, i.e., system topologies, are to be taken into account to determine the minimal and the maximal short-circuit current.

11.3.4 Meshed systems

Meshed systems are normally applied only for high voltage systems or in industrial supply systems for MV-level as well. With the consideration of consumer load and

(a)

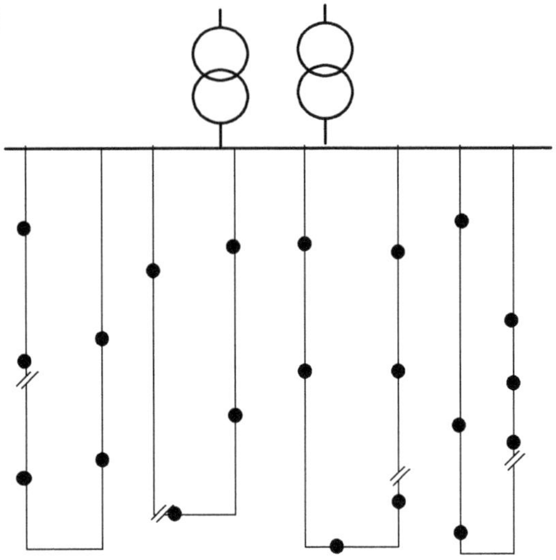

(b)

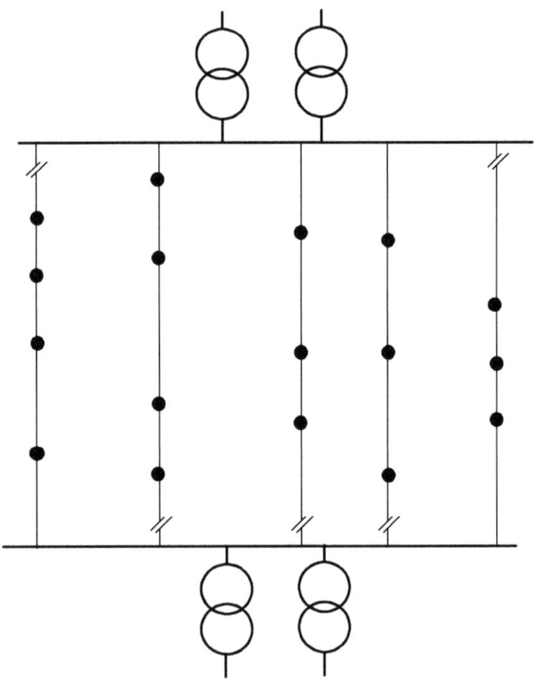

Figure 11.16 *General structures of ring-main systems: (a) Simple structure with one feeding busbar and (b) structure with two feeding busbars (feeding from opposite sides)*

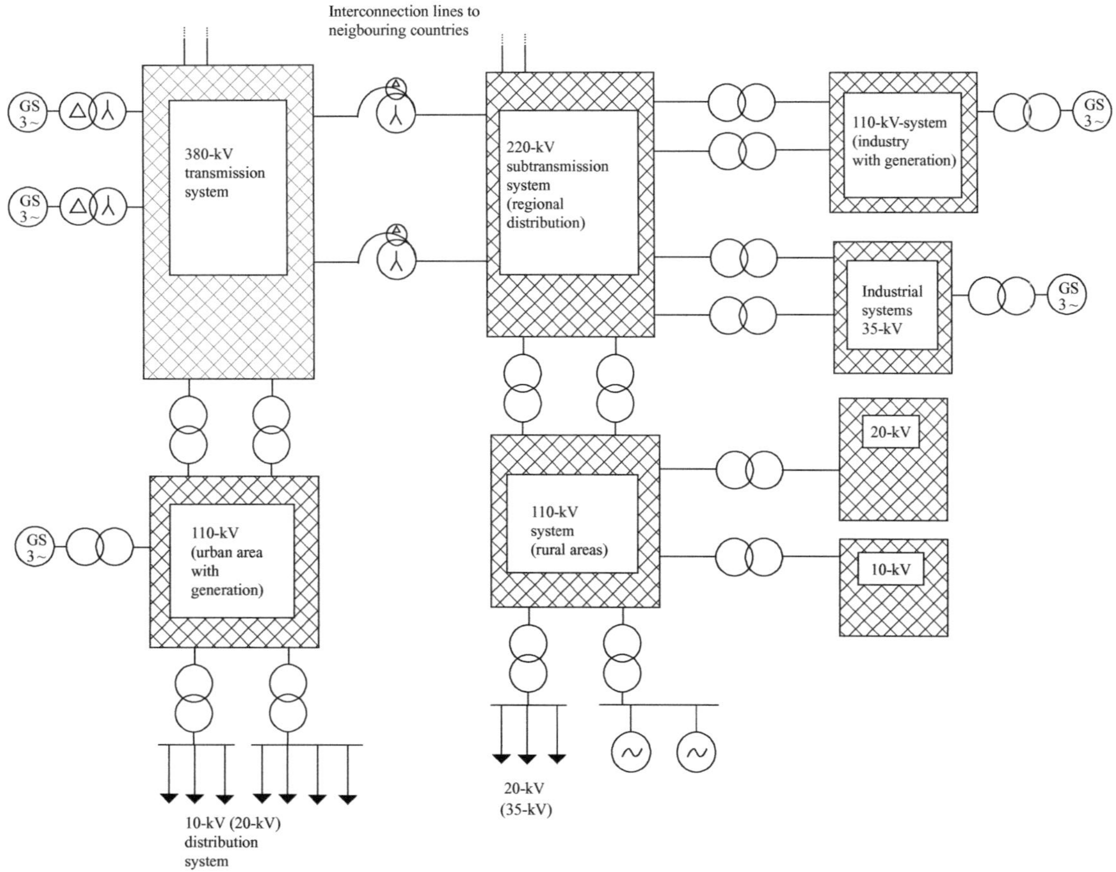

Figure 11.17 Principal structure of a high voltage system with different voltage levels

capability of power stations, the system is planned, constructed and operated in such a way as to allow the supply of consumers without overloading of any equipment and without violating the permissible voltage profile even under outage of one ($(n-1)$-criteria) or more equipment at the same time. The calculation of maximal short-circuit currents seems relatively simple, if all equipment are assumed to be in operation. As far as different operation schedules of the power stations are considered, different system topologies need to be considered. The calculation of the minimal short-circuit current, however, is much more difficult, as a close cooperation between planning and operation is required. A large number of different topologies in the meshed power system have to be analysed to ensure that the calculated short-circuit current is the minimal current. The principal structure of a meshed high voltage system is outlined in Figure 11.17.

A special type of meshed systems is applied to low voltage systems; the principal structure is outlined in Figure 11.18. The reliability of supply is comparatively high, as reserve in the case of outage of any line or infeed is provided through the remaining lines.

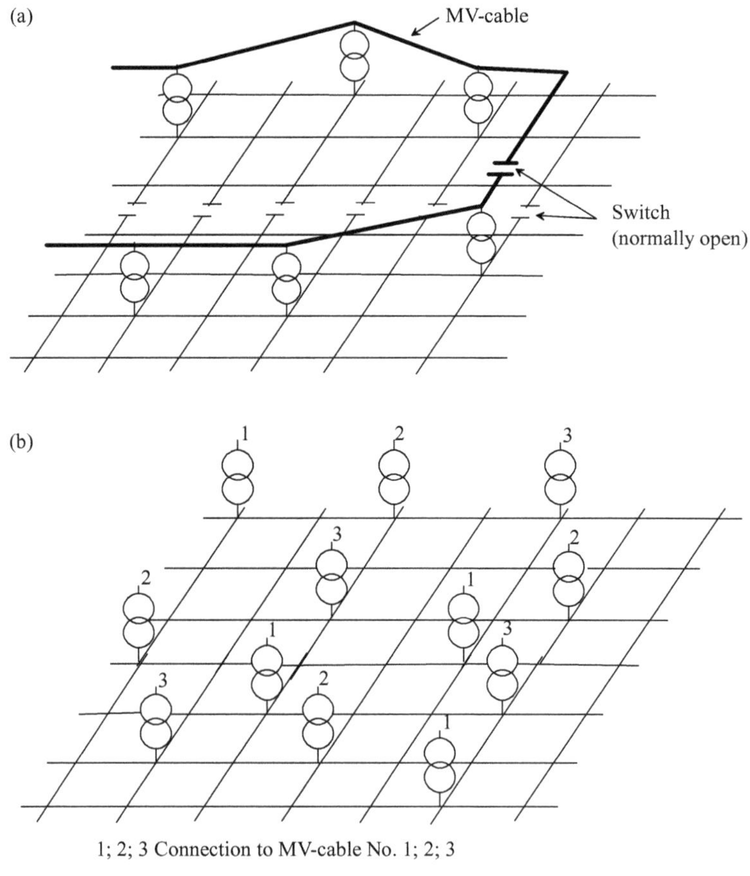

Figure 11.18 Principal structure of meshed low voltage system: (a) Single-fed meshed system and (b) meshed system with overlapping feeding

Chapter 12

Special problems related to short-circuit currents

12.1 Interference of pipelines

12.1.1 Introduction

Interference between overhead lines, communication circuits and pipelines is caused by asymmetrical currents, which may be due to short-circuits, asymmetrical operation or asymmetrical design of equipment, especially asymmetrical outline of overhead line towers with respect to pipelines and communication circuits. This interference is based on inductive, ohmic and capacitive coupling between the short-circuit path (e.g., overhead line) and the circuit affected by interference (e.g., pipeline). Normal operating currents, respectively voltages, cause magnetic as well as electric fields which are asymmetrical in the vicinity of overhead lines which may cause interference problems in the long-time range.

Short-circuit currents on overhead transmission lines and cables and short-circuit currents through earth cause interference in the short-time range only when the short-circuit is switched off after some seconds by the power system protection. Interference problems may arise in cable and overhead line systems to a different extent depending on the handling of the system neutral. The induced voltage in pipelines and communication circuits may endanger technical installations and safety of workers when defined limits will be exceeded. Within the context of this book, only the short-time interference, in the following called interference, is dealt with. Table 12.1 outlines the needs for the analysis of interference problems.

Interference problems may occur in most of the cases by inductive and ohmic coupling in power systems with low-impedance earthing consisting mainly of overhead lines, as can be seen from Table 12.1. In systems with isolated neutral or resonance earthing interference problems have to be regarded only when the fault current is not self-extinguishing (see Chapters 5 and 7). Capacitive coupling does not cause any

Table 12.1 Interference between power system, communication circuits and pipelines

Power system			Interference of communication circuits and pipelines by:		
Handling of neutrals	Operating condition		Inductive coupling	Capacitive coupling	Ohmic coupling
Low-impedance earthing	Without fault	OHTL	Only if circuits on same tower		No interference
		Cable	No interference problems		
	Short-circuit	OHTL	Present	No interference	Present
		Cable	Present in special cases	No interference	No interference
Isolated neutral; Resonance earthing	Without fault	OHTL	Only if circuits on same tower		No interference
		Cable	No interference		
	Short-circuit; Earth fault	OHTL	No interference	Present	No interference
		Cable	No interference		
	Double earth fault	OHTL	Present	No interference	Present
		Cable	No interference	No interference	No interference

severe problems in pipelines and communication circuits. It should be noted furthermore that interference of communication circuits is decreasing due to the decreasing installations of overhead communication circuits, which are replaced by wireless communication or by cable circuits, which can be protected easily against interference. The explanations on interference are therefore concentrated within this section to the interference of pipelines.

Regulations on the permissible values for voltages induced in pipelines and communication circuits and/or for touch voltages exist in various countries. The main aim is to protect any person likely to work on the pipeline or power circuit against electrocution hazard. According to an international survey carried out by CIGRE [20], the maximum permissible touch voltage is defined in different countries in different ways, ranging from 200 V up to 1500 V depending on the maximal fault duration time. In Germany, the maximal permissible touch voltage and the maximal permissible voltage pipeline-to-earth are both limited to 1000 V for a fault-duration of 0.5 s. Higher values are applied only in Australia (1500 V) and in Brazil (1700 V), whereas the Brazilian regulation defines the admissible value of a touch current, which is converted for comparison into the voltage limit. Within this survey, only two US-utilities have answered the questions on voltage limits. The limit for the touch voltage applied there is given to be 500 V, whereas the voltage pipeline-to-earth should be less than 5 kV. It is unclear and could not be clarified in the CIGRE-survey, why low value for touch voltage (500 V) as compared with other countries resulted in a comparatively high value for the pipeline-to-earth voltage (5 kV).

According to [21] the maximal permissible voltage pipeline-to-earth for short-time interference shall be below 1000 V. If ASME/IEEE-standard No. 80 is applied a maximal permissible touch voltage is defined in relation to the fault duration time for different body weight of the person involved. If the most severe restrictions are applied, i.e., 50 kg body weight and fault duration (clearing time) of 150 ms, the maximal permissible touch voltage U_{t50} is 350 V. This is the recommended limit as per IEEE-standard No. 80 item 6, where it is mentioned that the actual transferred voltage should be less than the maximum allowable touch voltage E_{touch} to ensure safety.

12.1.2 *Calculation of impedances for inductive interference*

In order to calculate the interference of pipelines the loop-impedances, coupling impedances and self-impedances of the line conductor, earth conductors and the pipeline itself are required. The loop-impedance of the pipeline with earth return is

$$\underline{Z}'_{\mathrm{P}} = R'_{\mathrm{P}} + \frac{\mu_0}{8} * \omega + j\left(\frac{\mu_0}{2\pi} * \omega * \left(\ln\frac{\delta}{r_{\mathrm{P}}} + \frac{\mu_{\mathrm{P}}}{4}\right)\right) \tag{12.1}$$

where r_{P} is the outer radius of the pipeline, μ_0 is the absolute permeability, μ_{P} is the fictitious relative permeability of the pipeline, δ is the depth of earth return path and ω is the angular frequency.

Equation (12.1) is composed of

R'_P	Resistance of the pipeline per unit length
$\frac{\mu_0}{8} * \omega$	Resistance of the earth return path per unit length
$\frac{\mu_0}{2\pi} * \omega * \ln \frac{\delta}{r_P}$	Outer reactance of the loop with earth return path per unit length
$\frac{\mu_0}{2\pi} * \omega * \frac{\mu_P}{4}$	Internal (inner) reactance of the conductor (pipeline) per unit length

The depth δ of the earth return path is given by Equation (12.2) with the resistivity of soil ρ (specific soil resistance) according to Table 12.2.

$$\delta = \frac{1.85}{\sqrt{\omega * (\mu_0/\rho)}} \tag{12.2}$$

Table 12.2 Resistivity of soil ρ for different types of soil conditions

Parameters	Type of soil						
	Alluvial soil – swamp soil	Clay	Limestone clay – farm soil	Wet sand	Wet gravel	Dry sand Dray gravel	Stony soil
Specific soil resistance ρ (Ωm)	30	50	100	200	500	1000	3000
$\sigma = 1/\rho$ (μS/cm)	333	20	10	5	2	1	0.33
Depth of earth return δ at 50 Hz (m)	510	660	930	1320	2080	2940	5100
Depth of earth return δ at 60 Hz (m)	465	600	850	1205	1900	2684	4655

The resistance R'_P of the pipeline can be calculated from the conductivity κ_P and the thickness d of the pipeline wall taking eddy currents and the dissipation of the current into the outer level of the pipeline wall into account.

$$R'_P = R'_{dc} * \left(x + 0.25 + \frac{3}{64 * x}\right) \tag{12.3}$$

The increase of the inner inductivity X'_i is given by

$$X'_i = R'_{dc} * \left(x - \frac{3}{64 * x} + \frac{3}{128 * x^2}\right) \tag{12.4}$$

with the parameter $x = d/(2 * \delta_P)$, the d.c. resistance of the pipeline wall R'_{dc}, the depth of current in the pipeline wall δ_P and d being the thickness of the pipeline wall:

$$\delta_P = \frac{1}{\sqrt{\omega/2 * (\kappa_P * \mu_P * \mu_0)}} \tag{12.5}$$

Comparing the loop-impedance as per Equation (12.1) with the inner inductance of the pipeline as per Equation (12.4) the fictitious relative permeability of the pipeline will be

$$\mu_P = 4 * R'_{dc} * \frac{x - 3/(64 * x) + 3/(128 * x^2)}{(\omega * \mu_0)/2\pi} \tag{12.6}$$

Coupling impedances [2] need to be calculated for the analysis of the interference problems. For the individual distances and impedances reference is made to Figure 12.1 indicating a typical interference problem between a 380-kV-line with two earth wires, counterpoise and a pipeline.

The coupling impedance $\underline{Z}_{LP}$ of the loop conductor and pipeline with earth return is obtained from

$$\underline{Z}'_{LP} = \frac{\mu_0}{8} * \omega + j\left(\frac{\mu_0}{2\pi} * \omega * \left(\ln \frac{\delta}{d_{LP\,min}}\right)\right) \tag{12.7}$$

The coupling impedance $\underline{Z}_{EP}$ of the loop earth conductor and pipeline with earth return is obtained from

$$\underline{Z}'_{EP} = \frac{\mu_0}{8} * \omega + j\left(\frac{\mu_0}{2\pi} * \omega * \left(\ln \frac{\delta}{d_{EP\,min}}\right)\right) \tag{12.8}$$

Coupling impedance $\underline{Z}_{LE}$ of the loop earth conductor and conductor with earth return is given by

$$\underline{Z}'_{LE} = \frac{\mu_0}{8} * \omega + j\left(\frac{\mu_0}{2\pi} * \omega * \left(\ln \frac{\delta}{d_{LE}}\right)\right) \tag{12.9}$$

In case a second earth conductor is installed, the coupling impedance $\underline{Z}_{EP2}$ of the loop second earth conductor and pipeline with earth return is calculated by

$$\underline{Z}'_{EP2} = \frac{\mu_0}{8} * \omega + j\left(\frac{\mu_0}{2\pi} * \omega * \left(\ln \frac{\delta}{d_{EP2\,min}}\right)\right) \tag{12.10}$$

The coupling impedance $\underline{Z}_{LE2}$ of the loop second earth conductor and conductor with earth return is obtained from

$$\underline{Z}'_{LE2} = \frac{\mu_0}{8} * \omega + j\left(\frac{\mu_0}{2\pi} * \omega * \left(\ln \frac{\delta}{d_{LE2}}\right)\right) \tag{12.11}$$

The coupling impedance $\underline{Z}_{E12}$ of the loop first and second earth conductor and conductor with earth return is obtained from

$$\underline{Z}'_{E12} = \frac{\mu_0}{8} * \omega + j\left(\frac{\mu_0}{2\pi} * \omega * \left(\ln \frac{\delta}{d_{E12}}\right)\right) \tag{12.12}$$

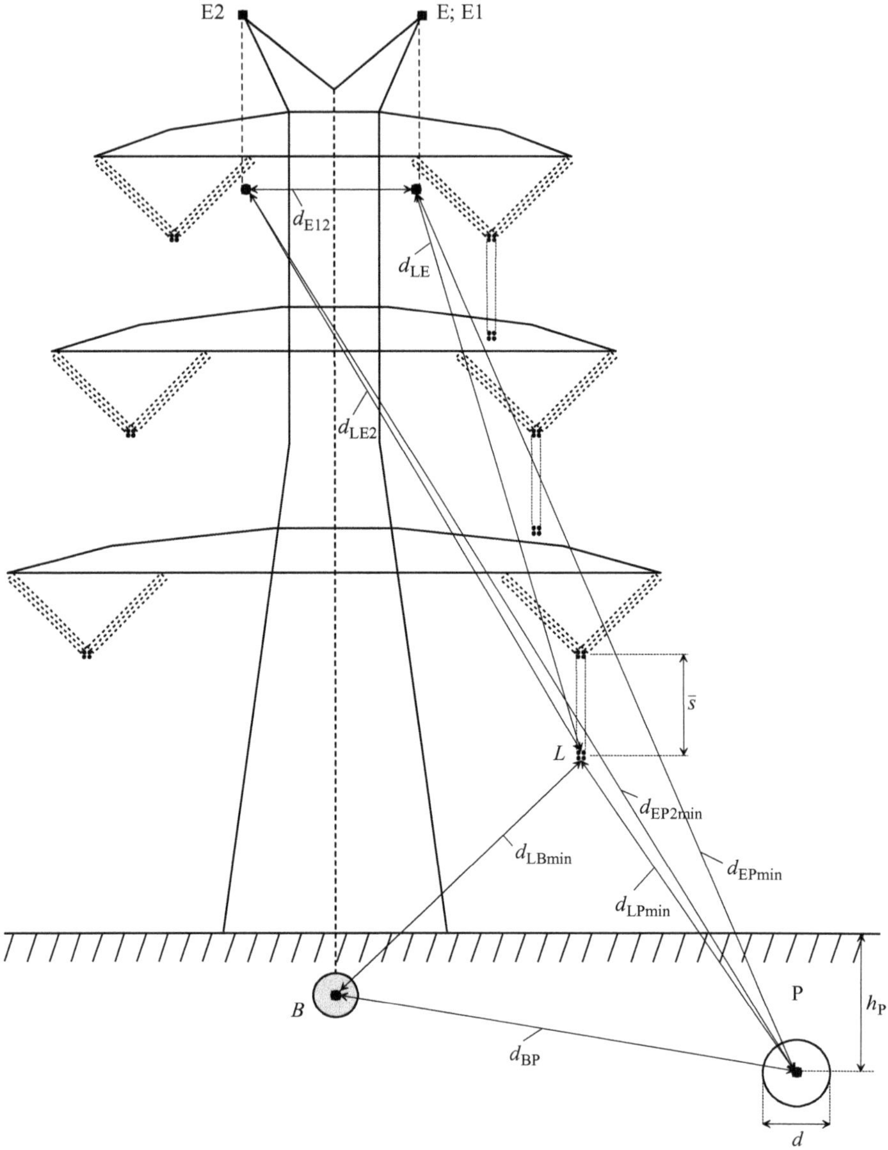

Figure 12.1 *Outline and distances of a high-voltage transmission-line tower. B: counterpoise; P: pipeline; L: conductor nearest to pipeline. E; first earth conductor (nearest to pipeline), also named E1; E2: second earth conductor*

Furthermore the loop-impedance $\underline{Z}_E$ of the earth conductor and earth return is given by

$$\underline{Z}'_E = R' + \frac{\mu_0}{8} * \omega + j\left(\frac{\mu_0}{2\pi} * \omega * \left(\ln\left(\frac{\delta}{r}\right) + \frac{\mu_r}{4}\right)\right) \tag{12.13}$$

Quantities as per Equations (12.7) to (12.13) are:

d_{LPmin}	Minimal distance between the pipeline and the lowest conductor nearest to the pipeline
d_{EPmin}	Minimal distance between the pipeline and the earth conductor
d_{LE}	Distance between the earth conductor and the lowest conductor nearest to the pipeline
d_{EP2min}	Minimal distance between the pipeline and the second earth conductor
d_{LE2}	Distance between the second earth conductor and the lowest conductor nearest to the pipeline
d_{E12}	Distance between the first and second earth conductor
r	Radius of earth conductor
R'	Resistance of earth wire per unit length.

Generally the minimal distances between the conductors and the pipeline have to be considered. This includes considerations on the conductor sag and the conductor swing under worst conditions. The mean effective height h_S of the conductor may be calculated from

$$h_S = h_L - 0.667 * \bar{s} \tag{12.14}$$

where h_L is the conductor height at the tower and $\bar{s}$ is the conductor sag.

In some cases a counterpoise parallel to the pipelines is used to reduce the induced voltage into the pipeline. The coupling impedances with the conductor are needed in these cases. The coupling impedance $\underline{Z}_B$ of the loop counterpoise and earth return is obtained from

$$\underline{Z}'_B = R'_B + \frac{\mu_0}{8} * \omega + j\left(\frac{\mu_0}{2\pi} * \omega * \left(\ln\left(\frac{\delta}{r_B}\right) + \frac{\mu_B}{4}\right)\right) \tag{12.15}$$

The coupling impedance $\underline{Z}_{LB}$ of the loop conductor and counterpoise with earth return is obtained from

$$\underline{Z}'_{LB} = \frac{\mu_0}{8} * \omega + j\left(\frac{\mu_0}{2\pi} * \omega * \left(\ln\frac{\delta}{d_{LB\,min}}\right)\right) \tag{12.16}$$

The coupling impedance $\underline{Z}_{BP}$ of the loop pipeline and counterpoise with earth return is calculated from

$$\underline{Z}'_{BP} = \frac{\mu_0}{8} * \omega + j\left(\frac{\mu_0}{2\pi} * \omega * \left(\ln\frac{\delta}{d_{BP}}\right)\right) \tag{12.17}$$

where r_B is the radius of counterpoise, μ_B is the fictitious relative permeability of counterpoise, R'_B the resistance of counterpoise per unit length, d_{LBmin} is the minimal distance between counterpoise and the lowest conductor nearest to the pipeline, d_{BP} is the distance between counterpoise and pipeline.

12.1.3 Calculation of induced voltage

Based on the impedance calculations as per Section 12.1.2 the induced voltage into the pipeline can be calculated for any configuration as follows:

$$\underline{U}_{iP} = -\underline{Z}'_{LP} * \underline{r}' * \underline{I}''_{kE} * l_{Pp} * w \tag{12.18}$$

where $\underline{Z}'_{LP}$ is the coupling impedance of the loop pipeline and conductor nearest to earth with earth return, $\underline{I}''_{kE}$ is the initial short-circuit current (asymmetrical) of the overhead line, l_{Pp} is the length of parallel exposure between pipeline and overhead line, $\underline{r}'$ is the screening factor as per Equation (12.18) and w is a probability factor taking into account that all worse conditions do not occur at the same time instant [21].

The screening factor depends on the presence of earth wires, counterpoises and any other compensation circuit capable to reduce the induced voltage into the pipeline. If only one earth wire is present as compensation circuit the screening factor $\underline{r}'$ is calculated based on the coupling and loop-impedances by

$$\underline{r}' = 1 - \frac{\underline{Z}'_{LE} * \underline{Z}'_{EP}}{\underline{Z}'_{E} * \underline{Z}'_{LP}} \tag{12.19}$$

If more than one earth conductor or additional compensation circuits are present, additional considerations for the screening factor are required. The total screening factor is then given by the difference of the individual factors of each earth conductor and/or compensation circuit taking into account the correction factor which represents the influence of each earth wire and compensation circuit on the current in the other wire [22]. In the case of two earth conductors the total screening factor $\underline{r}'_{tot}$ is given by

$$\underline{r}'_{tot} = 1 - \frac{\underline{Z}'_{LE} * \underline{Z}'_{EP}}{\underline{Z}'_{E} * \underline{Z}'_{LP}} * \underline{k}_2 - \frac{\underline{Z}'_{LE2} * \underline{Z}'_{EP2}}{\underline{Z}'_{E2} * \underline{Z}'_{LP}} * \underline{k}_1 \tag{12.20}$$

Correction factors k_1 and k_2 are calculated by

$$\underline{k}_1 = \frac{1 - (\underline{Z}'_{LE} * \underline{Z}'_{E12})/(\underline{Z}'_{E} * \underline{Z}'_{LE2})}{1 - (\underline{Z}'_{E12} * \underline{Z}'_{E12})/(\underline{Z}'_{E} * \underline{Z}'_{E2})} \tag{12.21a}$$

$$\underline{k}_2 = \frac{1 - (\underline{Z}'_{LE2} * \underline{Z}'_{E12})/(\underline{Z}'_{E2} * \underline{Z}'_{LE})}{1 - (\underline{Z}'_{E12} * \underline{Z}'_{E12})/(\underline{Z}'_{E} * \underline{Z}'_{E2})} \tag{12.21b}$$

Correction factors for other arrangements and numbers of conductors can be obtained, respectively calculated, as per [39]. Impedances as per Equations (12.19) to (12.21) are calculated according to Section 12.1.2

$\underline{Z}_{LP}$	Coupling impedance of the loop conductor and pipeline with earth return as per Equation (12.7)
$\underline{Z}_{EP}$	Coupling impedance of the loop first earth conductor and pipeline with earth return as per Equation (12.8)
$\underline{Z}_{LE}$	Coupling impedance of the loop first earth conductor and conductor with earth return as per Equation (12.9)
$\underline{Z}_{EP2}$	Coupling impedance of the loop second earth conductor and pipeline with earth return as per Equation (12.10)
$\underline{Z}_{LE2}$	Coupling impedance of the loop first earth conductor and conductor with earth return as per Equation (12.11)
$\underline{Z}_{E12}$	Coupling impedance of the loop first and second earth conductor with earth return as per Equation (12.12)
$\underline{Z}_{E}$	Loop-impedance of earth conductor and earth return as per Equation (12.13)
$\underline{Z}_{E2}$	Loop-impedance of second earth conductor and earth return as per Equation (12.13)

For three and more compensation circuits, e.g., two earth conductors and a counterpoise, the screening factor needs to be calculated either from the individual current distribution within the different compensation circuits or by means of the multiplication method or other methods as outlined in [3].

12.1.4 Characteristic impedance of the pipeline

The induced voltage $\underline{U}_{iP}$ is only an indication for the inductive interference but does not take into account the earthing conditions and the conductivity of the pipeline coating. In order to calculate step and touch voltages the voltage between pipeline and earth is to be calculated taking account of the earthing and conductivity conditions of the pipeline. Based on the analysis of the system equivalent, i.e., faulted phase conductor, presence of earth conductors, counterpoise, pipeline and earthing conditions, the pipeline must be represented by means of its characteristic impedance and its conductivity.

The conductivity of the pipeline against the surrounding earth is determined by the resistance R'_I of the pipeline coating and the resistance R'_{PE} of the bare (uncoated) pipeline in earth (see Table 12.3). The total resistance is given by

$$R'_{Ctot} = R'_I + R'_{PE} \tag{12.22}$$

The specific reactance due to the capacitance of the pipeline can be neglected for interference analysis as it is much smaller than the resistance. The resistance R'_{PE} of

Table 12.3 Resistance of pipeline coatings [23]

Type of coating	Thickness of coating (mm)	Specific coating resistance (k Ωm^2)
Bitumen	4–6	10
Polyethylene	2–3	100
Epoxy resin	0.3–0.6	10

the bare (uncoated) pipeline in earth is calculated by

$$R'_{PE} = \frac{\rho}{4\pi} * \left(2 * \ln\left(\frac{2 * l}{d}\right) + \ln\frac{\sqrt{(2 * h_P)^2 + (l/2)^2} + l/2}{\sqrt{(2 * h_P)^2 + (l/2)^2} - l/2}\right) \tag{12.23}$$

where h_P is the depth of pipeline under ground, l is the total length of pipeline, d is the outer diameter of pipeline and ρ is the resistivity of the soil as per Table 12.2.

It should be noted that the influence of the earth resistivity is comparatively low for high resistance of pipeline coating.

The characteristic impedance $\underline{Z}_W$, the propagation constant γ and the characteristic length L_K of the pipeline are required prior to the calculation of the voltage pipeline-to-earth.

$$\underline{Z}_W = \sqrt{\underline{Z}'_P * R'_{PE}} \tag{12.24}$$

$$\underline{\gamma} = \sqrt{\frac{\underline{Z}'_P}{R'_{PE}}} \tag{12.25}$$

$$L_K = \frac{1}{\mathrm{Re}\left\{\sqrt{\underline{Z}'_P / R'_{PE}}\right\}} \tag{12.26}$$

where $\underline{Z}'_P$ is the loop-impedance of pipeline and earth return per unit length as per Equation (12.1) and R'_{PE} is the resistance of the bare (uncoated) pipeline as per Equation (12.23).

12.1.5 Voltage pipeline-to-earth

The voltage pipeline-to-earth $\underline{U}_{PE}(x)$ along the exposure length is calculated based on the theory of line propagation for each individual location x of the pipeline. Using the abbreviations

$$\underline{I}^*_{kE} = \underline{r}'_{tot} * \underline{I}''_{kE} * \frac{\underline{Z}_{LP}}{\underline{Z}_P} * w \tag{12.27a}$$

and

$$\underline{b} = 0.5 * \sqrt{\underline{Z}'_{\text{P}} * R'_{\text{PE}}} * l \tag{12.27b}$$

where l is the length of pipeline, $\underline{Z}'_{\text{P}}$ is the loop-impedance of pipeline and earth return per unit length as per Equation (12.1), $\underline{R}'_{\text{PE}}$ is the resistance of the bare (uncoated) pipeline as per Equation (12.23), $\underline{r}_{\text{tot}}$ is the total reduction factor as per Equation (12.20), $\underline{Z}_{\text{LP}}$ the coupling impedance of the loop conductor and pipeline with earth return as per Equation (12.7), $\underline{I}''_{\text{kE}}$ is the short-circuit current through earth and w the probability factor.

The voltage pipeline-to-earth along the exposure length (parameter x) is calculated taking account of the earthing resistances R_1 and R_2 at the end of the pipeline, respectively at the end of section, under investigation by

$$\begin{aligned}\underline{U}_{\text{PE}}(x) = \underline{Z}_{\text{W}} * \underline{I}^*_{\text{kE}} * \{ & R_1(\underline{Z}_{\text{W}} + R_2)\text{e}^{\underline{b}+\underline{\gamma}x} + R_2(\underline{Z}_{\text{W}} - R_1)\text{e}^{-\underline{b}+\underline{\gamma}x} \\ & - R_1(\underline{Z}_{\text{W}} - R_2)\text{e}^{-\underline{b}-\underline{\gamma}x} - R_2(\underline{Z}_{\text{W}} + R_+)\text{e}^{\underline{b}-\underline{\gamma}x}\} \\ & *\{(\underline{Z}_{\text{W}}+R_1) * (\underline{Z}_{\text{W}} + R_2)\text{e}^{2\underline{b}} - (\underline{Z}_{\text{W}} - R_1) * (\underline{Z}_{\text{W}} - R_2)\text{e}^{-2\underline{b}}\}^{-1}\end{aligned} \tag{12.28}$$

Outside the exposure length (parameter y) the voltage pipeline-to-earth decays for each individual location y of the pipeline according to

$$\begin{aligned}\underline{U}_{\text{PE}}(y) = \underline{U}_1 * & \frac{R_{\text{A}} - \underline{Z}_{\text{W}}}{(R_{\text{A}} + \underline{Z}_{\text{W}})\text{e}^{2\underline{\gamma}l} + (R_{\text{A}} - \underline{Z}_{\text{W}})}\text{e}^{\underline{\gamma}l} \\ & + \frac{(R_{\text{A}} + \underline{Z}_{\text{W}})\text{e}^{2\underline{\gamma}l}}{(R_{\text{A}} + \underline{Z}_{\text{W}})\text{e}^{2\underline{\gamma}l} + (R_{\text{A}} - \underline{Z}_{\text{W}})}\text{e}^{-\underline{\gamma}l}\end{aligned} \tag{12.29}$$

where $\underline{U}_1$ is the voltage at the end of exposure length, R_{A} is the far-end impedance of the pipeline, l is the total length of pipeline outside exposure length, γ is the propagation constant and $\underline{Z}_{\text{W}}$ is the characteristic impedance.

The method described above assumes constant parameters $\underline{I}^*_{\text{kE}}$ and for the effective resistance R'_{Ptot} of the pipeline coating against earth and constant distance between the pipeline and the overhead line. In other cases the analysis has to be carried out for each individual section having constant or nearly constant parameters by separation of the exposure length into subdivisions for which constant parameters can be assumed.

In case of oblique exposures or crossing between overhead line and pipeline the exposure length has to be divided into different subsections having equal induced voltages. An average distance a_{o} between the overhead line and the pipeline has to be calculated

$$a_{\text{o}} = \sqrt{a_1 * a_2} \tag{12.30}$$

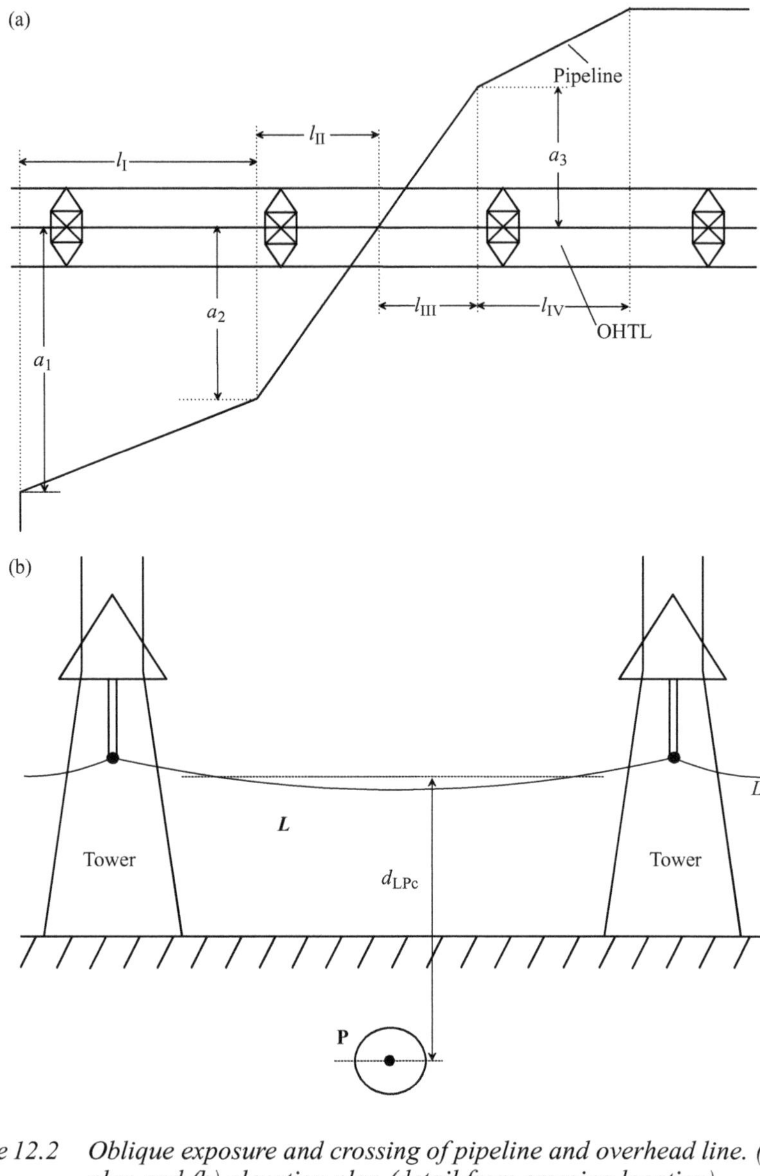

Figure 12.2 Oblique exposure and crossing of pipeline and overhead line. (a) Plot plan and (b) elevation plan (detail from crossing location)

The parameters a_1 and a_2 are the distances at the ends of the oblique exposure, see Figure 12.2 for details. The ratio a_1/a_2 should not exceed the value of three otherwise further subdivisions have to be selected. Distances of more than 1000 m between overhead line and pipeline can be neglected for short-time interference.

Crossings shall be handled similarly to oblique exposures with the restriction that the subdivisions in the vicinity of the crossing shall be selected as short as possible.

The distance between the overhead line and the pipeline at the location of crossing shall be set equal to the minimal height of the overhead conductor above the pipeline d_{LPc}, taking account of the average conductor sag, see Figure 12.2(b). The average distance a_{c} between the overhead line and the pipeline is given by

$$a_{\mathrm{c}} = \sqrt{a_2 * d_{\mathrm{LPc}}} \tag{12.31}$$

The results for each subdivision have to be superimposed for each location of the pipeline.

12.2 Considerations on earthing

12.2.1 General

The influence of neutral handling and to some extent of the earthing was outlined already in Chapters 5 and 7. The handling of neutrals, however, requires additionally the analysis of the earth itself, of earthing grids and rods and on step and touch voltages related to earthing. With respect to the main task of this book, i.e., calculation and analysis of short-circuit currents and their effects, the earthing in power systems are only dealt with in relation to the impact of short-circuit currents on earthing. Special problems such as corrosion of earthing material, influence of earthing on lightning and on fast-front overvoltages are not explained here. More details can be found in [5, 26].

Equipment and installations in power systems have to be designed and operated in such a way to avoid impermissible conditions with respect to the health of human beings and animals also taking into account reliable and sufficient operation of the technical installations. Earthing in power systems is one of the main items to ensure this safe and secure operation. Asymmetrical operation and short-circuits cause currents flowing through earth which may flow as well through the human body in the case of contact of the body with earth or with installations connected to earth. The earthing problem is determined by

- Resistance of human body
- Electrical conditions of the earth
- Current through earth
- Fault duration
- Earthing impedance

12.2.2 Resistance of human body

The impedance of the human body, which is mainly a resistance, is determined by the location of contacts to the electrical installations and depends on the touch voltage as well. Figure 12.3 indicates the resistance of the body if measured between the two hands, valid approximately 10 ms after initiation of currents through the body. As

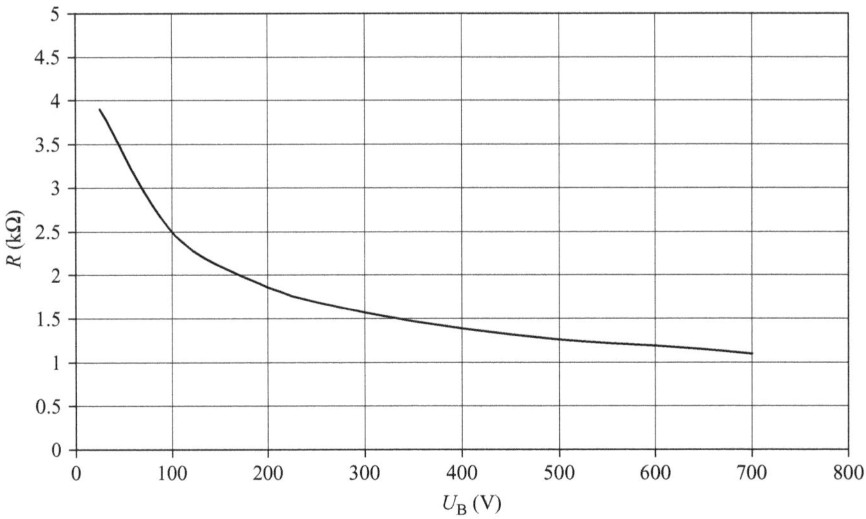

Figure 12.3 *Impedance of the human body (hand-to-hand) depending on the touch voltage*

can be seen the resistance is between 3.5 kΩ and 1.1 kΩ decreasing with increasing touch voltage. For other contacts the resistance is reduced, e.g.,

- One hand to breast reduction to 45 per cent of resistance as per Figure 12.3
- One hand to knee reduction to 70 per cent of resistance as per Figure 12.3
- Two hands to breast reduction to 23 per cent of resistance as per Figure 12.3
- Two hands to knee reduction to 45 per cent of resistance as per Figure 12.3

It should be noted in this respect that the current through the human body is the critical physical phenomena causing uncontrolled operation of the heart (ventricle flicker) or muscle convulsion. The critical current is a function of the exposure time [24]. The duration of the permissible touch voltage U_B therefore depends on the exposure time as well (fault duration) and is outlined in Figure 12.4.

The permissible touch voltage is set in most of the standards to $U_B = 65$ V, resulting in a negligible risk for ventricle flicker independent from the time of exposure. If this voltage limit cannot be guaranteed the exposure time must be limited.

12.2.3 Soil conditions

The resistivity of the soil depends on the type of soil (swamp soil, stony soil) as outlined in Table 12.2. The differences are determined by the humidity of the soil. The value for swamp soil is between 10,000 Ωm (humidity less than 10 per cent) and 30 Ωm (humidity 90 per cent). It is obvious that the resistivity of the soil varies in a wider range if surface electrodes will be used for earthing as the soil humidity

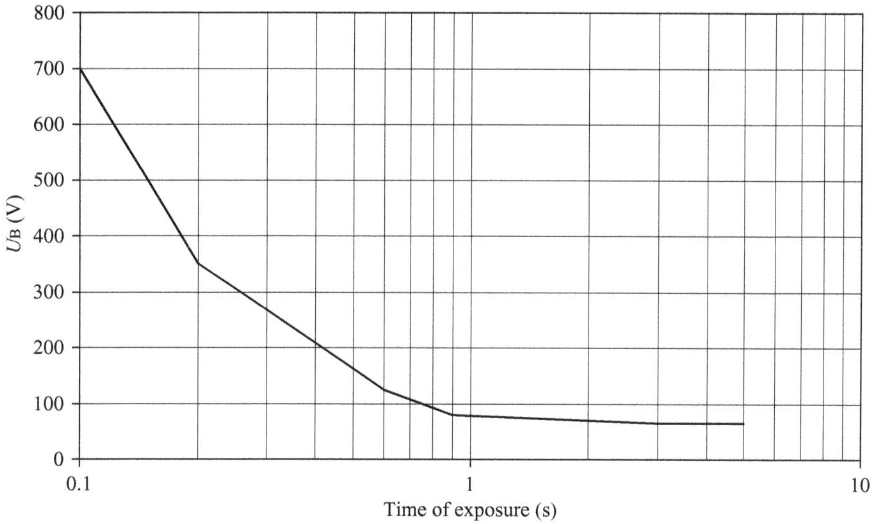

Figure 12.4 Permissible touch voltage depending on the time of exposure

varies in a wider range throughout the year as compared with deep-ground earthing by earthing rods. The resistivity for surface electrodes varies throughout the year in a range of ±30 per cent, in the case of deep-earthing only in a range of ±8 per cent of the average value, the highest value occurring in March and the lowest value in August (European countries). It is therefore recommended to use surface electrodes only in those installations where the soil humidity is nearly constant throughout the year and to use deep-ground earthing for other conditions.

12.2.4 Relevant currents through earth

Currents through earth are only existing in the case of asymmetrical short-circuits with earth connection. In most of the cases ($X_0/X_1 > 1$), the single-phase short-circuit current I''_{k1} is greater than the current through earth I''_{kE2E} in case of a double-phase short-circuit. Only in case of power systems (e.g., 115 and 132 kV) having high amount of cables, power stations with high rating closely connected to the system under investigation and when all transformer neutrals are earthed through low impedance, can the ratio X_0/X_1 be below One. In these cases the double-phase short-circuit current may exceed the single-phase short-circuit current.

Current through earth causes voltage drops U_E at the impedance of the earth, at the earthing itself and at the connection lines between the equipment (e.g., neutrals of transformers, transformer tank, overhead line tower) and earth. The currents to be taken into account as summarised in Table 12.4 depend on the type of neutral handling in the system.

Table 12.4 Currents through earth for the design of earthing installations

Valid for	Type of neutral handling			
	Isolated neutral	Low-impedance earthing	Earthing with current limitation	Resonance earthing
Voltage at earthing Touch voltage	I_{CE}	Current through earthing impedance	Current through earthing impedance	With reactor: $\sqrt{I_{rD}^2 + I_{res}^2}$ Without reactor: I_{Res}
Thermal stress of earthing and earthing connections	I''_{kEE}	I''_{k1} or I''_{kE2E}	I''_{k1} or I''_{kE2E}	I''_{kEE}
Earthing connections	I_{CE}	I''_{k1} or I''_{kE2E}	I''_{k1} or I''_{kE2E}	I''_{kEE}
Earthing voltage U_E Touch voltage U_B	$U_E \leq 125$ V: no measures required $U_E > 125$ V then $U_B \leq 65$ V or equivalent measures according to VDE 0141	$U_E > 125$ V then U_B according to VDE 0141 depending on fault duration or fast switch-off or equivalent measures according to VDE 0141	$U_E > 125$ V then U_B according to VDE 0141 depending on fault duration or fast switch-off or equivalent measures according to VDE 0141	$U_E > 125$ V then $U_B \leq 65$ V or equivalent measures according to VDE 0141

Table 12.5 Reduction factor for typical power system installations; distance of earth conductor to phase conductor $D \approx 20\ m$; $\rho = 100\ \Omega m$

Number of conductors	Type and cross-section	Reduction factor
1 Earth conductor	St 50–90	0.98
	Al/St 50/30	0.78
	Al/St 120/70	0.7
	Al/St 240/40	0.65
	Al/St 120/70; counterpoise Cu 120	0.52
2 Earth conductors	Al/St 95/55	0.5
Cable sheet	NA2XS(2)Y; 150Al; 20 kV	0.5
	Single-core cable; 110 kV	0.1
	N2XS(SL)2Y; 240Cu; 110 kV	0.27
	Cables in steel tube	<0.05

As mentioned already in Section 7.3, a part of the single- or double-phase earth-fault current is flowing through earth and the relevant impedances of the earthing installations, depending on the reduction factor of the earth conductor and/or of cable screens and sheets connected to earth. Detailed investigations are required to determine the reduction factor. For rough estimates the reduction factors for typical installations are given in Table 12.5.

12.2.5 Earthing impedance

The impedance, i.e., the resistance of the earthing installations is determined by the material of the earthing grid, electrodes and rods and by the presence of any connection of earth conductor, counterpoise, cable sheets and other earthed installations in the vicinity of the earthing installation. The earthing resistance is proportional to the resistivity of the soil and depends furthermore on the specific arrangements of the earthing installations as outlined in Table 12.6.

The equations as per Table 12.6 for the calculation of the resistance of earthing installation indicate that it is not recommendable to increase the number of meshes in an earthing grid in order to reduce the earthing impedance, as the effect is only marginal. Increasing the number of earthing rods on the same earthing installation is highly recommendable as the total earthing impedance is approximately reciprocal to the number of rods. Sufficient distance between the individual rods (at least more than the rod-length) shall be provided in this case. Material, cross-sections and laying of the earthing installations must comply with the relevant standards, e.g., ANSI C 33.8 (standard for safety grounding and bonding equipment).

Table 12.6 Resistance of earthing installations R_{EI} for different types and arrangement

Type of earthing			Earthing resistance
Single deep-ground rod	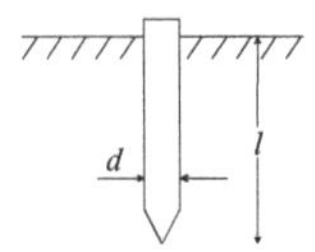		$R_{EI} = \frac{\rho}{2\pi * l} * \ln \frac{4l}{d}$
Multiple deep-ground rod	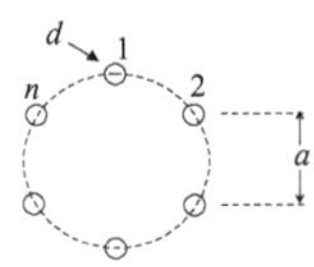	$a \geq l$ $k = 1.2$–1.5 $n = 5$: $k \approx 1.2$ $n = 10$: $k \approx 1.25$	$R_{EI} \approx k * \frac{1}{n} * \frac{\rho}{2\pi * l} * \ln \frac{4l}{d}$
Surface electrode	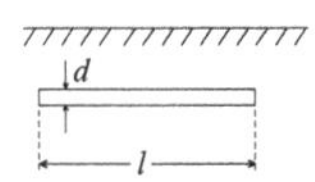		$R_{EI} = \frac{\rho}{\pi l} * \ln \frac{2l}{d}$
Crossed surface electrode	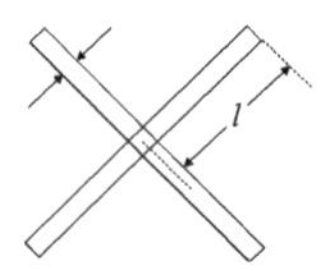		$R_{EI} \approx \frac{\rho}{2\pi * l} * \left(2.5 + \ln \frac{4l}{d}\right)$
Earthing grid: uniform resistivity of soil	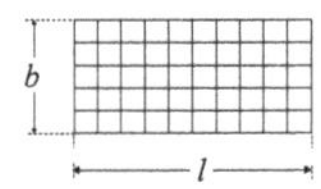	$D = \sqrt{\frac{4b * l}{\pi}}$ l_{tot} total length of earthing grid	$R_{EI} \approx \frac{\rho}{2 * D} + \frac{\rho}{l_{tot}}$
Earthing grid: two layers of resistivity	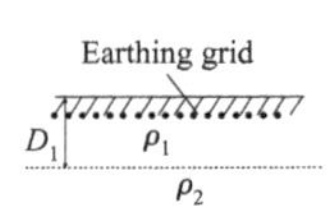	$D = \sqrt{\frac{4b * l}{\pi}}$ ρ_1: resistivity of surface layer ρ_2: resistivity of deep layer	$R_{EI} \approx \frac{\rho_2}{2 * D} + \frac{\rho_1 * D_1}{b * l}$

12.3 Examples

12.3.1 Interference of pipeline from 400-kV-line

The exposure of a 32″-pipeline with an overhead line (400 kV) as outlined in Figure 12.5 is analysed. The nearest distance of the pipeline is 27 m at tower No. 3, increasing to 160 m at tower No. 7 over a length of approximately 1600 m, decreasing to 120 m over a length of 1450 m and then crossing the overhead line at an angle of 90° between towers Nos 10 and 11. The elevation plan of the tower and the pipeline is given in Figure 12.6.

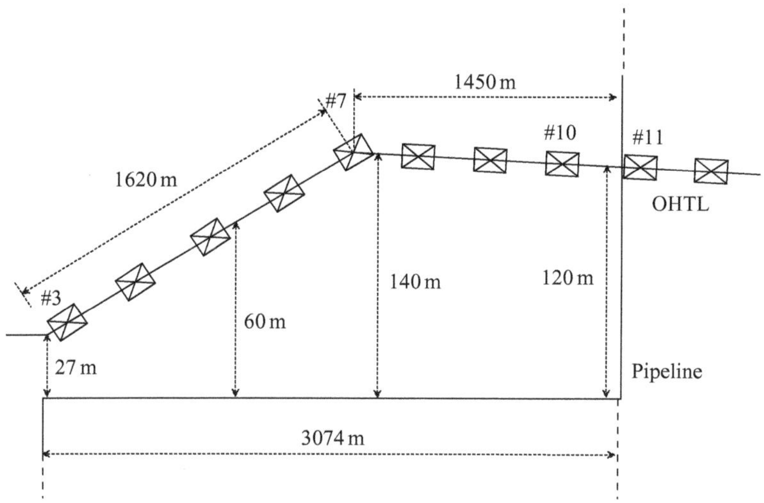

Figure 12.5 Plot plan of the exposure length pipeline and transmission line

The pipeline parameters are outlined below. The pipeline is not buried into earth, but laid directly on the surface. The total exposure length is 3074 m. At location $x = 0$ at tower No. 2 the pipeline is equipped with an isolating flange. The average sag of the line conductor and of the earth conductor is 10.2 m, the specific resistivity of the soil is 4 Ωm.

Type of pipeline	32″ steel pipe
Diameter	1524 mm
Thickness of wall	22.2 mm
Conductivity	5.56 Sm/mm^2
Relative permeability	200
Specific resistance of pipeline coating	20.3 Ω/km

The induced voltage of the pipeline was calculated with the procedure explained in the previous sections. The individual sections of the exposure length were chosen according to the distances between the transmission towers. The specific field strength and the specific induced voltage, both related to units of the short-circuit current, are given in Figure 12.7.

The analysis of the voltage pipeline-to-earth indicated a maximum value of $\underline{U}'_{PE} = 18.1$ V/kA which was obtained by superposition of the result from each section. The voltage $\underline{U}'_{PE}$ along the exposure length and a further 3 km outside the exposure section is outlined in Figure 12.8.

Short-circuit current calculation for the 400-kV-system under investigation indicated that the single-phase short-circuit current will always give the maximal asymmetrical short-circuit current. In order to cover future increase of short-circuit level, the maximal permissible short-circuit current in the system $I''_k = 40$ kA was taken as a basis for the assessment of interference problems. The voltage

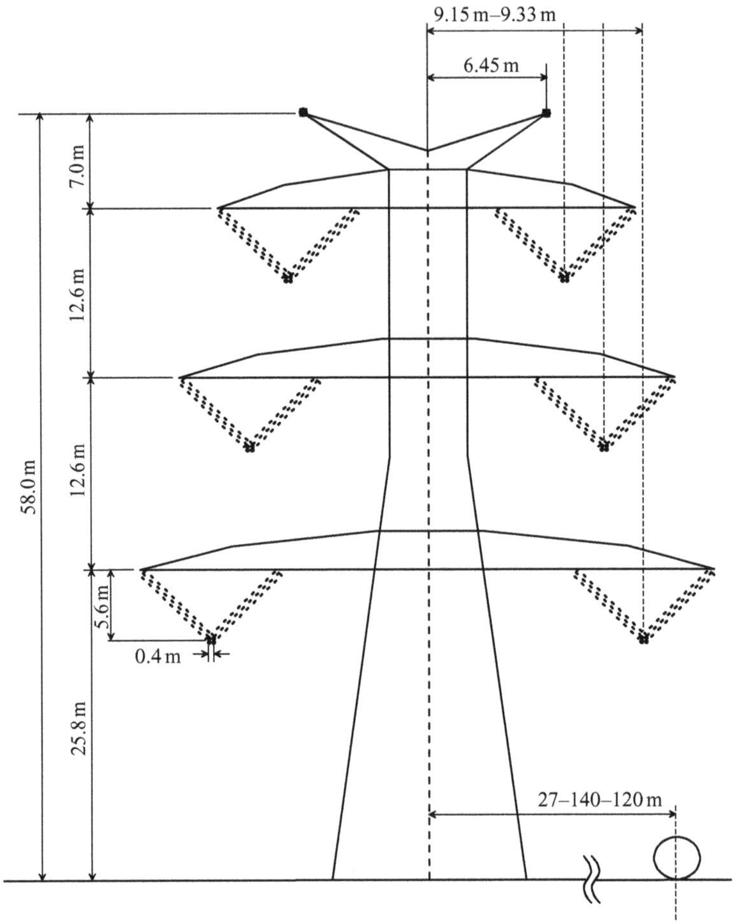

Figure 12.6 Elevation plan of the overhead transmission tower and the pipeline

pipeline-to-earth will reach $\underline{U}_{PE} = 724$ V in this case. With respect to [21] the voltage pipeline-to-earth will be below the maximal permissible voltage which is $\underline{U}_{PEmax} = 1000$ V. If ASME/IEEE-standard No. 80 is applied (50 kg body weight and fault duration 150 ms) the voltage pipeline-to-earth will be above the maximal permissible voltage $\underline{U}_{t50} \approx U_{PEmax} = 350$ V. Earthing at intermediate locations especially at the location where the voltage pipeline-to-earth is maximal must be done in order to reduce the voltage. More details on this example can be found in [27].

12.3.2 Calculation of earthing resistances

The resistance of an earthing grid within a switchyard of 80 m × 110 m; grid width 10 m × 10 m; $\rho = 100$ Ωm is

$$R_{EI} \approx \frac{\rho}{2 * D} + \frac{\rho}{l_{tot}} = \frac{100\ \Omega\text{m}}{2 * 105.9\ \text{m}} + \frac{100\ \Omega\text{m}}{1950\ \text{m}} = 0.532\ \Omega$$

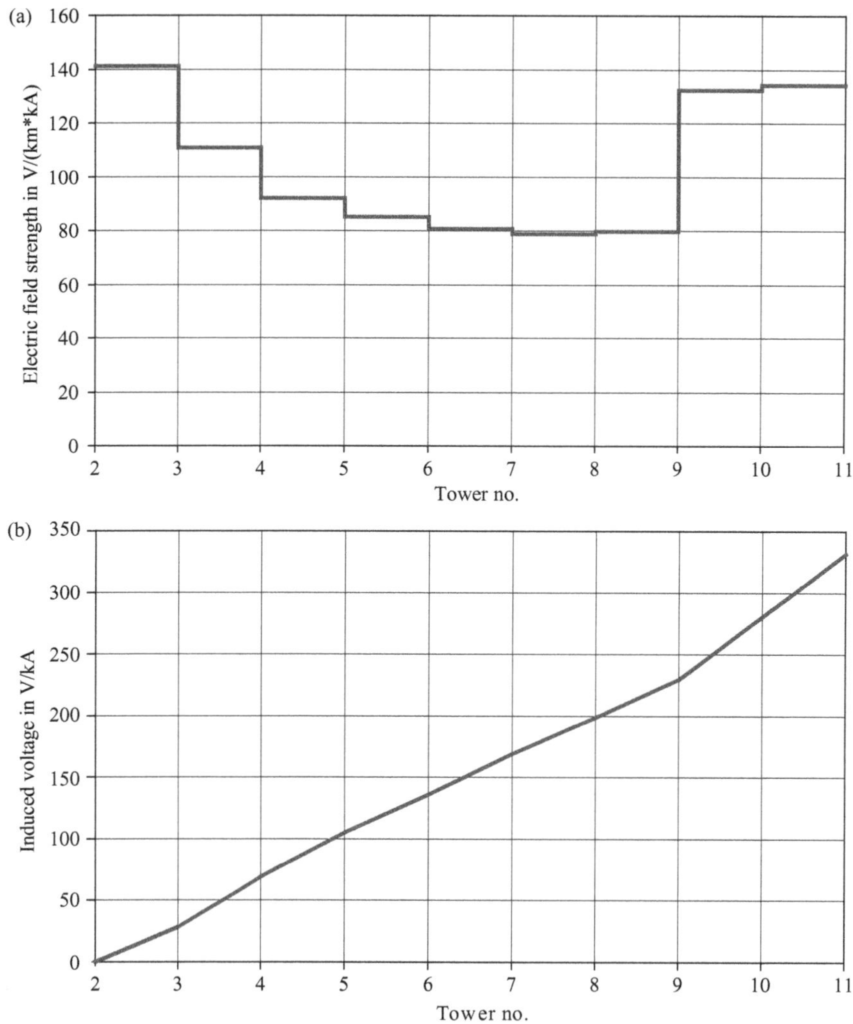

Figure 12.7 *Specific electric field strength (a) and specific induced voltage (b) of the pipeline between towers 2 and 11*

Effective length

$$D = \sqrt{\frac{4b * l}{\pi}} = \sqrt{\frac{4 * 80\ \text{m} * 110\ \text{m}}{\pi}} = 105.9\ \text{m}$$

Total length of the earthing grid $l_{tot} = 9 * 110\ \text{m} + 12 * 80\ \text{m} = 1950\ \text{m}$.

If the grid width is reduced to 5 m × 5 m, the total length of the earthing grid is $l_{tot} = 3550$ m. The resistance of the earthing grid is $R_{EI} = 0.5\ \Omega\text{m}$.

The same earthing grid (80 m × 110 m; grid width 10 m × 10 m) is now examined, except that the soil resistivity is assumed to be in two layers, the surface layer with a thickness $D_1 = 4$ m with $\rho_1 = 400\ \Omega\text{m}$ and the deeper layer with $\rho_2 = 600\ \Omega\text{m}$.

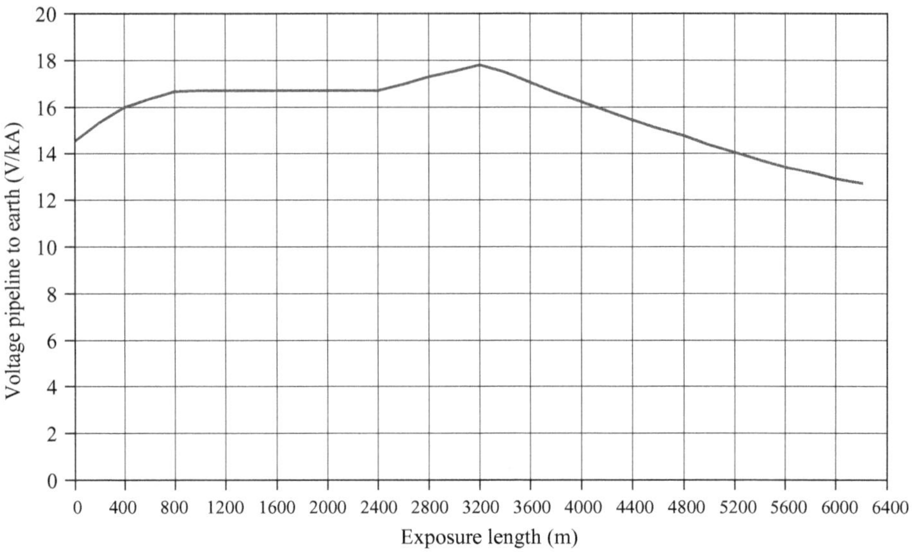

Figure 12.8 Voltage pipeline-to-earth along the exposure length (0–6400 m)

The earthing resistance is

$$R_{\mathrm{EI}} \approx \frac{\rho_2}{2 * D} + \frac{\rho_1 * D_1}{b * l} = \frac{400\ \Omega\mathrm{m}}{2 * 105.9\ \mathrm{m}} + \frac{100\ \Omega\mathrm{m} * 4\ \mathrm{m}}{80\ \mathrm{m} * 110\ \mathrm{m}} = 1.93\ \Omega$$

The resistance of an earthing with five deep-ground rods of length $l = 15$ m; diameter $d = 20$ mm; distance $a = 30$ m; $\rho = 100\ \Omega$m is given by

$$R_{\mathrm{EI}} \approx k * \frac{1}{n} * \frac{\rho}{2\pi * l} * \ln\frac{4l}{d} = 1.2 * \frac{1}{5} * \frac{100\ \Omega\mathrm{m}}{2\pi * 15\ \mathrm{m}} * \ln\frac{4 * 15\ \mathrm{m}}{0.02\ \mathrm{m}} = 2.04\ \Omega$$

whereas the earthing impedance of one rod is

$$R_{\mathrm{EI}} = \frac{\rho}{2\pi * l} * \ln\frac{4l}{d} = \frac{100\ \Omega\mathrm{m}}{2\pi * 15\ \mathrm{m}} * \ln\frac{4 * 15\ \mathrm{m}}{0.02\ \mathrm{m}} = 8.5\ \Omega$$

The earthing impedance of a surface electrode with length $l = 60$ m; diameter $d = 4$ mm; $\rho = 100\ \Omega$m is

$$R_{\mathrm{EI}} = \frac{\rho}{\pi l} * \ln\frac{2l}{d} = \frac{100\ \Omega\mathrm{m}}{\pi 60\ \mathrm{m}} * \ln\frac{2 * 60\ \mathrm{m}}{0.004} = 5.47\ \Omega$$

Chapter 13
Data of equipment

13.1 Three-phase a.c. equipment

A summary of relevant data of equipment is given in IEC 60909-2:1992. The data are based on a survey carried out by IEC TC 73. In some countries this document does not have the character of a standard.

13.1.1 System feeders

Impedances of power system feeders, respectively, their initial short-circuit power are difficult to determine as typical values, the structure of the power systems (cable or overhead line system), the voltage levels or the application task, i.e., for rural, urban or industrial power supply can vary in a wide range. Typical ranges of the initial short-circuit power (three-phase short-circuit) are given in Figure 13.1 for a power system with different voltage levels.

13.1.2 Transformers

Transformers are constructed with defined rated power with respect to their application. In low-voltage and to a certain extent in medium-voltage systems, transformers are build with standard rated power and standard impedance voltage. Transformers in high-voltage systems, sometimes also in medium-voltage systems, have to meet special application conditions, such as the internal standard of the utility with special defined values for rated power, impedance voltage and ohmic losses. It should be noted in this respect that the minimisation of the impedance voltage is limited by the minimal insulation thickness of the windings. Typical values for the impedance voltage of transformers are outlined in Figure 13.2. Figure 13.3 gives data for the short-circuit losses (ohmic losses). The relation of the impedance voltage (%-value)

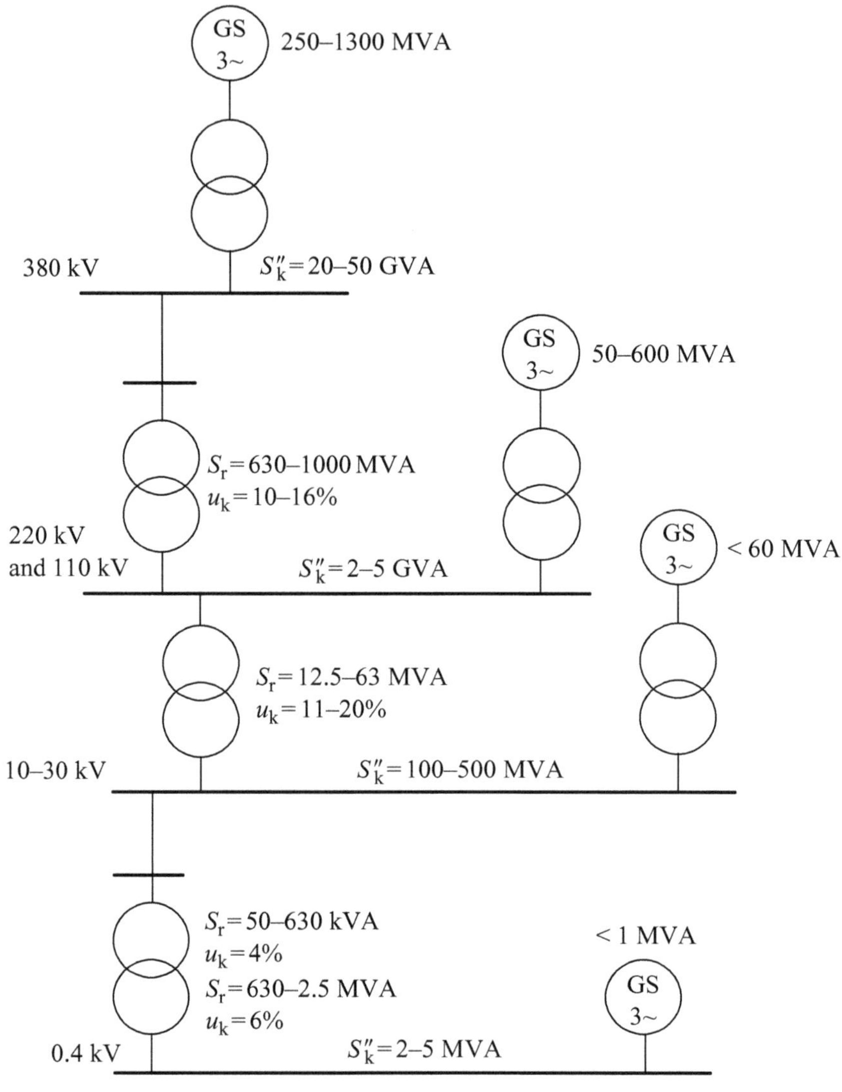

Figure 13.1 Principal structure of a power supply system and typical values of initial short-circuit power of public supply system [7]

to the rated power (MVA-value) as per IEC 60909-2:1992 is given below:

$$u_{\mathrm{kr}} = 8 + 0.92 * \ln(S_{\mathrm{rT}}) \tag{13.1}$$

Further data are included in Table 13.1.

The impedance voltage of auto-transformers is lower than that of full-winding transformers.

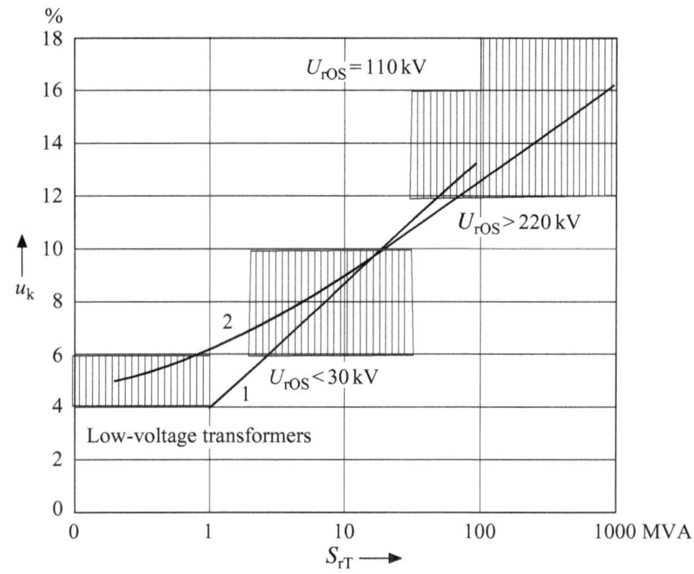

Figure 13.2 Typical values for the impedance voltage of two-winding transformers

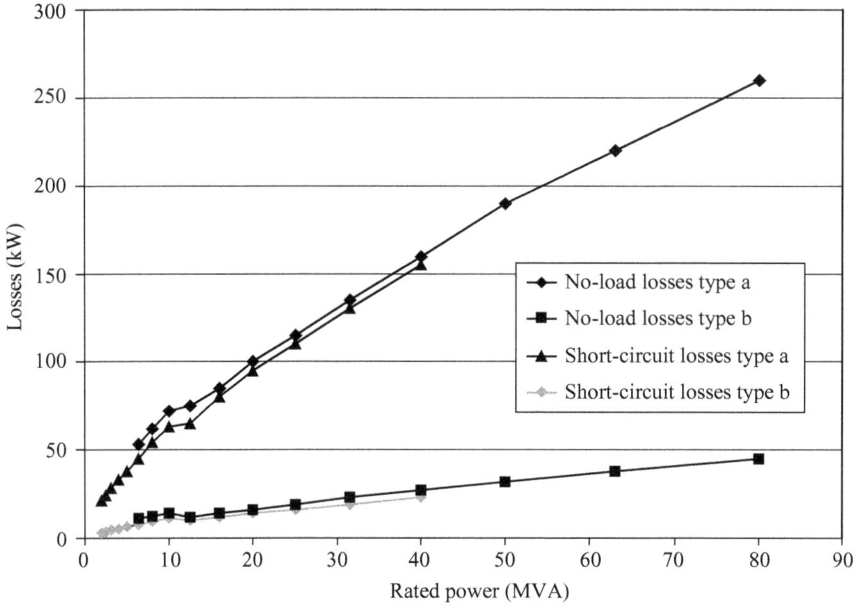

Figure 13.3 Typical values for the ohmic losses, no-load losses and no-load current of two-winding transformers

Table 13.1 Data of transformers

Voltage levels	S_r (MVA)	u_{kr} (%)	u_{Rr} (%)
MV/LV $U_n < 1$ kV	0.05–0.63 0.63–2.5	4 6	1–2 1–1.5
MV/MV $U_n = 1$–66 kV	2.5–25	6–9	0.7–1
HV/MV $U_n > 66$ kV	25–63	10–16	0.6–0.8

The ratio of positive to negative sequence impedance of transformers depends on the vector group and is typically in the range of

Vector group YNd	$X_0/X_1 \approx 0.8$–1.0
Vector group Yzn	$X_0/X_1 \approx 0.1$
Vector group Yyn (three-limb core)	$X_0/X_1 \approx 3.0$–10.0
Vector group Yyn (five limb core and three single-phase transformers)	$X_0/X_1 \approx 10.0$–100.0
Vector group YNynd	$X_0/X_1 \approx 1.5$–3.2

13.1.3 Generators

The parameters such as rated power, rated voltage, power factor and subtransient reactance and synchronous reactance are needed to calculate the impedance of generators and by this the contribution to the short-circuit current.

Rated voltages within one power range may vary depending on the construction type of the generator. The subtransient reactance of synchronous generators is typically in the range of 10–30 per cent depending on the rated power as mentioned in IEC 60909-2. The synchronous reactance is between 100 and 300 per cent, whereas salient pole generators normally have lower values than turbo generators. Typical values are summarised in Table 13.2.

Power factor of generators with rated power below 20 MVA is approximately $\cos\varphi_r = 0.8$ and increases for high rated machines (>1000 MVA) to $\cos\varphi_r = 0.85$ on average. The ratio of saturated to unsaturated reactance x_{dsat}/x_d is between 0.8 and 0.9, whereas in the case of rated power below 100 MVA, the ratio can be between 0.65 and 1.0. Zero-sequence reactance of synchronous generators are $x_0 \approx (0.4$–$0.8)x''_d$ depending on the winding arrangement.

Table 13.2 Typical data of synchronous generators (**average values**)

Rated power S_{rG} (MVA)	Rated voltage U_{rG} (kV)	Subtransient reactance x''_d (%)	Synchronous reactance x_d (%)
Below 4	0.48–11.5	10–25	110–**170**–230
4–20	2.2–13.8	8–16	120–**180**–210
20–200	6–22	10–20	160–**205**–260
Above 200	20–27	18–30	220–**230**–240
Synchronous motors	Up to 11.5	12–25	

13.1.4 Overhead lines

Impedances of overhead lines depend on the geometrical arrangement of phase conductors, on the tower outline and on the number and type of conductors. The zero-sequence impedance furthermore depends on the earth resistivity, on the arrangement of earth conductors and on the design of the earthing system, including conductive installations in earth, respectively, connected to earth, e.g., pipelines, counterpoise and cable sheaths. The calculation of impedances of overhead lines in the positive-sequence and the zero-sequence component are outlined below based on the tower outline as per Figure 13.4. The equations are valid for overhead lines which are symmetrically constructed and operated. All circuits are assumed to be in operation except as noted.

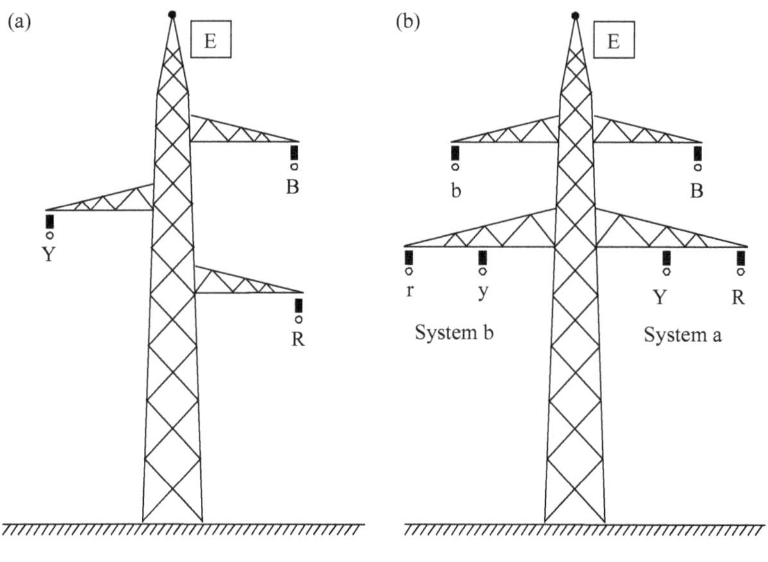

Figure 13.4 Tower outline of high-voltage transmission lines. (a) Single-circuit line and (b) double-circuit line

The distances between phase wires RYB are named d_{RY}, d_{YB} and d_{BR} for the single-circuit line (and similar for the double-circuit line) and the average distances for the calculation of impedances are

$$D = \sqrt[3]{d_{RY}d_{YB}d_{BR}} \tag{13.2a}$$

$$D_{mRy} = \sqrt[3]{d_{Ry}d_{Yb}d_{Br}} \tag{13.2b}$$

$$D_{mRr} = \sqrt[3]{d_{Rr}d_{Yy}d_{Bb}} \tag{13.2c}$$

$$D_{ab} = \sqrt{\sqrt[3]{d_{Rr}d_{Yy}d_{Bb}} * \sqrt[3]{d_{Ry}d_{Rb}d_{Yb}}} \tag{13.2d}$$

$$D_{aE} = \sqrt[3]{d_{Rq}d_{Yq}d_{Bq}} \tag{13.2e}$$

Bundle-conductors of n conductors with radius r in a circular arrangement on the radius r_T must be represented by an equivalent radius r_B

$$r_B = \sqrt[n]{nr * r_T^{n-1}} \tag{13.3}$$

The impedance in the positive-sequence component is calculated for a single-circuit line ($\mu_0 = 4\pi 10^{-4}$ H/km)

$$\underline{Z}'_{1I} = \frac{R'_1}{n} + j\omega\frac{\mu_0}{2\pi}\left(\ln\frac{D}{r_B} + \frac{1}{4n}\right) \tag{13.4}$$

The impedance of the double-circuit line is

$$\underline{Z}'_{1II} = \frac{R'_1}{n} + j\omega\frac{\mu_0}{2\pi}\left(\ln\frac{D * D_{mRy}}{r_B * D_{mRr}} + \frac{1}{4n}\right) \tag{13.5}$$

The calculation of the impedance of the zero-sequence component has to take account of the earth conductor. The impedance of the zero-sequence component of a single-circuit line without earth conductor is given by

$$\underline{Z}'_{0I} = \frac{R'_1}{n} + 3\omega\frac{\mu_0}{8} + j\omega\frac{\mu_0}{2\pi}\left(3 * \ln\frac{\delta}{\sqrt[3]{r_B D^2}} + \frac{1}{4n}\right) \tag{13.6a}$$

and for operation with one earth conductor (index E)

$$\underline{Z}'_{0IE} = \frac{R'_1}{n} + 3\omega\frac{\mu_0}{8} + j\omega\frac{\mu_0}{2\pi}\left(3 * \ln\frac{\delta}{\sqrt[3]{r_B D^2}} + \frac{1}{4n}\right) - 3\frac{(\underline{Z}'_{aE})^2}{\underline{Z}'_E} \tag{13.6b}$$

The loop-impedance $\underline{Z}'_E$ of the arrangement earth conductor and earth return, see Equation (12.13), is

$$\underline{Z}'_E = R'_E + \omega\frac{\mu_0}{8} + j\omega\frac{\mu_0}{2\pi}\left(\ln\frac{\delta}{r_E} + \frac{\mu_r}{4}\right) \tag{13.7a}$$

The coupling impedance $\underline{Z}'_{aq}$ of phase conductor and earth conductor (see Equation (12.11), where $\underline{Z}'_{aE}$ is named $\underline{Z}'_{LE2}$)

$$\underline{Z}'_{aE} = \omega\frac{\mu_0}{8} + j\omega\frac{\mu_0}{2\pi} * \ln\frac{\delta}{D_{aE}} \tag{13.7b}$$

The distance D_{aq} is given by Equation (13.2e) and the depth δ of the earth return path (see Equation (12.2)) is

$$\delta = \frac{1.85}{\sqrt{\mu_0(\omega/\rho)}} \tag{13.8}$$

The resistivity of the soil ρ is between 30 Ωm (Swamp soil) and 3000 Ωm (Stony soil) as outlined in Table 12.2.

The impedance of the zero-sequence component of a double-circuit overhead line without earth conductor is given by

$$\underline{Z}'_{0\text{II}} = \frac{R'_1}{n} + 3\omega\frac{\mu_0}{8} + j\omega\frac{\mu_0}{2\pi}\left(3 * \ln\frac{\delta}{\sqrt[3]{r_B D^2}} + \frac{1}{4n}\right) + 3\underline{Z}'_{ab} \tag{13.9a}$$

and in the case of operation with earth conductor

$$\underline{Z}'_{0\text{IIE}} = \frac{R'_1}{n} + 3\omega\frac{\mu_0}{8} + j\omega\frac{\mu_0}{2\pi}\left(3 * \ln\frac{\delta}{\sqrt[3]{r_B D^2}} + \frac{1}{4n}\right) + 3\underline{Z}'_{ab} - 6\frac{(\underline{Z}'_{aE})^2}{\underline{Z}'_E} \tag{13.9b}$$

where $\underline{Z}'_{aE}$ is the coupling impedance of phase conductor and earth conductor according to Equation (13.7b), $\underline{Z}'_E$ is the impedance of the loop phase conductor and earth return according to Equation (13.7a) and $\underline{Z}'_{ab}$ is the coupling impedance between the systems a and b according to Equation (13.10).

$$\underline{Z}'_{ab} = \omega\frac{\mu_0}{8} + j\omega\frac{\mu_0}{2\pi} * \ln\frac{\delta}{D_{ab}} \tag{13.10}$$

The relative permeability μ_r relevant for overhead lines is

Conductors from Cu or Al	$\mu_r = 1$
Conductors from Al/St, cross-section ratio > 6	$\mu_r \approx 1$
Conductors from Al/St, one layer of Al only	$\mu_r \approx 5$–10
Conductors from Steel (St)	$\mu_r \approx 25$

Typical values for the impedances of MV-overhead lines are summarised in Table 13.3. Table 13.4 shows the impedances of HV-overhead lines.

Table 13.3 Typical values of impedance of the positive-sequence component of MV-overhead lines

Conductor	U_n (kV)	Resistance (Ω/km)	Reactance (Ω/km)
50 Al	10–20	0.579	0.355
50 Cu	10–20	0.365	0.355
50 Cu	10–20	0.365	0.423
70 Cu	10–30	0.217	0.417
70 Al	10–20	0.439	0.345
95 Al	20–30	0.378	0.368
150/25 Al/St	110	0.192	0.398
2 * 240/40 Al/St	220	0.06	0.3
4 * 240/40 Al/St	380	0.03	0.26

Table 13.4 Typical values of impedances of the positive- and zero-sequence component of HV-overhead lines ($\rho_E = 100\ \Omega m$)

Conductor	Earth wire	U_n (kV)	Positive-sequence impedance (Ω/km)	Zero-sequence impedance (Ω/km)	
				One circuit	Two circuits
AlSt 240/40	St 50	110	$0.12 + j0.39$	$0.31 + j1.38$	$0.5 + j2.2$
	Al/St 44/32			$0.32 + j1.26$	$0.52 + j1.86$
	Al/St 240/40			$0.22 + j1.1$	$0.33 + j1.64$
	Al/St 44/32	220	$0.12 + j0.42$	$0.3 + j1.19$	$0.49 + j1.78$
	Al/St 240/40			$0.22 + j1.1$	$0.32 + j1.61$
Al/St 2 * 240/40	Al/St 240/40		$0.06 + j0.3$	$0.16 + j0.98$	$0.26 + j1.49$
Al/St 4 * 240/40	Al/St 240/40	400	$0.03 + j0.26$	$0.13 + j0.091$	$0.24 + j1.39$

A detailed list of impedances of overhead lines of different voltage levels is given in IEC 60909-2:1992.

The capacitances of overhead lines are only needed for special problems, i.e., in case of isolated neutral or if the system is operated with resonance earthing (see Chapter 5), or in case of double-circuit faults. The capacitance of the positive-sequence component for single-circuit line is given by ($\varepsilon_0 = 8.854 * 10^{-12}$ F/m)

$$C'_{1I} = \frac{2\pi\varepsilon_0}{\ln(D/r_B)} \tag{13.11a}$$

The capacitance of the positive-sequence component for a double-circuit line is given by

$$C'_{1II} = \frac{2\pi\varepsilon_0}{\ln(DD_{mRy}/r_B D_{mRr})} \tag{13.11b}$$

Similar to the calculation of the impedance of the zero-sequence component, the earth conductor has to be considered for the calculation of capacitances of the zero-sequence component. In the case of operation without earth conductor the capacitance of the single-circuit line is calculated by

$$C'_{0I} = \frac{2\pi\varepsilon_0}{3*\ln(2h/\sqrt[3]{r_B D^2})} \tag{13.12a}$$

$$C'_{0IE} = \frac{2\pi\varepsilon_0}{3*(\ln(2h/\sqrt[3]{r_B D^2}) - (\ln((h+h_q)/D_{aq}))^2/\ln(2h_q/r_q))} \tag{13.12b}$$

The capacitances of the double-circuit line are calculated by

$$C'_{0II} = \frac{2\pi\varepsilon_0}{3*\ln\left((2h*\sqrt{4h^2+D^2_{mRy}})/(\sqrt[3]{r_B D^2}*\sqrt[3]{D_{mRr}D^2_{mRy}})\right)} \tag{13.13a}$$

$$C'_{0IIE} = 2\pi\varepsilon_0 \Big/ \left\{3*\ln\frac{2h*\sqrt{4h^2+D^2_{mRy}}}{\sqrt[3]{r_B D^2}*\sqrt[3]{D_{mRr}D^2_{mRy}}} - 2\frac{(\ln((h+h_E)/D_{aE}))^2}{\ln(2h_E/r_E)}\right\} \tag{13.13b}$$

To take account of the conductor sag, the average height h of the conductor is used

$$h = \sqrt[3]{h_R h_Y h_B} = \sqrt[3]{h_r h_y h_b} \tag{13.14}$$

The influence of the tower on the capacitance in the zero-sequence component is considered by an increase of 6 per cent (overhead lines with nominal voltage 400 kV), up to 10 per cent (overhead lines with nominal voltage 60 kV).

It should be observed that the capacitance in the positive-sequence component is given by

$$C_1 = 3C_L + C_E \tag{13.15a}$$

and the capacitance in the zero-sequence component is identical to the line-to-earth capacitance (see Section 13.2)

$$C_0 = C_E \tag{13.15b}$$

13.1.5 Cables

Impedances of cables differ very much depending on the type and thickness of insulation, the cable construction, cross-section of conductor, screening, sheaths and armouring and on the type of cable laying, i.e., flat formation or triangle formation. Sheaths and armouring have especially in low voltage cables a strong influence on the impedance. The installation of other conductive installations, e.g., pipelines and screening, armouring and sheaths of other cables have a strong influence on the zero-sequence impedance, which therefore can only be given for simple arrangements. Reference is made to [1], [2], [8], [9] and to data-sheets of manufacturers.

Due to the high permittivity ε_r of the insulation and the small distance between phase conductor and sheeth, identical to earth potential, the capacitances of cables are significantly higher as compared with overhead lines. Figures 13.5 and 13.6 indicate values for the capacitances and the capacitive loading currents of MV- and HV-cables.

Figure 13.7 indicates typical values of reactances (positive-sequence system) of cables of different construction.

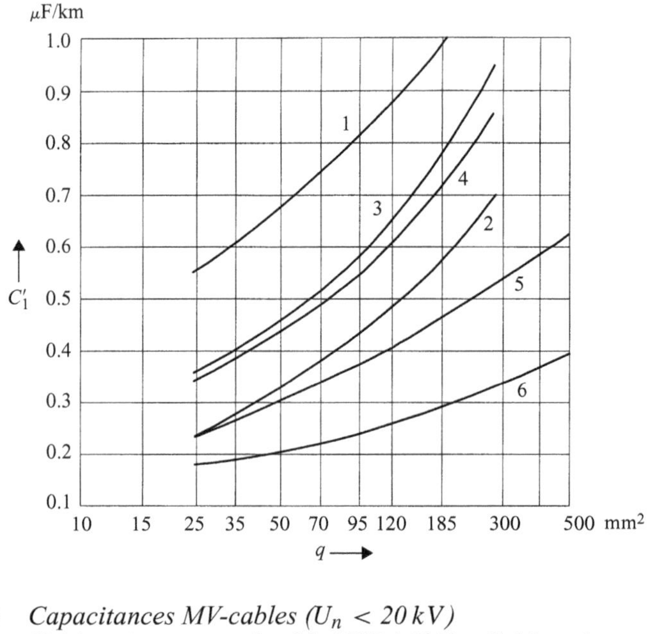

Figure 13.5 Capacitances MV-cables (U_n < 20 kV)
1) Mass-impregnated cable NKBA 1kV
2) Mass-impregnated cable NKBA 6/10 kV
3) Three-core cable NEKEBY 10 kV
4) PVC-cable NYSEY 10 kV
5) VPE-cable N2XSEY 10 kV
6) VPE-cable N2XSEYBY 20 kV

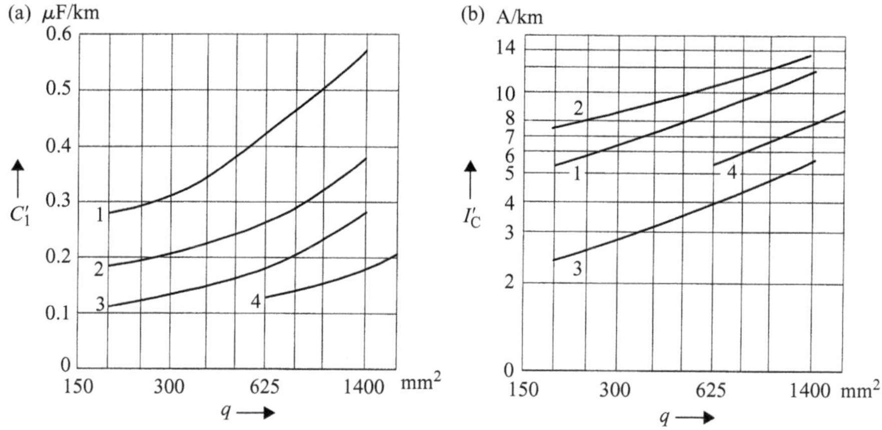

Figure 13.6 *Capacitances C_1' (a) and capacitive loading current I_c' (b) of HV-cables*
1) Single-core oil-filled cable 110 kV *2) single-core oil-filled cable 220 kV*
3) VPE-cable 110 kV *4) VPE-cable 220 kV*

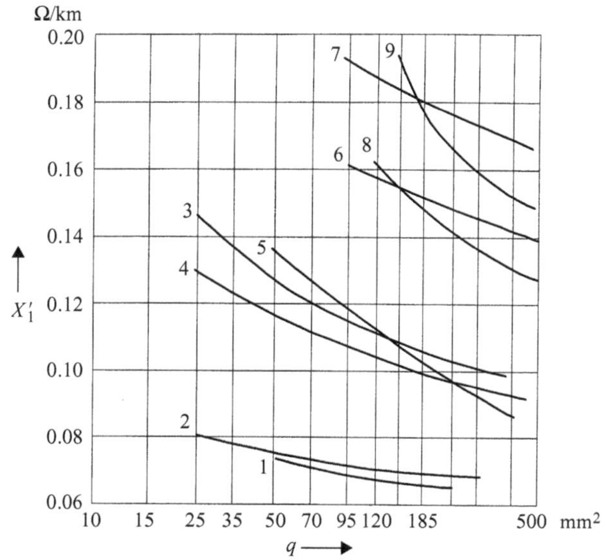

Figure 13.7 *Reactance (positive-sequence system) of three-phase cables ($U_n \leq 110$ kV)*
1) 0,6/1 kV, 4-conductor, NKBA
2) 0,6/1 kV, 4-conductor, NA2XY
3) Three-core cable with armouring 10 kV
4) PVC-cable NYFGby, 10 kV
5) VPE-cable NA2XSEY, 10 kV
6) Single-core oil-filled cable (triangle formation) 110 kV
7) Single-core oil-filled cable (flat formation) 110 kV
8) VPE-cable (triangle formation) 110 kV

13.1.6 Reactors and resistors

Short-circuit limiting reactors are constructed for all voltage levels, from low voltage up to 750 kV. The reactors are manufactured with oil-insulated windings and as air-insulated core-type reactors. Figure 13.8 shows a short-circuit limiting reactor (air-insulated core-type; 10 kV; 630 A; 6 per cent).

Petersen-coils are constructed as reactor with fixed reactance, with tap-changer and with continuous controllable reactors. The control range normally is limited to 1 : 2.5 for tap-changer control and 1 : 10 for continuous control. Standards as per

Figure 13.8 Arrangement of a short-circuit limiting reactor

Source: Mohndruck

IEC 60289:1998 are applicable. The values as per below need to be specified:

Rated voltage	U_r (phase voltage) or $U_r/\sqrt{3}$ depending on application
Rated current	I_r (fixed reactance) or maximal current to be controlled
Rated frequency	50 or 60 Hz; for traction systems other frequencies are used
Operating method	Continuous operation or short-time operation (e.g., 2 h or some minutes)
Control range	Minimal and maximal adjustable current

Figure 13.9 shows a Petersen-coil adjustable in steps by tap-changer (oil-insulated).

Earthing resistors are designed individually for the special applications. They are typically made from stainless steel, cast steel, NiO–Cr- or CuO–Ni-alloy. The maximal permissible temperature, the temperature coefficient and the assumed cost determine the selection of material. Stainless steel is an advantage compared with cast steel due to low temperature coefficient. Low-impedance resistors are mainly arranged from meandered wire elements, high-impedance resistors are arranged from

Figure 13.9 Adjustable Petersen-coil 21 kV/$\sqrt{3}$; 4 MVAr; $I_r = 70.1$–330 A; adjustable in 64 steps, 4.13 A each

Source: SGB Starkstromgerätebau

steel plate grid or steel fabrics. Some characteristic parameters of resistor elements are given in Table 13.5.

Rating and design of resistors are based on ANSI/IEEE 32:1972 and EN 60529, creeping distances are determined in accordance to IEC 60815, insulation must comply with IEC 60071 and high-voltage testing shall be carried out based on

Table 13.5 Characteristic parameters of resistor elements

Material	CuNi 44 or NiCr 8020	Cast steel	CrNi-alloy steel
Arrangement	Wire elements	Cast element	Steel grid and fabric
Resistance at 20°C (Ω)	1500–0.5	0.2–0.01	0.75–0.04
Rated current (A)	<20	25–125	25–250
Thermal time constant (s)	20–90	240–600	120
Temperature coefficient (K^{-1})	0.004	0.075	0.05
Maximal temperature (°C)		400	760

Figure 13.10 Earthing resistor made from CrNi-alloy steel fabric for indoor installation 3810 Ω, 5 A for 10 s, 170 kV BIL, IP 00

Source: Schniewindt KG

Figure 13.11 *Earthing resistor made from meandering wire for outdoor installation 16 Ω, 400 A for 10 s, 75 kV BIL, IP 20*

Source: Schniewindt KG

IEC 60060. Earthing resistors can be suitably designed for indoor and outdoor installations. Figures 13.10 and 13.11 show two different types of resistors.

13.1.7 Asynchronous motors

Data of asynchronous motors are included in IEC 60781:1989 (mentioned as A in Table 13.6) , IEC 60909-2:1992, in [10] (mentioned as B in Table 13.6) and in IEC 60909-1:1991 (mentioned as C in Table 13.6). Table 13.6 outlines the relevant data of asynchronous motors for LV- and MV-application.

13.2 d.c. equipment

Data of d.c. equipment are not documented in a similar way as for a.c. equipment. Literature also presents less information. All data listed hereafter are based on manufacturers' data, information from calculation examples and incomplete data from literature. They should be used for preliminary information only.

13.2.1 Conductors

Resistance of conductors in d.c. auxiliary installations is calculated from the cross-section of the cable or busbar and the material constant. The specific resistance of

Table 13.6 Data of asynchronous motors

S_{rM} (kVA)	P_{rM} (kW)	U_{rM} (kV)	I_{rM} (A)	η_{rM}	$\cos\varphi_{rM}$	I_{anM}/I_{rM}	R_S (mΩ)	I_0 (A)	Pair of poles	Ref.
14.9	11	0.38	22.5	0.89	0.83	8.5	370	9	2	B
24	18.5	0.38	36.6			6.0	[a]1002		2	C
24	18.5	0.38	36.5	0.89	0.85	6	236.7	13	2	B
	20	0.4		0.93	0.85	6				A
29.5	22	0.38	45				[a]813		3	C
29.5	22	0.38	45	0.9	0.83	6	160	16	3	B
40	30	0.38	60.8	0.9	0.84	6.5	91.67	22.6	2	B
	40	0.4		0.93	0.85	6				A
57.8	45	0.38	88	0.92	0.85	6.7	50.34	31.0	2	B
	50	0.41		0.94	0.84	6				B
69.6	55	0.38	106	0.92	0.86	1.3	36.67	38	2	B
90	75	0.38	137	0.94	0.88	6.8	22.97	43	2	B
191.5	160	0.38	291	0.95	0.88	6.3	7	74	2	B
S_{rM} (MVA)	P_{rM} (MW)	U_{rM} (kV)	I_{rM} (A)	η_{rM}	$\cos\varphi_{rM}$	I_{anM}/I_{rM}	R_S (Ω)	I_0 (A)	Pair of poles	Ref.
0.197	0.16	6	19				[a]30.39		1	C
0.259	0.175	6	25				[a]32.2		6	C
0.218	0.18	6	21			5.7	[a]28.94		2	C
0.27	0.225	6	19				[a]22.97		2	C
0.281	0.23	6	27				[a]21.38		1	C
0.299	0.25	6	29	0.94	0.89	5.3	1830	6.6	1	B
0.353	0.3	6	34				[a]16.43		1	C
0.374	0.32	6	36				[a]16.04		1	C
0.467	0.4	6	45	0.955	0.89	5.33	720	11.8	2	B
0.54	0.46	6	52				[a]9.52		1	C
0.685	0.55	6	66	0.95	0.85	5.3	0340	21.5	3	B
0.837	0.63	6	80.5	0.94	0.8	5.2	806.1	27.4	6	B
0.842	0.7	6	81				[a]7.13		3	C
1.697	1.4	6	163	0.948	0.87	5	169	44	2	B
2.09	1.8	6	201	0.968	0.89	5.2	98	50	3	B
2.4	2.1	6	231	0.971	0.9	5.1	100	58.6	2	B
3.07	2.65	6	296	0.96	0.9	5	73.26	60	1	B
5.245	4.5	6	504	0.975	0.88	4.7	27.3	94	2	B
6.85	6	6	659	0.973	0.9	5.5	11	138	2	B
11.64	10	6	1120	0.977	0.88	4	9	154	2	B

[a] Motor impedance instead of resistance.

materials at temperature of 20°C shall be used in accordance with IEC 61660-3:2000 as given below:

$$\text{Copper} \quad \rho = \frac{1}{54}\frac{\Omega\text{mm}^2}{\text{m}}$$

$$\text{Aluminium} \quad \rho = \frac{1}{34}\frac{\Omega\text{mm}^2}{\text{m}}$$

Resistance for other temperatures has to be calculated as given in Table 3.12.

Inductance of d.c. conductors installations depends on the arrangement of the conductors and can only be calculated for simple layout as mentioned in Table 3.12.

13.2.2 Capacitors

Capacitors in d.c. auxiliary installations are installed up to some 10 mF for smoothing of the d.c. voltage. Typical values of the d.c. resistance and the a.c. resistance, respectively, are summarised together with other relevant data in Tables 13.7 and 13.8.

According to information received from manufacturers, the inductance of capacitors is in the range of nano-Henry and can be neglected as compared with the inductance of the connecting cables.

13.2.3 Batteries

The detailed data of batteries as requested in IEC 61660-1:1997 are not available from data sheets of manufacturers, as some data, such as voltage of loaded and unloaded

Table 13.7 Typical values of MKP-capacitors; self-healing dry insulation; different make of capacitor can and fuse

Capacitor can	Capacity (μF)	Nominal voltage (V)	Resistance (mΩ)
Rectangular	12.000	900	0.8
	9.000	1000	0.5
	4.000	1900	0.5
Round	500	900	2.5
	200	1100	3
Prismatic	1250	440	<2
(Internal fuse)	1.600	690	<2
Prismatic	490	440	<7.5
(External fuse)	250	250	<3

Table 13.8 Typical values of MKP-capacitors; resin insulation; round can

Capacity (μF)	Nominal voltage (V)	Resistance (mΩ)
500	420	0.6
1100	420	0.5
2000	420	0.5
500	500	0.8
750	500	0.6
1500	500	0.5
500	640	0.6
1000	640	0.6
1800	640	0.5

Table 13.9 Resistance of loaded batteries (data from several manufacturers)

2-V-batteries											
Capacity	Ah	100	200	300	400	500	600	800	1000	2000	3000
R_B; manufacturer 1	mΩ	0.9	0.8	0.75	0.7	0.65	0.63	0.6	0.55	0.5	0.45
R_B; manufacturer 2	mΩ	—	0.4	0.35	0.32	0.3	—	0.2	0.15	0.08	0.05
12-V-batteries											
Capacity	Ah	40	55	65	75	80	90	100	150	200	
R_B; manufacturer 3	mΩ	9.5	5.8	5.8	5.5	5.5	5.2	4.3	4.0	3.8	—
R_B; manufacturer 4	mΩ	9.7	8.5	—	6.3	6.3	—	5	4.0	3.6	—

battery, depend on the layout and the operational requirements, e.g., required voltage tolerance and voltage drops of connecting cables, of the whole battery plant.

A battery cell with $U_{nB} = 2.0$ V is taken as an example. The voltage of the loaded battery is 2.23 V/cell, which is contrary to IEC 61660-1:1997 stating a value of $E_{Bge} = 1.115 * U_{nB}$ instead of $E_{Bge} = 1.05 * U_{nB}$. When the minimal voltage at the consumer inside a 220-V-installation shall be not less than $U_{min} = 0.9 * U_n$ the minimal permissible voltage at the battery plant is $E_{Bmin} = 191.4$ V taking account of a voltage drop of 3 per cent at the connecting circuits. It is therefore necessary to install 108 battery cells with a minimal voltage of $E_{Bun} = 1.772$ V/cell. If the minimal voltage shall be set to $E_{Bun} = 1.833$ V/cell as recommended by some manufacturers, only 105 cells need to be installed. The maximal voltage at the battery plant is $E_{Bge} = 240.8$ V (108 cells) and $E_{Bge} = 234.2$ V (105 cells), respectively. The number of cells and the minimal permissible voltage depend on the loading time of the battery, the required capacity and the discharge time and vice versa.

Values of the inductance of 12-V-batteries are in the range of $L_{B} = 1$–$10\ \mu H$ and $L_{BZ} = 0.17$–$1.7\ \mu H/cell$, respectively [11]. These values are in the same range as inductances of conductors and cannot be neglected.

Values of the internal resistance of batteries are in the range of $R_{B} = 0.05$–$70\ m\Omega/cell$. Batteries with high capacity have small value of resistance and batteries with low capacity have high value of resistance. Typical values are outlined in Table 13.9. The values of the resistance differ very much especially for 2-V-batteries.

It should be noted, that batteries with very low internal resistance are made from lead-grid cathode and grid-anodes, and batteries with high internal resistance are made from grid-anodes and lead-sheaths cathode sometimes also with additional stretched copper grid.

Symbols, superscripts and subscripts

A detailed explanation of the quantities and symbols is included in the text for each equation, table and figure. It cannot be avoided that some symbols and subscripts are used for different physical quantities. It should be noted that e.g., using the symbol 'J' for the current density cannot be mixed up with the symbol 'J' for the second mechanical moment, as both symbols will not occur in the same equations and even not in a similar context.

Symbols for quantities

A	Aperiodic current
C	Capacitance
D	Geometric mean distance, geometric factor
E	Young's modulus
$\underline{\mathbf{E}}$	Matrix of unity
E	Voltage
F	Mechanical force
G	Conductance
I	Current
J	Current density
K	Factor
L	Inductance
M	Mutual impedance
P	Active power
Q	Reactive power
Q	Thermal heat
R	Resistance
S	Apparent power
T	Total time, time constant
U	Voltage
X	Reactance
Y	Admittance

Z	Impedance
Z	Section moduli
a	Distance
$a; a^2$	Complex operational phasor
c	Voltage factor
d	Distance, diameter
d	Damping factor
f	Frequency
h	Conductor height
k	Factor
$l; \ell$	Length
m	Mass
n	Number
p	Control range
q	Cross-section
q	Factor of plasticity
q	Factor
r	Radius
r	Reduction factor
s	Conductor sag
t	Time instant
t	Transformation ratio
v	Detuning factor
w	Probability factor
α	Temperature & permittivity coefficient
ε	Permittivity
τ	Time constant
ω	Angular frequency
φ	Phase angle
ρ	Resistivity
μ	Factor according to IEC 60909
μ	Permeability
λ	Factor according to IEC 60909
κ	Factor according to IEC 60909
δ	Earth-fault factor (a.c.)
δ	Decay coefficient (d.c.)
δ	Depth of earth return path
γ	Impedance angle
γ	Propagation constant
σ	Mechanical stress

Quantities (Example U)

U	Capital letter used for r.m.s-value
$\underline{U}$	Underlined capital letter indicates phasor (vector)

$\hat{U}$	Peak value
$\lvert\underline{U}\rvert$	Complex value
u	%/MVA-value
u	Instantaneous value
$u(t)$	Time course
$\bar{U}$	Average value
$\mathbf{U}$	Vector (matrix)
$\underline{\mathbf{U}}$	Complex vector (matrix)

Superscripts (Example U)

$\underline{U}^*$	Conjugate–complex value
u'	p.u.-value
U'	Transient value
U''	Subtransient value

Order of subscripts (Example U)

First order	Component (R, Y, B or 0, 1, 2)	U_{Y}
Next order	Type of operation (n; r; b)	U_{Yb}
Next order	Indication of equipment	U_{YbT}
Next order	Number of equipment	U_{YbT4}
Next order	Special condition	U_{YbTmax}
Next order	Index	U_{YbTmaxi}
U_{YbTmaxi}	Indicates voltage of phase **Y** ***b***efore short-circuit, equipment is ***T***ransformer number **4** ***max***imal value for alternative ***i***	

In most of the cases, positive-sequence system is used without subscript '1'

Subscripts, components, systems

0; 1; 2	Zero-, positive-, negative-sequence systems
ac	a.c. system
dc	d.c. system
R; Y; B	Phases of three-phase a.c. system

Subscripts, type of operation

0	No-load value, eigenvalue
an	Locked rotor, starting
b	Before (prior to) fault, beginning
b	Breaking
e	End

k	Short-circuit value
k1	Single-phase short-circuit
k2	Double-phase short-circuit
k3	Three-phase short-circuit
m	Highest value (IEC 60038)
n	Nominal value
r	Rated value
s; sat	Saturated value
δ	At temperature δ

Subscripts, indication of equipment

B	Battery
B	Counterpoise
Br	Branch
C	Capacitor
D	Reactor, rectifier
E	Earth, earth wire
F	Field
G	Generator
J	Joint for connection
KW	Power station
L	Line; load
M	Motor
M	Mutual value
N	Network (a.c. system)
P	Pipeline
Q	System feeder
R	Resistance
S	Special earthing impedance
T	Transformer
To	Overhead-line tower
X	Reactance
Y	Common branch (d.c. system)
Z	Battery cell

Subscripts, special conditions

20	Value at 20°C
f	Fictitious
ge	Loaded
max	Maximal value
min	Minimal value

o	Without tap-changer
res	Residual
s	With tap-changer
th	Thermal
tot	Total
un	Unloaded

Other subscripts

0; 1	Condition at beginning or ending respectively
1; 2; 3	Different sides of a transformer
a; *b*; *c*	Index
B	Base or reference
c	Equivalent
d	Direct axis
HV; MV; LV	High-, medium-, low-voltage side of a transformer
I	Current
i; *j*	Index
Mec	Mechanical
m	Main conductor
p	Pole pair
q	Quadrature axis
res	Equivalent resistance
s	Subconductor
U	Voltage
W	Characteristic impedance
I	One circuit in operation
II	Two circuits in operation

References

1 Schlabbach, J.: Electrical power system engineering (Elektroenergieversorgung). 2nd revised edition, VDE-Verlag. Berlin, Offenbach/Germany, 2003. ISBN 3-8007-2662-9.
2 Oeding, D. and Oswald, B.: Electrical systems and power stations (Elektrische Kraftwerke und Netze). 6th edition, Springer-Verlag. Berlin, Heidelberg, New York, 2004. ISBN 3-5400-0863-2.
3 Schlabbach, J.: Neutral handling (Sternpunktbehandlung). Systems engineering, Vol. 15. VWEW-Energieverlag, Frankfurt/Germany, 2002. ISBN 3-8022-0677-0.
4 CCITT: Directives CCITT, Vol. V, Chapter 5. ITU, Geneva/Switzerland.
5 Niemand, T. and Kunz, H.: Earthing in power systems (Erdungsanlagen). Systems engineering, Vol. 6. VWEW-Energieverlag, Frankfurt/Germany, 1996. ISBN 3-8022-0362-3.
6 Arbitration agency of VDEW: Technical recommendation No. 1 – Induced voltages in telecommunication circuits. VWEW-Energieverlag, Frankfurt/Germany, 1987.
7 Schlabbach, J., Blume, D., and Stephanblome, T.: Voltage quality in electrical power systems. Power and Energy Series, No. 36. IEE-publishers, Stevenage, UK, 2001. ISBN 0-85296-975-9.
8 ABB: Switchgear Manual. 9th edition, Cornelsen-Girardet, Essen/Germany, 1993. ISBN 3-464-48234-0.
9 VDEW: Cable-book (Kabelhandbuch). 6th edition, VWEW-Energieverlag, Frankfurt/Germany, 2001. ISBN 3-8022-0665-7.
10 Scheifele, J.: Contribution of asynchronous motors to short-circuit currents (Beitrag von Drehstrom-Asynchronmotoren zum Kurzschlussstrom). Ph.D. thesis, Technical University of Darmstadt/Germany, 1984.
11 Gretsch, R.: Design of electrical installations in automobiles (Ein Beitrag zur Gestaltung der elektrischen Anlage in Kraftfahrzeugen). Dr.-Ing. habil. thesis, University Erlangen-Nürnberg/Germany, 1979.
12 VDEW: Distributed generation in LV-systems (Eigenerzeugungsanlagen am Niederspannungsnetz). 4th edition, VWEW-Energieverlag, Frankfurt/Germany, 2001. ISBN 3-8022-0790-4.

13 Balzer, G., Nelles, D., and Tuttas, C.: Short-circuit current calculation acc. VDE 0102 (Kurzschlußstromberechnung nach VDE 0102). VDE-technical reports, Vol. 77. VDE-Verlag, Berlin, Offenbach/Germany, 2001. ISBN 3-8007-2101-5.
14 Pistora, G.: Calculation of short-circuit currents and voltage drop (Berechnung von Kurzschlussströmen und Spannungsfällen). VDE-technical reports, Vol. 118. VDE-Verlag, Berlin, Offenbach/Germany, 2004. ISBN 3-8007-2640-8.
15 Gröber, H. and Komurka, J.: Transformation of zero-sequence voltage through transformers (Übertragung der Nullspannung bei zweiseitig geerdeten Transformatoren). Technical report of FGH, Mannheim/Germany, 1973.
16 Balzer, G.: Double-side earthing of transformers (Beidseitige Sternpunktbehandlung von Transformatoren). In VDE: Neutral handling in 10-kV- to 110-kV-system. ETG-Report, Vol. 24. VDE-Verlag, Berlin, Offenbach/Germany, 1988, pp. 172–187.
17 SPEZIELEKTRA: Resonant earthing controller EZR2 (Erdschlußkompensationsregler EZR2). Operation manual N 9/4.88, Spezielektra, Linz/Austria, 1988.
18 Schäfer, D., Schlabbach, J., Gehrmann, A., and Kroll, R.: Increase of displacement voltage in MV cable systems with resonant earthing (Erhöhung der Verlagerungsspannung in Mittelspannungs-Kabelnetzen mit Erdschlußkompensation). *Elektrizitätswirtschaft*, Vol. 93 (1994), VWEW-Energieverlag, Frankfurt/Germany, pp. 1295–1298.
19 Fiernkranz, K.: MV-systems with isolated neutrals or resonant earthing (Mittelspannungsnetze mit isoliertem Sternpunkt oder Erdschlußkompensation). ETG-Report, Vol. 24. VDE-Verlag, Berlin, Offenbach/Germany, 1988.
20 Kouteynikoff, P. and Sforzini, A.: Results of an international survey of the rules limiting interference coupled into metallic pipes. CIGRE Committee 36. *ELECTRA*, Geneva, Switzerland, 1986.
21 Arbitration agency of VDEW: Technical recommendation No. 3 – Measures for construction of pipelines in the vicinity of HV/AC three-phase installations (German). VWEW-Energieverlag, Frankfurt/Germany, 1982.
22 Arbitration agency of VDEW: Technical recommendation No. 5 – Principles of calculation and measurement of reduction factor of pipelines and earth wires (German). VWEW-Energieverlag, Frankfurt/Germany, 1980.
23 Arbitration agency of VDEW: Technical recommendation No. 7 – Measures for the installation and operation of pipelines in the vicinity of three-phase high-voltage installations (German). VWEW-Energieverlag, Frankfurt/Germany, 1985.
24 IEC 60479-1: Effects of currents passing through the human body. International electrotechnical commission, Geneva, Switzerland, 1994.
25 Jenkins, N., Allan, R., Crossley, P., Kirschen, D., and Strbaq, G.: Embedded generation. Power and Energy Series, No. 31. IEE Publishers, Stevenage, UK, 2000. ISBN 0-85296-774-8.
26 Schlabbach, J. (Ed.) and Metz, D.: Power system engineering (Netzsystemtechnik). VDE-Verlag, Berlin, Offenbach/Germany, 2005. ISBN 3-8007-2821-4.

27 Schlabbach, J.: Short-time interference of pipelines. Report on research and development No. 14. University of Applied Sciences, Bielefeld/Germany, 2000. ISBN 3-923216-52-1.
28 Cory, B. and Weedy, B.: Electrical power systems. 4th edition. John Wiley & Sons Ltd., Chichester, England, 1998. ISBN 0-471-91659-5.
29 Funk, G.: System of symmetrical components (Symmetrische Komponenten). Elitera-Verlag, Berlin, Germany, 1976. ISBN 3-8708-7087-7.
30 Kories, R. and Schmidt-Walter, H.: Electrotechnical handbook (Taschenbuch der Eletrotechnik). 6th revised edition, Verlag Harri Deutsch, Frankfurt/Germany, 2004. ISBN 3-8171-1734-5.
31 Metz, D., Naundorf, U., and Schlabbach, J.: Handbook of electrotechnical equations (Kleine Formelsammlung Elektrotechnik). 4th edition, Fachbuchverlag Leipzig, Germany, 2003. ISBN 3-446-22545-5.
32 Wildi, T.: Units and conversion charts. IEEE Press, New York, USA, 1990. ISBN 0-87942-273-4.
33 Oeding, D. and Schünemann, H.: Calculation of short-circuit currents in HV-systems using %/MVA-system (Berechnung der Kurzschlussströme in Hochspannungsnetzen mit %/MVA-system). BBC-News, Mannheim/Germany, 1965.
34 Koglin, H.: The decaying d.c. component of short-circuit currents (Der anklingende Gleichstrom beim Kurzschluss in Elektroenergieversorgungsnetzen). Ph.D. thesis, University of Darmstadt/Germany, 1971.
35 Pitz, V. and Waider, G.: Impedance correction factors of network transformers for short-circuit current calculation (Impedanzkorrekturfaktoren für Netztransformatoren bei der Kurzschlussstromberechnung). *Elektrie* 47 (1993), pp. 301–304.
36 Scheifele, J. and Waider, G.: Maximal short-circuit currents through linear optimisation (maximale Kurzschlussströme durch lineare Optimierung). *etzArchiv* 10 (1998), pp. 275–280.
37 Oeding, D. and Waider, G.: Maximal partial short-circuit currents of power stations (Maximale Teilkurzschlussströme von Kraftwerksblöcken ohne Stufenschalter). *etzArchiv* 10 (1988), pp. 173–180.
38 Balzer, G. and Deter, O.: Calculation of thermal effects of equipment due to short-circuit currents (Berechnung der thermischen Kurzschlussbeanspruchung von Starkstromanlagen). *etzArchiv* 7 (1985), pp. 287–290.
39 Tuttas, C.: Approximation of reduction factors of complex conductor arrangement (Berechnung des Reduktionsfaktors komplizierter Leiteranordnungen). AEG-report 54, Frankfurt, Germany, 1981, pp. 153–157.
40 Hosemann, G., Nietsch, C., and Tsanakas, D.: Short-circuit stress in d.c. auxiliary systems. Cigre-report 23-104, Geneva, Switzerland, 1992.
41 Tsanakas, D.: Substitution function for the calculation of mechanical and thermal stress due to short-circuits in d.c. installations (Ersatzfunktion für die Bestimmung der mechanischen und thermischen Kurzschlussbeanspruchung in Gleichstromanlagen). *etzArchiv* 10 (1988), pp. 355–360.

42 Nietsch, C.: Calculation of short-circuit currents in d.c. installations. University of Erlangen/Germany, EV-report F254, 1989.
43 Tsanakas, D., Meyer, W., and Safigianni, A.: Dynamical short-circuit stress in d.c. installations (Dynamische Kurzschlussbeanspruchung in Gleichstromanlagen). *Archiv für Eletrotechnik* 74 (1991), pp. 305–313.
44 Tsanakas, D. and Papadias, A.: Influence of short-circuit duration on dynamic stresses in substations. *IEEE Transactions PAS 102* (1983), pp. 492–501.
45 Hosemann, G. and Tsanakas, D.: Dynamic short-circuit stress of busbar structures with stiff conductor. Parametric studies and conclusions. *Electra* 68, Geneva, Switzerland, 1980, pp. 37–64.
46 Meyer, W.: Additional calculation acc. IEC 60865-1 for the determination of short-circuit stress of lines (Ergänzung des Berechnungsverfahrens nach IEC 60865-1 zur Ermittlung der Kurzschlussbeanspruchung von Leitungsseilen mit Schlaufen im Spannfeld). EE-report, University of Erlangen/Germany, 2002.
47 Meyer, W.: Mechanical short-circuit stress of rigid conductors in IEC 865-1 – Information on the norm (Mechanische Kurzschlussbeanspruchung von biegesteifen Leitern in IEC 865-1 – Hintergründe zur Norm). EE-report, University of Erlangen/Germany, 2002.
48 Tsanakas, D.: Dynamic stress in high-voltage structures by short-circuits of short-duration. CIGRE-Symposium *High currents in power systems*. Proc. Report 500-01, Brussels, Belgium, 1985.
49 Rüger, W. and Hosemann, G.: Mechanical short-circuit effects of single-core cables. IEEE Trans. PD, 4 (1989) pp. 68–74.
50 Nietsch, C. and Tsanakas, D.: Short-circuit currents in d.c. auxiliary installations (Kurzschlussströme in Gleichstrom-Eigenbedarfsanlagen). *Elektrie* 46 (1992), pp. 18–22.
51 Hosemann, C., Zeitler, E., Miri, A., and Stein, N.: The behaviour of droppers in HV substations under short-circuit. Proceedings of the 5th International symposium on short-circuit currents in power systems. Proc. Report 3.2, Zlin, Czech Republic, 1992.
52 Tsanakas, D., Meyer, W., and Nietsch, C.: Short-circuit currents of motors in d.c. auxiliary installations in power plants and substations. *Electromotion '99*, Patras/Greece, pp. 489–496.
53 Stein, N., Meyer, W., and Miri, A.: Test and calculation of short-circuit forces and displacements in HV substations with strained conductors and droppers. *ETEP* 10 (2000), pp. 131–138.
54 Herold, G. and Kunz, M.: Fast analytical short-circuit current calculation of rectifier fed auxiliary subsystems. *ETEP* 13 (2003), pp. 151–159.
55 Pitz, V., Köster, H.-J., *et al.*: Short-circuit mechanical effects on outdoor HV substations with wide bundling. CIGRE-Session 2004. Proc. Report B3-107, Paris, France.
56 Wessnigk, K. and Griesbach, P.: Digital calculation of short-circuit current in battery-fed d.c. installations (Digitale Berechnung des zeitlichen Kurzschlussstromverlaufs in batteriegespeisten Gleichstromnetzen). *Elektrie* 43 (1989), pp. 379–381.

57 Albert, K., Apelt, O., Bär, G., and Koglin, H.-J. (Ed.): Electrical power supply (Elektrischer Eigenbedarf). VDE-Verlag, Berlin, Offenbach/Germany, 1993. ISBN 3-8007-1586-4.

IEC-standards, EN-norms and other standards and norms mentioned within the context of this book are listed in Tables 1.1 and 1.2.

Standards are only to be applied based on their actual issues, available from VDE-Verlag GmbH, Bismarckstr. 33, D-10625 Berlin, Beuth-Verlag GmbH, Burggrafenstr. 6, D-10787 Berlin or the national standard organisation. English versions of the norms are also available through BSI in London/UK.

The English text of this book was neither checked by DIN (Deutsches Institut für Normung e.V., Berlin) nor by VDE (Verband der Elektrotechnik Elektronik Informationstechnik e.V., Frankfurt).

Index

%/MVA system semirelative units 33–4
conversion of quantities with p.u. and ohm systems 35
example 40
correction factor 34
HV three-phase examples 93
impedance based on measurement example 41–2
impedance calculations for a three-winding transformer 37–40
impedance results of various equipment 91
impedances calculation HV three-phase example 92
a.c. systems, impedance: *see* impedance calculations for a.c. equipment
a.c. systems, mechanical effects: *see* conductors, a.c., mechanical effects of short circuits
a.c. systems, thermal effects
about thermal effects 195
see also conductors, a.c., thermal effects
admittance/reactance/impedance 12
American National Standards Institute, web address 4
asynchronous motors
data of 281–2
factor q for short-circuit breaking current 162–3
impedance calculations 55, 63–4
auxiliary installations
supply of a power station calculations example 94–6
see also batteries in d.c. auxiliary installations; d.c. auxiliary installations, effects of short circuits; d.c. auxiliary installations, short-circuit current calculations; d.c. motors in auxiliary installations

batteries in d.c. auxiliary installations
example data 283–5
impedance calculations 60, 64, 65
short-circuit currents 170–2
breaking current with short-circuits: *see* short-circuit breaking current
British Standards Institute, web address 4
busbar/switchgear short-circuit current limitation measures
busbar sectionaliser in single busbar switchgear 233–4
Ip-limiter usage 234–6
medium and low voltage limiting equipment 234–6
multiple busbar operation 232–6
time curves of short-circuit currents 235

cables
capacitances 276–7
impedances 276
reactances 276–7
see also conductors
capacitors in auxiliary installations
factor κ 169–71
impedance calculations 59, 64, 65
short-circuit currents 169–70
typical values of capacity and resistance 283–4
causes of short circuit currents 1–3
communication circuit interference: *see* pipelines/communication circuits, interference with
complex calculations 11–14

conductors, a.c., mechanical effects of short circuits
about mechanical effects 201–2
conductor oscillation influence 207–9
bending stress and dynamic force on supports 208–9
mechanical natural frequency 207–8
electromagnetic forces 202–4
fixing clamps calculations example 217–18
forces and stresses in rigid conductor support calculations 204–7
conductors, a.c., thermal effects of short circuits
bare conductors 198–9
cable calculation example 216–17
cable screening, armouring and sheath 199–201
maximal permissible short-circuit current density 201
maximal permissible temperature 200–1
heat dissipation considerations, factors *m* and *n* 197–8
IEC 60865–1 197
IEC 60909–0 197
impregnated paper-insulated cables 199–200
maximum permissible temperature and current density table 196
overhead line conductors type Al/St 198
rated short-time current density 199
thermal equivalent short-time current 195–7
conductors, d.c. equipment, impedance/resistance calculations 58, 64, 65, 281–3
converter fed drive, impedance calculations 56
correction factors
d.c. auxiliary installations, short-circuit current calculations, σ 179
example 190–1
with HV three-phase systems examples 76, 91–4
impedance correction factor K_G 154–6
examples 91–4
impedances of equipment and short-circuit current example 154
%/MVA- or p.u.-system 34, 152–4
current limitation earthing power system 102–5
earth-fault factor δ 103, 104
d.c. auxiliary installations, effects of short circuits
bending stress on conductors 213–14
bending stress of sub-conductors 211
calculation example 223–4
equivalent circuit diagrams and short-circuit current time course 166
force calculation for main- and sub-conductors 212–15
calculation example 222–3
forces on supports 214–15
mechanical natural frequency of main and sub-conductors 211
standard approximation function (simplified approach) 209
electromagnetic effect example 220
substitute rectangular function approach 209–16
calculation example 220–3
thermal effects calculation example 218–19
thermal short circuit strength 215–16
vibration period and natural mechanical frequency of conductors 212
calculation example 220–1
d.c. auxiliary installations, short-circuit current calculations
about auxiliary installations 165–8
correction factor σ 179
equivalent circuit diagram 180
example
battery short-circuit current 186
capacitor short-circuit current 185–6
correction factors and corrected parameters 190–1
data and parameters 183–4
equivalent diagram 182
impedances of cable and busbar conductors 184–5
partial short-circuit currents 191–3
rectifier short-circuit current 186–8
total short-circuit current 193–4
maximal short-circuit current calculations 168
minimal short-circuit current calculations 168
time function calculations 167
total short-circuit current 178–82
typical time curves 181
see also batteries in d.c. auxiliary installations; capacitors in auxiliary installations; d.c. motors in auxiliary installations; impedance calculations

for d.c. equipment; rectifiers in auxiliary installations
d.c. motors in auxiliary installations
factor κ 175–8
IEC 61000–1:1997 175
IEC 61660–1:1997 175, 178
impedance calculations 62, 65, 66
with independent excitation, short-circuit currents 174–8
factors k 175–8
short-circuit current calculations, example 188–9
definitions and terms 30–2
Deutsches Institute für Normung, web address 4
double earth-faults
about double-earth faults 139
fault in a 20-kV-system example 146–8
impedances 139–40
peak short-circuit current 143
power system configurations 140–2
steady-state short-circuit current 143
symmetrical short-circuit breaking current 143
symmetrical short-circuit current 139–40
see also earth currents from short circuits

earth currents from short circuits
about short-circuits through earth 143–5
short-circuit at overhead-line tower 145–6
short-circuit inside a switchyard 144–5
single-phase short-circuit in a 110-kV-system example 148–9
earth faults: *see* double earth-faults; earth currents from short circuits
earth-fault factor δ
and current limitation earthing 103, 104
and low impedance neutral earthing effects 102
earthing considerations
about earthing 257
earthing grid resistance calculation example 264–6
earthing impedance 261–2
resistance of different types of earthing 262
human body resistance 257–9
relevant currents through earth 259–61
design information 260
soil conditions and resistivity 258–9
earthing rods 259
see also neutral earthing
earthing resistors
about earthing resistors 279–81
characteristic parameters 280–1
electromagnetic forces, on a.c. conductors 202–4
equivalent circuit diagrams for short circuits 24–9, 43

factors
about factors 151
factor k for d.c. motor short-circuit currents 175–8
factor κ for peak short-circuit current 77–80, 156–8
with d.c. motors 175–8
factor q for short-circuit breaking current of asynchronous motors 162–3
factor λ for steady-state short-circuit current 82–3, 160–2
factor μ for symmetrical short-circuit breaking current 81–2, 158–60
factors m and n, heat dissipation 197
IEC 60909–1:1991–10 151, 162
impedances of equipment and short-circuit current example 152–3
induced voltage calculation, screening factor 252
isolated neutral power system, overvoltage factor 108
see also correction factors
faults, as causes of short circuit currents 1–3
fuses, and short circuit currents 2

generators
power factor data 270
rated voltage data 270–1
short circuit current limitation, subtransient reactance influence 239–40
sutransient reactance data 270–1
synchronous reactance data 270–1
zero-sequence reactance data 270

human body resistance 257–9
HV a.c. three-phase system short circuit current calculations
about HV short circuits 67–8
about parameter calculations 70–2
assumptions and methods 68–71
auxiliary supply of a power station calculations example 94–6
decaying (aperiodic) component 80–1

HV a.c. three-phase system short circuit current calculations (*continued*)
far-from generator short-circuit 67–9
IEC 60909–0 71, 72–83
impedance correction factor examples 91–4
initial symmetrical short-circuit current 72–4
double-phase 73
far-from-generator 73
line-to-earth 73
three-phase 73
inside power plant short-circuit currents 72–6
correction factors 76, 91–4
equivalent circuit diagrams 74–5
locations around transformers 74–6
line-to-earth (single-phase) short-circuit example 87–8
maximal and minimal short-circuit currents 71–2
meshed 110-kV-system example 89–91
minimal short-circuit currents 85–6
motor influences 84–5
near-to-generator short-circuit example 86–7
peak short-circuit currents 77–80
calculation example 88–9
equivalent frequency method 80
factor κ 77–80
meshed networks 79–80
non-meshed fed 77–8
ratio R/X method 80
single-fed 77
steady-state short-circuit current 82–3
factor λ 82–3
subtransient reactance 67
symmetrical short-circuit breaking current 80–2
factor μ 81–2
types of short circuit 67–9

IEC 60038 6, 226–7
IEC 60050 6, 7
IEC 60071 7
IEC 60265 7
IEC 60282 7
IEC 60479 7
IEC 60781 2, 5, 9, 131, 132
IEC 60865 2, 5, 9, 197
IEC 60896 7
IEC 60909 5, 6, 9, 71, 72–83, 131, 132, 139, 151, 162, 197
IEC 60949 7
IEC 60986 7, 8
IEC 61000 175
IEC 61071 7
IEC 61660 5, 9, 175, 178
IEC document list 5–9
impedance
impedance/admittance/reactance 12
impedances of a three-winding transformer in MVA example 37–40
measurement for symmetrical components 20–4
in %/MVA-system for auxiliary supply 96
in %/MVA-system based on measurement example 41–2
impedance calculations for a.c. equipment
about impedance for a.c. systems 45
asynchronous motors 55, 63–4
calculation examples 63–4
converter fed drive 56
for HV a.c. three-phase equipment under short circuit examples 91
overhead transmission lines 53, 63–4
power-station unit 52, 63–4
short-circuit limiting reactor 54, 63–4
synchronous generators 51, 63–4
system feeder 46, 63
system loads 57, 63–4
three-winding transformer 48–50, 63–4
two-winding transformer 47, 63
impedance calculations for d.c. equipment
about impedance for d.c. systems 50
batteries 60, 64
calculation examples 64–6
capacitors 59, 64, 65
conductors 58, 64, 65
motors, d.c. 62, 65, 66
rectifiers 61, 63–4, 65–6
impedance correction factor K_G 154–6
examples 91–4
impedance matrix, transformation of 19–20
inductive interference calculations 247–52
interference: *see* pipelines/communication circuits, interference with
International Electrotechnical Commission, web address 4
Ip-limiters 234–6
isolated neutral power system 105–8
capacitive earth-fault current 106–7
earth faults 105–8
earth-fault arcs 107–8
overvoltage factor 108

limitation of short-circuit currents
about limitation measures 225–6
coupling of power system at busbars with low short-circuit level 230–1
distribution of feeding locations 228–9
earthing impedances in neutral of transformers 238
and generator subtransient reactance 239–40
meshed systems 241–4
nominal system voltage selection 226–7
IEC 60038:1987 recommended voltages 226–7
operating subsystems separately 227–8
radial systems 241
restructuring power systems 231
ring-main systems 241–2
short-circuit limiting reactors 238
see also busbar/switchgear short-circuit current limitation measures; transformers
limiting value, definition 31
low-impedance neutral earthing power system 98–102
earth-fault factor δ 102, 103, 104
single-phase short-circuit current 98–100
single/three-phase short circuit relation 100–1
low-voltage (LV) systems, short circuit current calculations
about short-circuit currents in LV systems 131
calculation methods 132
IEC 60781 131, 132
IEC 60909–0 131, 132
impedance calculations 132–3
approximations for generators and rectifiers 133
initial symmetrical short-circuit current 133
maximal short-circuit current calculation example 135–6
minimal short-circuit currents 134–5
calculation example 135, 137
motor influences 134
peak short-circuit current 133
steady-state short-circuit current 134
symmetrical short-circuit breaking current 133
types of faults 131

matrix equations 14–17
cyclic-symmetrical matrix 15–16
see also transformations/transformation matrix
maximal and minmal short-circuit currents 3
meshed systems, limitation of short-circuit currents 241–4
motors, influence for HV a.c. three-phase systems 84–5
motors, d.c.: *see* d.c. motors

neutral earthing
about neutral earthing 97–8
characteristics of different fault types summary 99
design of an earthing resistor connected to an artificial neutral example 124
limitation of single-phase short-circuit current by earthing through impedance example 123–4
transformer HV-side and LV-side considerations 116–19
compensation windings 116–18
voltages at neutral of a unit transformer example 126–9
see also current limitation earthing power system; isolated neutral power system; low-impedance neutral earthing power system; resonance earthing power systems (Peterson-coil)
nominal value, definition 31
norms
about technical standards and norms 4–5
international documents and norms 5–6
norms as referred in standards 6–9

ohm-system physical quantities voltage/current/impedance 32
conversion of quantities with p.u. and %/MVA systems 35
example 40
HV three-phase correction factor examples 94
overhead transmission lines
capacitances 274–6
conductor sag considerations 275
coupling impedance 273
distance between phase wires 272
impedance calculations/data 53, 63–4, 271–6
impedance in the positive-sequence component for a single and double circuit line 272

overhead transmission lines (*continued*)
impedance of the zero-sequence component 272
loop impedance 272
relative permeability for 273
resistivity of soil 273
thermal effects 198
typical impedance values 274

parallel and series connection 27, 30
peak short-circuit currents 2–3
double earth faults 143
factor κ 77–80, 156–8, 175–8
low-voltage (LV) systems 133
see also HV a.c. three-phase system short circuit current calculations
Peterson-coil: *see* resonance earthing power systems (Peterson-coil)
phasor diagrams 11–14
pipelines/communication circuits, interference with
about pipeline interference 245–7
characteristic impedance of the pipeline 253–4
pipeline coating resistance 254
coupling types summary table 246
induced voltage calculation 252–3
coupling impedances 253
screening factor 252
inductive interference calculations 247–52
coupling impedances 249–51
earth return path considerations 248–9
interference of pipeline from 400-kV-line example 262–6
pipeline-to-earth voltage 254–7
oblique exposure 255–6
safety aspects 247
Potier's reactance 160–1
power system short-circuit current limitation: *see* limitation of short-circuit currents
power-station unit, impedance calculation 52, 63–4
p.u.-system relative units 33–4
conversion of quantities with ohm and MVA systems 35
example 40
correction factor 34

radial systems, limitation of short-circuit currents 241
rated data, definition 32
rated value, definition 31
reactance/admittance/impedance 12
reactors, short-circuit current limiting: *see* resonance earthing power systems (Peterson-coil)
rectifiers in auxiliary installations
impedance calculations 61, 63–4, 65–6
short-circuit currents 172–4
relative quantities/p.u.-system 33–4
resonance earthing power systems (Peterson-coil) 108–16, 278–9
20-kV-system example 124–5
about resonance earthing 108–11
alternate earthing of parallel transformer neutrals 118, 121
basic principles 109–11
capacitive earth-fault current and residual current calculation example 125–6
current limits 111
displacement voltage calculations 112–15
fault currents in MV-system from fault in HV-system 118, 121
impedance calculations 54, 63–4
increase of displacement voltage example 119–23
capacitive asymmetry in a 20-kV-system 120, 122
resonance curve for detuning factors in a 20-kV-system 122
Peterson-coils 278–9
tuning 115–16
resonance frequency/tuning 110–11
usage worldwide 108–9
see also neutral earthing
ring-main systems, limitation of short-circuit currents 241–2
r.m.s. value of short-circuit currents 2–3
RYB-system
representation of a line example 42–4
see also symmetrical components

safety, interference with pipelines and communication circuits 247
screening factor, pipelines/communication circuits 252
semirelative quantities and %/MVA system: *see* %/MVA system semirelative units
series and parallel connection 27, 30
short-circuit breaking current
about breaking current value 2–3
double earth faults 143

factor μ for symmetrical breaking 81–2, 158–60
low-voltage (LV) systems current calculations 133
three-phase systems calculations 80–2
short-circuit current limitation: *see* limitation of short-circuit currents
short-circuit currents
about short circuit currents 1–3
equivalent circuit diagrams 24–9
importance of 1–3
maximal and minimal 3
typical time course 2–3
see also busbar/switchgear short-circuit current limitation measures; d.c. auxiliary installations, short-circuit current calculations; earth currents from short circuits; HV a.c. three-phase system short circuit current calculations; limitation of short-circuit currents; low-voltage (LV) systems, short circuit current calculations
short-circuit limiting reactors: *see* resonance earthing power systems (Peterson-coil)
soil conditions and resistivity 248, 258–9
standards
about technical standards and norms 4–5
American National Standards Institute 4
British Standards Institute 4
definitions and terms 30–2
Deutsches Institute für Normung 4
international documents and norms 5–6
International Electrotechnical Commission 4
norms as referred in standards 6–9
VDE-Verlag 4
see also IEC
substitute rectangular function approach: *see* d.c. auxiliary installations, effects of short circuits
subtransient reactance of generators 67, 239–40
switchgear
and short circuit currents 2
see also busbar/switchgear short-circuit current limitation measures
symmetrical components
example 34–7
impedance measurement of 20–4
interpretation of the system of 18–19
and representation of a line in the RYB-system example 42–4
transformation of impedances 19–20
transformation matrix 14–17
see also ohm-system physical quantities voltage/current/impedance; %/MVA system semirelative units; p.u.-system relative units
synchronous generators, impedance calculation 51, 63–4
system feeders
impedance calculations 46, 63
ranges of initial short-circuit power 267, 268
system loads, impedance calculations 57, 63–4

terms and definitions 30–2
thermal effects of short circuits: *see* conductors, a.c., thermal effects of short circuits
thermal equivalent short-time current for conductors 195–7
three-phase systems: *see* HV a.c. three-phase system short circuit current calculations
time course of short circuit currents 2
transformations/transformation matrix 14–17
complex rotational phasors 17
cyclic-symmetrical matrix 15–16
delta-star-transformation 31
reverse transformation 17
star-delta-transformation 30
transformation of impedances 19–20
transformers
impedance calculations in %/MVA 37–40
and impedance measurement for symmetrical components 20–3
ratio of positive to negative sequence impedance 270
short-circuit current limitation measures 236–8
earthing impedances in the neutrals 238–9
impedance voltage considerations 236–8
loadflow and short-circuit analysis results 237
three-winding, impedance calculations 21–4, 48–50, 63–4

transformers (*continued*)
  two-winding, impedance calculations 47, 63
  typical values two-winding transformer
    impedance voltage 267–70
    ohmic losses, no-load losses and no-load current 267, 269
  see also neutral earthing
transmission lines: *see* overhead transmission lines

VDE documents 5–9
VDE-Verlag, web address 4
vectors/vector diagrams 11–15
  example 34–7

web addresses, standards institutions 4

www.ingramcontent.com/pod-product-compliance
Ingram Content Group UK Ltd.
Pitfield, Milton Keynes, MK11 3LW, UK
UKHW020238290926
15470UKWH00002B/613

9 780863 415142